Android 4.X
应用与开发实战手册(第2版)
——适用 Android 4.X~2.X

黄彬华　著

清华大学出版社

北京

内 容 简 介

本书以 Android 4.X 进行开发示范，通过大量图示与 step by step 方式，详细介绍了使用 Android 开发智能手机、平板电脑应用程序的方法，读者无须强记即可灵活运用各项开发技巧。本书还介绍了如何将应用程序上传到 Google Play(原 Android Market)供全球 Android 移动设备用户下载，以及如何将 AdMob 广告板置入应用程序，即使应用程序免费也可以获利。此外，本书所有范例都已录制成长达 450 分钟的视频，是帮助初学者学习和教师教学的最佳强化工具，读者可通过 www.tupwk.com.cn 下载。

本书语言简洁，内容丰富，既可作为 Android 初学者的入门教材，也可作为 Android 应用开发人员的自学参考手册。

Android 4.X App 開發教戰手冊：第二版|適用 Android 4.X~2.X
黃彬華
碁峯資訊股份有限公司，2012.08
ISBN 978-986-276-581-4
本书为碁峯资讯股份有限公司授权清华大学出版社于中国大陆（港澳台除外）地区发行中文简体版。未经出版者书面许可，不得以任何方式复制或抄袭本书内容。
北京市版权局著作权合同登记号　图字：01-2013-4599

本书封面贴有清华大学出版社防伪标签，无标签者不得销售。
版权所有，侵权必究。侵权举报电话：010-62782989　13701121933

图书在版编目(CIP)数据

Android 4.X 应用与开发实战手册——适用 Android 4.X～2.X / 黄彬华 著. —2 版. —北京：清华大学出版社，2013.7
ISBN 978-7-302-32213-9

Ⅰ. ①A… Ⅱ. ①黄… Ⅲ. ①移动终端—应用程序—程序设计—手册 Ⅳ. ①TN929.53-62

中国版本图书馆 CIP 数据核字(2013)第 084571 号

责任编辑：王　定　于　平
装帧设计：牛艳敏
责任校对：蔡　娟
责任印制：何　芊

出版发行：清华大学出版社
网　　址：http://www.tup.com.cn, http://www.wqbook.com
地　　址：北京清华大学学研大厦 A 座　　邮　编：100084
社 总 机：010-62770175　　邮　购：010-62786544
投稿与读者服务：010-62776969, c-service@tup.tsinghua.edu.cn
质 量 反 馈：010-62772015, zhiliang@tup.tsinghua.edu.cn

印 刷 者：北京鑫丰华彩印有限公司
装 订 者：三河市李旗庄少明印装厂
经　　销：全国新华书店
开　　本：185mm×260mm　　印　张：26.5　　字　数：661 千字
版　　次：2013 年 7 月第 1 版　　印　次：2013 年 7 月第 1 次印刷
印　　数：1～4000
定　　价：58.00 元

产品编号：051257-01

FOREWORD

　　Android 是由 Google 公司和开放手机联盟领导并开发的一种基于 Linux 的自由及开放源代码的操作系统，主要用于移动设备(如智能手机和平板电脑)。Android 操作系统最初由 Andy Rubin 开发，主要支持手机，随后，逐渐扩展到平板电脑及其他领域(如电视、数码相机、游戏机等)。经过几年的快速发展，Android 已是今非昔比。数据显示，到 2012 年底，Android 占据全球智能手机操作系统市场 76%的份额，在中国市场占有率为 90%。

　　本书以 Android 4.X 进行开发示范，通过大量图示与 step by step 方式，详细介绍了使用 Android 开发智能手机、平板电脑应用程序的方法，读者无须强记即可灵活运用各项开发技巧。本书还介绍了如何将应用程序上传到 Google Play(原 Android Market)供全球 Android 移动设备用户下载，以及如何将 AdMob 广告板置入应用程序，即使应用程序免费也可以获利。此外，本书所有范例都已录制成长达 450 分钟的视频，是帮助初学者学习和教师教学的最佳强化工具，读者可通过 www.tupwk.com.cn 下载。

　　本书语言简洁，内容丰富，既可作为 Android 初学者的入门教材，也可作为 Android 应用开发人员的自学参考手册。

　　本书由姬朝阳、吉振涛进行繁体版转简体版工作，并对书中的程序代码做了进一步调试和运行，在此表示衷心感谢。在本书的编辑过程中，我们还对繁体版中的一些术语做了转换，使其更符合大众的阅读习惯，但由于水平有限，疏漏之处在所难免，欢迎广大读者批评指正。

FOREWORD

  智能手机(smart phone)之所以比传统手机(feature phone)更智能化，是因为智能手机可以自由新建或删除各种应用程序(App)；传统手机则仅限于出厂时内置的应用程序，很难自由扩增。智能手机可以安装各种应用程序，从而将手机功能发挥到极致。例如，使用手机定位导航、翻译各国语言、收发 E-mail、浏览网页、拨打网络电话、使用 facebook、录音录像与享受游戏乐趣等，不再只是拨打电话与发送短信而已。

  智能手机的功能几乎等同于迷你笔记本电脑，让手机与计算机之间的区分越来越模糊。平板电脑的添加，更让智能手持设备(笔者将智能手机与平板电脑归类为智能手持设备)逐渐形成了除个人电脑外的一大势力，而且智能手持设备有下列优点。

- 携带方便：智能手机体积小不在话下，平板电脑无论 7 寸或 10 寸也都可以轻易地放在提包内，方便随时使用。
- 操作简单：掌上电脑可以说专为普通人设计，不像操作一般计算机需要较专业的知识，例如硬件配置、网络设置等。
- 系统稳定：除了山寨机外，市场上的掌上电脑几乎都是品牌机，由知名厂商设计，所以系统稳定性高。

  除了智能手机与平板电脑外，许多其他移动设备也准备嵌入该操作系统。按照这种趋势发展下去，以后开发 Android 应用程序，不一定都是为了手机或平板电脑，也可能是开发其他设备的应用程序，例如电子书设备、移动电视等。为了让更多人能够进入到移动设备的世界，甚至能够编写出方便自己或他人日常生活的应用程序，笔者决定将在大学与教育训练中心所讲授的数据汇总起来，以深入浅出的方式说明 Android 应用程序的编写技巧，让读者可以很轻松地开发出实用的应用程序。

黄彬华

CONTENTS

第1章 Android 导论与新版功能介绍 ······ 1
- 1.1 认识 Android ······ 2
  - 1.1.1 Android 属于 Linux 移动平台 ······ 2
  - 1.1.2 Android 历史 ······ 3
  - 1.1.3 版本更新过程 ······ 3
  - 1.1.4 开放手机联盟介绍 ······ 7
- 1.2 Android 成功的原因 ······ 8
  - 1.2.1 开放源代码与采用 Apache 授权方式 ······ 8
  - 1.2.2 Android 向 Java 招手 ······ 9
- 1.3 Google Play 介绍与获利实例 ······ 10
  - 1.3.1 Google Play 介绍 ······ 10
  - 1.3.2 Android 应用程序能否获利 ······ 11
- 1.4 Android 新版功能介绍 ······ 12

第2章 开发工具下载与安装 ······ 15
- 2.1 开发 Android 应用程序所需的工具 ······ 16
- 2.2 JDK 下载与安装 ······ 16
  - 2.2.1 JDK 下载 ······ 16
  - 2.2.2 JDK 安装 ······ 18
- 2.3 Eclipse 下载与安装 ······ 19
  - 2.3.1 Eclipse 下载 ······ 19
  - 2.3.2 Eclipse 安装 ······ 21
- 2.4 ADT 与 Android SDK 安装 ······ 22
  - 2.4.1 第一次安装 ······ 22
  - 2.4.2 旧版升级成新版 ······ 30
- 2.5 Eclipse 编码设置成 UTF-8 ······ 31
  - 2.5.1 整个 workspace 都改成 UTF-8 ······ 31
  - 2.5.2 单一项目改成 UTF-8 ······ 32

第3章 Android 项目与系统架构 ······ 35
- 3.1 设置 Android 模拟器 ······ 36
- 3.2 创建与管理 Android 项目 ······ 40
  - 3.2.1 创建 Android 项目 ······ 40
  - 3.2.2 运行 Android 项目 ······ 43
  - 3.2.3 移除 Android 项目 ······ 46
  - 3.2.4 导入 Android 项目 ······ 47
  - 3.2.5 打开范例程序 ······ 48
- 3.3 DDMS 的使用 ······ 49
- 3.4 Android 系统架构介绍 ······ 51
- 3.5 项目目录架构与 Manifest 文件介绍 ······ 53
  - 3.5.1 Android 项目目录架构 ······ 54
  - 3.5.2 Manifest 文件 ······ 57
- 3.6 应用程序本地化 ······ 58
  - 3.6.1 Android 支持的地区与语言 ······ 59
  - 3.6.2 创建支持多国语言的应用程序 ······ 60

第4章 UI 设计基本概念 ······ 63
- 4.1 Android UI 设计基本观念 ······ 64
- 4.2 Widget 组件介绍 ······ 66
  - 4.2.1 与用户间的交互——以 Button 事件处理为例 ······ 67
  - 4.2.2 TextView 与 EditText ······ 71
  - 4.2.3 CheckBox、RadioButton 与 ToggleButton ······ 76
  - 4.2.4 RatingBar ······ 79
  - 4.2.5 SeekBar ······ 81
  - 4.2.6 ImageView ······ 84
  - 4.2.7 WebView ······ 85

4.3	页面配置与 layout 组件介绍 …… 89
	4.3.1 LinearLayout ……………… 89
	4.3.2 RelativeLayout …………… 91
	4.3.3 TableLayout ……………… 93
	4.3.4 ScrollView ………………… 96
4.4	设置 UI 样式——使用 style 与 theme ……………………… 98
	4.4.1 定义 style ………………… 98
	4.4.2 继承 style ………………… 99
	4.4.3 应用 theme ……………… 100
	4.4.4 继承 theme ……………… 101
4.5	触控与手势 ……………………… 102
	4.5.1 触控事件处理 …………… 102
	4.5.2 手势 ……………………… 104

第 5 章 UI 高级设计 ……………… 109

5.1	Menus …………………………… 110
	5.1.1 Options Menu ……………… 110
	5.1.2 Context Menu …………… 114
	5.1.3 Submenu ………………… 117
5.2	对话窗口与日期选择器 ………… 119
	5.2.1 AlertDialog ……………… 120
	5.2.2 DatePickerDialog 与 TimePickerDialog ………… 122
	5.2.3 CalendarView …………… 129
5.3	Spinner 与 AutoComplete- TextView ………………………… 132
	5.3.1 Spinner …………………… 132
	5.3.2 AutoCompleteTextView …… 136
5.4	Gallery 与 GridView …………… 137
	5.4.1 Gallery …………………… 137
	5.4.2 GridView ………………… 140
5.5	ListView ………………………… 143
	5.5.1 ListActivity ……………… 144
	5.5.2 ListView ………………… 146
5.6	自定义 View 组件与 2D 绘图 …………………………… 149
5.7	补间动画 ………………………… 153

5.8	Drawable 动画 …………………… 159

第 6 章 Activity 生命周期与平板电脑 设计概念 …………………… 163

6.1	Activity 生命周期 ……………… 164
6.2	Activity 间传递数据 …………… 168
6.3	通知信息 ………………………… 172
6.4	Broadcast ………………………… 175
	6.4.1 单纯接收 Broadcast ……… 175
	6.4.2 自行发送与接收 Broadcast …… 178
6.5	Service 生命周期 ……………… 182
	6.5.1 调用 startService()打开 Service ……………………… 183
	6.5.2 调用 bindService()连接 Service ……………………… 189
6.6	平板电脑 UI 设计概念 ………… 194
	6.6.1 Fragment 生命周期 ……… 195
	6.6.2 Activity 画面拆分 ……… 201
	6.6.3 ActionBar ………………… 208
	6.6.4 Tabs ……………………… 213

第 7 章 数据访问 …………………… 217

7.1	Android 数据访问概论 ………… 218
7.2	Assets …………………………… 218
7.3	Resources ………………………… 220
7.4	Shared Preferences ……………… 223
7.5	Internal Storage ………………… 228
7.6	External Storage ………………… 231

第 8 章 移动数据库 SQLite ………… 239

8.1	SQLite 数据库概论与数据 类型 ……………………………… 240
	8.1.1 SQLite 数据库概论 ……… 240
	8.1.2 SQLite 数据类型 ………… 240
8.2	使用命令行创建数据库 ………… 242
8.3	SQL 语法 ……………………… 243
	8.3.1 创建数据表 ……………… 244
	8.3.2 DML 语法 ………………… 245

8.4	Android 应用程序访问 SQLite 数据库	247
8.5	SQLite 新增功能	250
8.6	SQLite 查询功能	255
	8.6.1 输入欲查询的数据	256
	8.6.2 数据浏览	259
8.7	SQLite 修改与删除功能	264

第 9 章 Google 地图 ··· 271

9.1	申请 Google 地图的 API 密钥	272
9.2	在 Google 地图上呈现自己位置	275
	9.2.1 显示与缩放 Google 地图	276
	9.2.2 呈现自己位置	277
9.3	在 Google 地图上指定位置	281
9.4	标记的使用	284
9.5	LocationListener 与 LocationManager	287
9.6	以地名/地址查询位置	293
9.7	导航功能	296

第 10 章 传感器应用 ··· 301

10.1	传感器介绍	302
10.2	加速度传感器	303
10.3	方位传感器	308
	10.3.1 调用 getOrientation()取得方位信息	309
	10.3.2 通过 Sensor.TYPE_ORIENTATION 取得方位信息	313
10.4	接近传感器	315
10.5	光线传感器	317

第 11 章 多媒体与相机功能 ··· 321

11.1	Android 多媒体功能介绍	322
11.2	播放 Audio 文件	324
	11.2.1 播放资源文件	324
	11.2.2 播放外部文件或网络数据流	328
11.3	播放视频文件	332
	11.3.1 简易视频播放器	332
	11.3.2 MediaPlayer 播放视频文件	334
11.4	录制音频文件	339
11.5	拍照功能	346
11.6	录像功能	352

第 12 章 手机实用功能开发 ··· 359

12.1	手机铃声设置	360
12.2	手机音量与振动的设置	369
12.3	来短信与来电处理	376
12.4	查询联系人数据	382

第 13 章 AdMob 广告制作 ··· 389

13.1	AdMob 简介	390
13.2	注册 AdMob 账户	391
13.3	AdMob 广告实现	396

第 14 章 将应用程序发布至 Google Play ··· 399

14.1	如何将应用程序发布至 Google Play	400
14.2	产生并签名应用程序	400
	14.2.1 使用 Eclipse + ADT 产生并签名应用程序	400
	14.2.2 签名应用程序注意事项	403
14.3	申请 Android 开发人员账号	407
14.4	使用开发人员管理界面发布应用程序	409
	14.4.1 应用程序首次发布	409
	14.4.2 应用程序改版	412

附录 导入范例程序错误时的解决方法 ··· 413

第 1 章
Android导论与新版功能介绍

本章学习目标：

- 认识 Android
- Android 成功的原因
- Google Play 介绍与获利实例
- Android 新版功能介绍

认识 Android

2011 年对移动设备而言是具有重大意义的一年,因为该年仅智能手机(smart phone)的销售量(约 4.9 亿)就远超过台式计算机(desktop)、笔记本电脑(notebook)与上网本(netbook)的销售总量(约 3.5 亿),参见表 1-1[①]。除此之外值得注意的是平板电脑(pad 或 tablet)的成长率高达 274.2%,为所有产品之首。这样的结果导致全球最大 PC 制造商 HP 曾一度计划出售 PC 事业[②],可见移动设备的来势汹汹。

表 1-1

种 类	2011(百万)	2011/2010 成长率
Smart phones	487.7	62.7%
Pads	63.2	274.2%
Netbooks	29.4	-25.3%
Notebooks	209.6	7.5%
Desktops	112.4	2.3%

至于 2011 年各智能手机的操作系统销售市场占有率与成长率则如表 1-2 所示,Android 手机无论市场占有率 48.8%或是成长率 244.1%都居所有智能手机之首,所以值得我们深入研究。

表 1-2

操作系统	2011(百万)	市场占有率	2011/2010 成长率
Android	237.8	48.8%	244.1%
iOS	93.1	19.1%	96.0%
Symbian	80.1	16.4%	-29.1%
BlackBerry	51.4	10.5%	5.0%
bada	13.2	2.7%	183.1%
Windows Phone	6.8	1.4%	-43.3%
Others	5.4	1.1%	14.4%
Smart phones	487.7	100.0%	62.7%

1.1.1 Android 属于 Linux 移动平台

Android(机器人)是一种专门为了移动设备,例如:移动电话、平板电脑以及上网本而设计的操作系统,所以该系统以达到文件精简、运行性能佳而且省电为目的。Android 主要以 Linux 为核心(Linux kernel,目前采用 3.0.1 版)与 GNU[③]软件(GNU software)为基础;也就是说,Android 是属于 Linux 操作系统的一种。Linux 是一个相当成熟且稳定的操作系统,无论安全性、多任

---

① 参见 http://www.canalys.com/newsroom/smart-phones-overtake-client-pcs-2011。
② 参见 http://zh.wikipedia.org/wiki/%E6%83%A0%E6%99%AE%E5%85%AC%E5%8F%B8。
③ http://en.wikipedia.org/wiki/GNU。

务处理能力甚至软硬件的支持度,都非常优秀。不过 Android 并不完全兼容于传统的 Linux 系统,例如,它没有 X Window 系统,也没有完全支持 GNU 函数库,所以无法将所有 Linux/GNU 的应用程序都移植到 Android 上。

1.1.2 Android 历史

在 2005 年的时候,有传言说 Google 想要扩张事业版图到手机业务,甚至想要成为手机制造商,推出专门以提供位置服务(例如卫星导航、地图等服务)为主的自有品牌手机。如果 Google 真有这样打算的话,那么要达成此目的最大的问题就是 Google 没有自己的手机操作系统。结果在 2005 年 7 月,Google 并购了一家位于加州、成立仅 22 个月的小公司,名为 Android。Google 宣称该公司仅开发手机上的软件,但有可靠消息指出该公司不仅开发手机专用软件,还致力于开发手机操作系统。

过了近两年半,Google 于 2007 年 11 月 5 日公开发表他们所研发的,以 Linux 系统为核心的移动平台,名为 Android。当时 Google 不仅发表他们新的移动平台,更宣布开放该平台源代码。同时也促成 OHA(Open Handset Alliance,开放手机联盟)的成立。

Google、HTC(宏达国际电子股份有限公司)、T-Mobile(美国电信公司)更是于 2008 年 9 月 23 日共同发布了全球第一个基于 Android 平台的手机——T-Mobile G1①(也称为 HTC Dream),如图 1-1 所示,该手机包含 GPS 定位功能、310 万像素数字相机以及一系列的 Google 应用程序,打开了 Google 在移动通信领域上的征途。

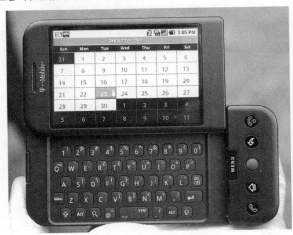

图 1-1

1.1.3 版本更新过程②

表 1-3 详细列出了 Android 系统各版本对应的 API 层级(API level)、发布时间与重大更新项目,让读者更了解 Android 系统的整个演变过程。

---

① http://www.htc.com/www/press.aspx?id=66338&lang=1028。

② http://en.wikipedia.org/wiki/Android_version_history。

表 1-3

Android 系统版本	API 层级	发布日期与重大更新
1.0	1	2008 年 9 月 23 日发布
1.1	2	2009 年 2 月 9 日发布，并修复之前的 Bug
1.5 (Cupcake) 架构在 Linux Kernel 2.6.27	3	2009 年 4 月 30 日发布，主要的更新如下： 1. 支持上传视频到 YouTube，图片可以直接上传到 Picasa 2. 支持立体声蓝牙耳机，同时改善自动配对性能 3. 采用最新 WebKit 技术的浏览器，支持复制/粘贴和页面中搜索 4. GPS 性能大大提高 5. 加入屏幕虚拟键盘，支持智能选字 6. 主屏幕可以增加应用程序或目录 7. 应用程序自动随着手机页面旋转 8. 短信、Gmail、日历、浏览器的操作界面大幅改善，可以批量删除邮件 9. 来电照片显示
1.6 (Donut) 架构在 Linux Kernel 2.6.29	4	2009 年 9 月 15 日发布，主要的更新如下： 1. 改进 Android Market 2. 整合摄影、拍照、浏览操作界面，并可复选相片并删除 3. 改进响应时间与应用程序整合 4. 改善整体搜索功能(包含语音搜索) 5. 支持手势 6. 支持 CDMA 网络 7. 文字转语音系统(text-to-speech) 8. 应用程序耗电查看 9. 支持更多种类的屏幕分辨率
2.0/2.0.1/2.1 (Eclair) 架构在 Linux Kernel 2.6.29	5 (2.0) 6 (2.0.1) 7 (2.1)	2009 年 10 月 26 日发布 2.0，2009 年 12 月 3 日发布 2.0.1，2010 年 1 月 12 日发布 2.1，主要的更新如下： 1. 改善硬件速度 2. 支持更多种类的屏幕分辨率 3. 操作界面重整 4. 新的浏览器操作界面和支持 HTML5 5. 新的联系人功能 6. 更佳的白色/黑色背景比率 7. 改进 Google Maps 3.1.2 8. 支持 Microsoft Exchange 9. 支持内置相机闪光灯 10. 数位变焦 11. 改进虚拟键盘 12. 蓝牙 2.1

(续表)

Android 系统版本	API 层级	发布日期与重大更新
2.2 (Froyo) 架构在 Linux Kernel 2.6.32	8	2010 年 5 月 20 日发布,主要的更新如下: 1. 整体速度和性能改进 2. 加强软件实时编译的速度 3. 集成 Chrome 的 V8 JavaScript 引擎应用到浏览器 4. 增加对 Microsoft Exchange 的支持(安全政策、日历同步功能、auto-discovery, GAL look-up) 5. USB 分享器和 WiFi 热点功能 6. 自动和批量更新 Android Market 的应用程序 7. 可以快速在多国语言键盘与对应字典间作切换 8. 通过蓝牙功能实现语音拨号以及分享联系人数据 9. 支持在浏览器上传文件 10. 支持软件可以安装在扩展的内存中 11. 支持 Adobe Flash 10.1
2.3/2.3.3 (Gingerbread) 架构在 Linux Kernel 2.6.35	9 (2.3) 10 (2.3.3)	2010 年 12 月 6 日发布 2.3,2011 年 2 月 9 日发布 2.3.3,主要的更新如下: 1. 支持 extra-large 屏幕尺寸与分辨率 2. 支持 SIP VoIP 通信 3. 支持 WebM/VP8 视频播放以及 AAC 音频编码 4. 支持复数镜头 5. 支持 NFC(近距离通信) 6. 支持更多声音特效,例如 Bass 音 7. 新增下载管理功能,方便长时间下载 8. 同步资源回收(concurrent garbage collection)功能,改善运行性能 9. 强化 UI(User Interface,用户界面)设计 10. 强化剪贴板功能(system-wide copy-paste functionalities) 11. 加强支持原生码开发(native code development) 12. 原生码支持更多传感器(sensor),例如:gyroscopes(陀螺仪)与 barometers(气压计) 13. 强化电源、应用程序管理功能 14. 增强音效、图形画面与用户输入,使游戏开发更方便
3.0/3.1/3.2 (Honeycomb) 架构在 Linux Kernel 2.6.36	11 (3.0) 12 (3.1) 13 (3.2)	2011 年 2 月 22 日发布。这个版本的 Android 系统是专门为平板电脑设计(tablet-only)的。主要的更新如下: 1. Google eBooks 上提供数百万本书 2. 支持平板电脑大屏幕、高分辨率 3. 新版 Gmail 4. Google Talk 视频功能 5. 3D 加速处理 6. 网页版 Market(Web store)详细分类显示,按个人需求 Android 分别设置安装应用程序 7. 新的短信通知功能

(续表)

Android 系统版本	API 层级	发布日期与重大更新
3.0/3.1/3.2 (Honeycomb) 架构在 Linux Kernel 2.6.36	11 (3.0) 12 (3.1) 13 (3.2)	8. 专为平板电脑设计的用户界面(重新设计的通知栏与系统栏) 9. 加强多任务处理的界面 10. 重新设计适用大屏幕的键盘及复制粘贴功能 11. 多个标签的浏览器以及私密浏览模式 12. 快速切换各种功能的相机 13. 增强的图库与快速翻阅的联系人界面 14. 更有效率的 Email 界面 15. 支持多核心处理器 16. 3.2 优化 7 寸平板显示
4.0/4.0.3 (Ice Cream Sandwich) 架构在 Linux Kernel 3.0.1	14 (4.0) 15 (4.0.3)	2011 年 10 月 19 日发布，主要的更新如下： 1. 提升硬件的性能以及系统的优化，提升系统流畅度 2. 每一页都有 System Bar 3. 界面以新的标签页形式展示，并且将应用程序和其他内容的图标分类 4. 更方便地在主界面创建活页夹，并且使用拖放的操作方式 5. 在日历中也可以使用多点触控，进行缩放和拖曳操作 6. Gmail 支持缩放操作，并支持左右拖曳进行查看 7. 增加屏幕截图功能 8. 改进虚拟键盘出现的错误操作 9. 在屏幕锁状态下也可以拍照和观看信息 10. 改进复制/粘贴功能 11. 更好的语音功能 12. 脸部解锁 13. 网页浏览器具有新增标签页功能，最大可同时打开 16 个标签页 14. 浏览器的书签可以与其他装置的 Chrome 进行同步 15. 全新、现代化 Roboto 字体 16. 内置网络流量监控功能，用户可以对流量进行设置，超出设置上限时，手机会自动关闭上网功能，并且可以随时查看已使用和未使用的流量，并且以报表形式呈现 17. 随时可以关闭正在使用的应用程序 18. 提升相机功能 19. 内置图片处理软件 20. 新的图库软件 21. 可以在公共场合快速地与其他 NFC 设备通信 22. 新的启动屏幕，Home 画面右下角类似 Tray 的图标，内有多个程序可运行 23. 支持硬件加速的功能 24. 支持 WiFi Direct 功能 25. 支持 1080p 视频播放和录制

01 Android 导论与新版功能介绍

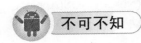 不可不知

Android 的吉祥物如图 1-2 所示。

图 1-2

除了吉祥物之外,Android 1.5 版开始都有一个代号,而且官方网站也提供对应的图片,如图 1-3 所示。

Cupcake(Android 1.5)

Donut(Android 1.6)

Eclair(Android 2.1)

Froyo(Android 2.2)

Gingerbread(Android 2.3)

Honeycomb(Android 3.0)

Ice Cream Sandwich(Android 4.0)

图 1-3

1.1.4 开放手机联盟介绍

开放手机联盟(OHA,Open Handset Alliance)是一个商业性的联盟,其目的在于共同制订

7

Android 开放源代码的移动设备标准,将与其他来自于 Symbian OS(Nokia)、iOS(Apple)、BlackBerry(RIM)、Windows Mobile OS(Microsoft)、Palm OS(Palm)的移动平台竞争。截至目前为止共有 80 多位成员[①],分类如表 1-4(未将所有成员列出)。

表 1-4

成员种类	成员名称
手机制造商(Handset Manufacturers)	HTC、Acer、Motorola、LG、Samsung
电信运营商(Mobile Operators)	T-Mobile、NTT DoCoMo、中国移动(China Mobile Communications)
软件开发商(Software Companies)	Google、eBay、Ascender
半导体制造商(Semiconductor Companies)	Intel、Nvidia、Texas Instruments
商品化公司(Commercialization Companies)	Aplix、Borqs、L&T Infotech

1.2 Android 成功的原因

Android 操作系统能够成功打入移动市场的原因分析如下。

1.2.1 开放源代码与采用 Apache 授权方式

Android 除了操作系统外,还包含许多移动设备所需使用到的软件,Google 将 Android 的源代码公开出来,而且大部分都采用 Apache 2.0 授权方式[②](Apache Software License, 2.0),这样一来对企业有非常大的好处,说明如下。

(1) 免费使用 Android 系统:移动设备安装 Android 操作系统,不像安装其他操作系统那样,需要支付费用给操作系统厂商[③];这样一来,整体成本大幅下降,在现今微利时代,更显重要,对移动设备厂商的获利更具保障。

(2) 可以按需求修改 Android:Google 公开 Android 相关软件的源代码,厂商可以自行下载研究[④],而且根据 Apache 授权方式,厂商可以按照自己的需求,修改或重制其内容,开发适合自己路线的产品[⑤]。例如:Nookcolor 就是将 Android 应用在电子书阅读器上[⑥]。

(3) 不需公开源代码[⑦]:Apache 授权方式与一般 GPL 授权方式不同,修改以 Apache 授权的源代码而产生的新程序代码,不必再将此新程序代码的内容公开出来。GPL 授权方式要求公开修改后的源代码,虽然有利于软件的发展,但却不利于企业营利,因为一旦公开源代码,商

---

① http://www.openhandsetalliance.com/oha_members.html。
② http://source.android.com/source/licenses.html。
③ 例如手机若安装 Windows Mobile 操作系统,则手机制造商需支付权利金给微软公司。
④ Android 源代码下载说明网页 http://source.android.com/source/download.html。
⑤ http://www.apache.org/licenses/LICENSE-2.0 的 "2. Grant of Copyright License"。
⑥ http://www.barnesandnoble.com/。
⑦ http://en.wikipedia.org/wiki/Apache_License。

业对手就可迅速反击。Apache 授权方式不要求更改后再公开源代码，企业会更愿意投入大量心力去开发相关软件。

1.2.2　Android 向 Java 招手

　　智能移动设备之所以被称为智能，并不是简单拥有一个智能的操作系统，还必须辅以大量的应用程序(也就是俗称的软件)，才会让用户觉得该移动设备功能强大，从而愿意使用。所以如何吸引大量程序设计师来 Android 平台上开发各式各样的应用程序让用户使用，也是决定一个操作系统成败的关键。Google 在发展 Android 平台时，必定思考过该选择哪种程序语言来开发 Android 的应用程序。自行研发新的程序语言？新程序语言往往不稳定且需要经过长时间的调试、宣传才可能被开发者接受，这样一来不仅成本过高，而且会拖延 Google 进军移动市场的时间。最好的方法是找一个不仅在市场上已经非常成熟稳定，而且受到大家欢迎的程序语言来当作开发 Android 应用程序的主要程序语言。所以 Google 最后选上 Java 程序语言来开发 Android 应用程序也就不令人意外了。

　　Java 程序语言有许多优点[1]，所以几乎年年稳坐全球最受欢迎程序语言的冠军宝座[2]。Google 采取的策略是 Android 的应用程序直接以 Java 程序语言编写(到目前为止也只支持 Java 语言)，而且支持程度几乎包括 Java SE 5.0 版所有函数库。这样一来，全球众多的 Java 程序开发者很有可能都会投入 Android 应用程序的开发，所以在很短的时间内，Android 将会有数以万计的应用程序供移动设备用户下载[3]，因此目前才会有如此多的用户喜爱 Android 平台的移动设备。看来 Android 应用程序采用 Java 程序语言开发的策略也已奏效。

不可不知

　　虽然 Android 应用程序是以 Java 写成的，但 Android 平台上没有 JVM(Java Virtual Machine)，所以 Java 的 class(Java byte code)或 jar 文件不能在 Android 上运行。Android 有一个类似 JVM 的功能，称作 Dalvik VM(Virtual Machine)。Java 的 class 文件必须先编译成 dex 文件(Dalvik Executable)，然后由 Dalvik VM 运行[4]。Dalvik VM 是一个特别为 Android 量身定做的 VM，可以在 CPU 或内存配置都比 PC 差的移动设备上仍能达到高运行性能但低耗电量。除此之外，因为不直接使用 JVM 来运行应用程序，Google 可以避开 Java 版权的问题[5]，并且不需要遵循以 Oracle[6]为主导的 Java 标准。

---

[1] Java 优点：跨平台，纯面向对象程序语言易于模块化、易于分布式计算，安全性高，有庞大的群体致力于 Java 技术的研究并开发源代码。
[2] 请参见 TIOBE 网站，http://www.tiobe.com/index.php/content/paperinfo/tpci/index.html。
[3] Android Market 从 50 000 个增长到 100 000 应用程序只花了 3 个月不到的时间，请参见 http://www.engadget.com/2010/07/15/android-market-now-has-100-000-apps-passes-1-billion-download-m/。
[4] 可参见本书"3.4 Android 系统架构介绍"的"Android Runtime(Android 运行环境)"图解说明。
[5] 请参见 http://www.betaversion.org/~stefano/linotype/news/110/，这篇文章对 Google 如何避免 Java 版权问题，有精辟说明。
[6] Java 的创始公司 Sun 已于 2009 年 4 月 20 日被 Oracle 并购，请参见 http://www.oracle.com/us/corporate/press/018363。

1.3 Google Play 介绍与获利实例

1.3.1 Google Play 介绍

Google Play 原名为 Android Market。要探讨 Google Play，可以分成 Android 应用程序用户与开发者两个角度来说明。

1. 应用程序用户

用户可以通过移动设备至 Google Play 下载并安装各式各样开发好的应用程序，如图 1-4 所示。

2011 年 2 月 2 日开始，Android 用户也可以通过 PC 上的浏览器直接将 Google Play[①] 上的应用程序安装在已注册的 Android 移动设备上[②]，如图 1-5 所示。

图 1-4

图 1-5

目前用户可以下载的应用程序已经超过 30 万个，应用程序数量统计可参见表 1-5[③]：

表 1-5

统 计 日 期	应用程序总数量
2009/3	2 300
2009/12	20 000
2010/8	80 000
2011/5	200 000
2011/7	250 000
2011/11	310 000

---

① 网址为 https://play.google.com/store。
② 详细的下载说明请参见 http://briian.com/?p=7430。
③ 参见 http://zh.wikipedia.org/wiki/Android_Market。

截至目前，Google Play 免费软件比例已接近 7 成，是所有应用软件商店中最高的，如表 1-6 所示，这对用户来说是相当有利的。

表 1-6

商 店 名 称	免费软件比例
Google Play	67%
Windows Phone Marketplace	61%
Samsung Apps	39%
Apple App Store	37%
BlackBerry App World	26%
Nokia Ovi Store	26%

2. 应用程序开发者

Android 开发者可以将开发好的应用程序发布到 Google Play，而且一生仅需缴纳一次费用 (25 美金)给 Google，比上传至 App Store 便宜很多[①]。应用程序是否收费，要看该程序的开发者在发布时是否设置过要收取费用；也就是说，应用程序是否收费由开发者自行决定。如果应用程序要收费，Google Play 采用"三七分账"方式，Google 会收取应用程序售价的 30% 当作使用销售平台的费用，开发者则获得售价的 70%。如何发布应用程序到 Google Play，将在第 14 章详加说明。

1.3.2　Android 应用程序能否获利

编写 Android 应用程序并发布到 Google Play 是否可以获利？一直以来都是开发者关心的话题。下面有一个实际案例，是一个开发者在自己的网站上公布他发布到 Google Play 的应用程序获利情况[②]。说明如下。

(1) 应用程序：Advanced Task Manager，有 0.99 美元的付费版和展示广告的免费版(图 1-6 是免费版，广告在最下面)。

(2) 开发者：Arron La。

(3) 收益。

- 付费版：2009/2~2010/8，共 50 000 美金(纯收益，已扣除付给 Google 30%的费用)。
- 免费版：2009/11~2010/8，共 29 000 美金。
- 总收益：79 000 美金。

[①] iPhone 应用程序发布至 Apple 公司的 App Store，必须申请成为 iPhone developer，年费 99 美元。
[②] http://www.mobilewebgo.com/android-revenues-80000-total-and-monthly-revenues-advanced-task-manager。

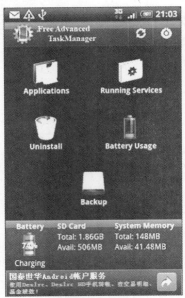

图 1-6

虽然将应用程序发布到 Google Play 不一定就会像 Arron La 一样获利,但随着 Android 移动设备的市场占有率越来越大,获利的可能性就越来越高。除此之外,想要获利还可以采取另外一个方法,那就是应用程序不收取任何费用,但是加上 AdMob 的广告,只要用户点击该广告,照样可以获利,上述实例中的免费版应用程序也有获利,来源就是广告收益。开发者可以在应用程序中插入 AdMob 广告,AdMob 会按照广告被点击的情况,分享广告收益给开发者。广告的制作十分简单,将在第 13 章详加说明。

Android 新版功能介绍

Android 4.0 版在 2011 年 10 月 19 日发布,主要整合了手机(Android 2.x)与平板电脑(Android 3.x)API 的版本,所以无论手机或平板电脑都可以使用相同的 UI 组件与功能。其他新增功能说明如下。

1. System Bar

无论目前运行什么应用程序,都会显示 System Bar,如图 1-7 的最下面部分,虚拟按键由左至右按照顺序为 Back、Home、Recent Apps,方便用户退回上一页,直接回到主画面,或显示最近使用过的应用程序。

2. New Lock Screen Actions

用户不需要解除屏幕锁就可以直接拍照或将状态栏向下拨动以查看所有通知信息,如图 1-8 所示。用户也可以在未解锁状态直接控制音乐播放。

Android 导论与新版功能介绍 01

图 1-7

图 1-8

3. Control over network data

Android 4.0 版开始内置网络流量监控功能，用户可以对网络流量进行上限设置，超过上限时，移动设备会自动关闭上网功能。用户也可以随时查看已使用的网络流量，系统也会以报表形式呈现各个应用程序访问网络的情况，如图 1-9 所示。

4. Face Unlock

要使用脸部解锁功能，用户必须先拍摄自己的脸，之后进行屏幕解锁时，移动设备会先自动打开视频相机，以进行脸部识别，比较成功即可解锁，如图 1-10 所示。即使比较不成功，仍可通过输入密码或图形解锁方式进行解锁。

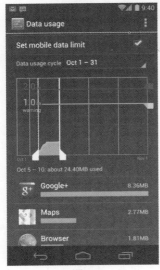

图 1-9

图 1-10

5. Android Beam for NFC-Based Sharing

Android Beam 是 NFC(Near Field Communication)技术的应用。两个具备 NFC 功能的移动设备一旦靠近，就可以分享 YouTube 视频、联系人、地图、网页、应用程序等信息，不需要另外设置，也不必配对，如图 1-11 所示。

图 1-11

6. Wi-Fi Direct[①]

Android 4.0 版开始支持 Wi-Fi Direct 功能，用户的移动设备不需要通过 AP 或 Internet，就可以直接与其他同样支持 Wi-Fi Direct 的设备进行点对点连接而相互分享数据，如图 1-12 所示。

图 1-12

---

① WiFi Alliance 录制一段简明易懂的动画说明 WiFi direct 的概念，参见 http://www.youtube.com/watch?v=je2lWjfpywQ。

第 2 章

开发工具下载与安装

本章学习目标：

- 开发 Android 应用程序所需的工具
- JDK 下载与安装
- Eclipse 下载与安装
- ADT 与 Android SDK 安装
- Eclipse 编码设置成 UTF-8

2.1 开发 Android 应用程序所需的工具[①]

"工欲善其事，必先利其器"，想要快速、无碍地开发 Android 应用程序(Android Application，简称 Android App)，必须选择适当的开发工具。到底开发 Android 应用程序需要哪些开发工具？Android 开发官方网站建议最好安装下列相关工具[②]：

- JDK(Java Development Kit)——开发 Android 应用程序时需要使用 JDK 的工具，例如开发 Google Map 相关应用程序需要使用 JDK 的 keytool。
- Eclipse——集成开发环境(Integrated Development Environment，IDE)，包含开发一般 Java 应用程序所需的工具，如程序内容编辑器、调试工具等。
- ADT(Android Development Tools)Plugin for Eclipse——ADT 是 Google 专门为了利用 Eclipse 开发 Android 应用程序而设计的插件。
- Android SDK(Software Development Kit)——Android SDK 会将 Java 的 class 文件编译成 dex 文件后交由 Dalvik VM 执行；除此之外 Android SDK 还包含开发 Android 应用程序时所需用到的 Android 模拟器(emulator)，与专属调试工具 LogCat。

2.2 JDK 下载与安装

2.2.1 JDK 下载

建议至少要下载 JDK 5 以后的版本，在此下载最新版本——JDK 7。可至 Java 官方网站 http://java.oracle.com/下载[③]。请按照下列步骤下载：

STEP 1 单击 Software Downloads 的 Java SE 连接，如图 2-1 所示。

STEP 2 按下 JDK 的 Download 按钮，如图 2-2 所示。

STEP 3 选择 Accept License Agreement，并单击适合平台[④]的 JDK 下载连接，如图 2-3 所示。

[①] 大部分开发者都熟悉微软 Windows 操作系统，所以之后开发工具的下载与安装步骤都以微软 Windows 操作系统当作示范环境。若欲下载其他平台的开发工具，请自行参见各个开发工具官方网站上的说明。
[②] http://developer.android.com/sdk/index.html。
[③] 会自动跳转到 http://www.oracle.com/technetwork/java/index.html。
[④] Windows x86 适用于 32 位版本，例如 Windows XP。Windows x64 适用于 64 位版本，例如 Windows 7。

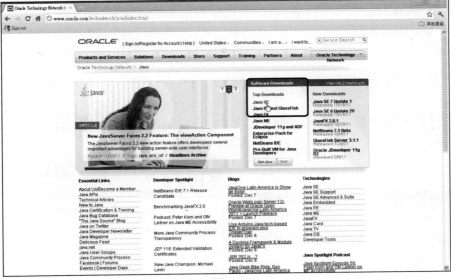

图 2-1

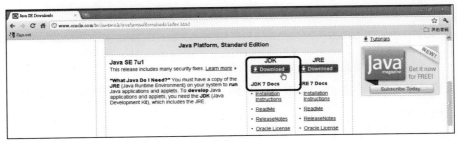

图 2-2

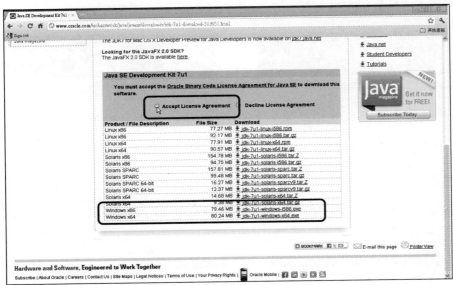

图 2-3

2.2.2　JDK 安装

请按照下列步骤安装 JDK：

STEP 1　双击 JDK 安装文件即可启动安装程序，如图 2-4 所示。

STEP 2　进入 JDK 安装程序欢迎画面，如图 2-5 所示，按下 Next 按钮继续。

图 2-4

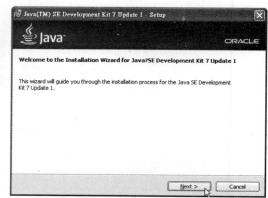

图 2-5

STEP 3　选择需要安装的功能，如图 2-6 所示。在此使用默认，不需要更改，直接按下 Next 按钮继续。

STEP 4　设置 JRE 安装路径，如图 2-7 所示。如不更改，直接按下 Next 按钮继续；如欲更改，请按下 Change 按钮。

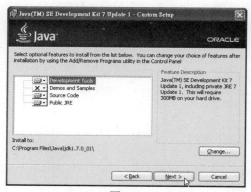

图 2-6

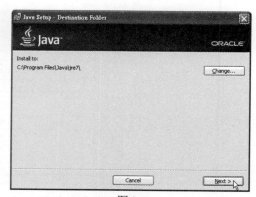

图 2-7

STEP 5　看到图 2-8 画面代表安装完成，按下 Finish 按钮结束。

开发工具下载与安装 02

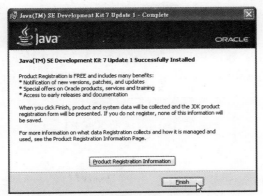

图 2-8

2.3 Eclipse 下载与安装

2.3.1 Eclipse 下载

建议至少下载 Eclipse 3.4 以后的版本，在此我们下载 Eclipse 3.7.1 版[①]。请至 Eclipse 官方网站 http://www.eclipse.org/下载。下载步骤说明如下：

 进入 Eclipse 网站首页后单击 Downloads 连接，如图 2-9 所示。

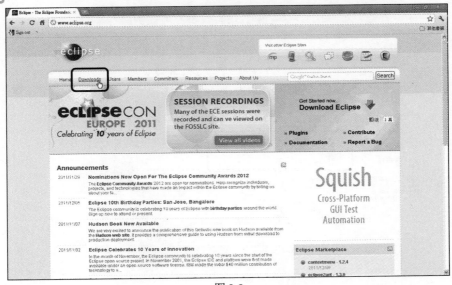

图 2-9

---

① 不建议下载 Eclipse 3.6.0 版(2010 年 6 月发布)，因为该版在处理自定义的 xml 文件时会出现 NullPointerException 异常事件，而且确定会在开发 Android 应用程序时发生(尤其在编辑 strings.xml 文件时)。有兴趣请参见 Eclipse Bugs 网站，Bug ID 为 318108 — "https://bugs.eclipse.org/bugs/show_bug.cgi?id=318108"。

STEP 2　Android 官方网站建议安装 Eclipse Classic 版本[①]，不过笔者测试过另外两个版本 Eclipse IDE for Java Developers、Eclipse IDE for Java EE Developers 也都可以正常运行，参见图 2-10。建议 Eclipse 版本的 bit 最好与 JDK 相同，例如下载的 JDK 是 64 bit 版本，Eclipse 也请下载 64 bit 版本，否则可能无法启动 Eclipse。

图 2-10

STEP 3　单击下载连接开始下载，如图 2-11 所示。

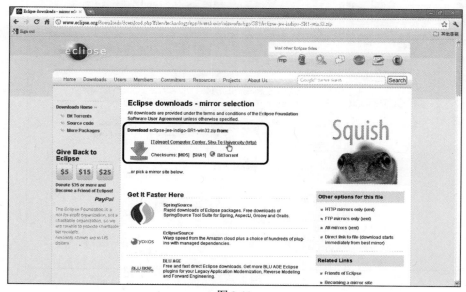

图 2-11

---

① 请参见 http://developer.android.com/sdk/eclipse-adt.html 的 Preparing Your Development Computer 部分。

2.3.2 Eclipse 安装

Eclipse 没有 Install.exe、Setup.exe 等安装文件，只要直接将下载的 Eclipse zip 压缩文件解压缩即可。解压缩后会出现 eclipse 目录，其中的 eclipse.exe 即为启动的运行文件，如图 2-12 所示。

图 2-12

Eclipse 启动后会弹出 Workspace Launcher 窗口，如图 2-13 所示。该窗口主要说明之后开发的项目将会放在指定的路径下，如果不更改默认路径，可直接按 OK 按钮继续；如欲更改请按右边 Browse 按钮并指定路径。如果勾选 Use this as the default and do not ask again，以后启动 Eclipse 开发工具就不会再弹出此窗口。

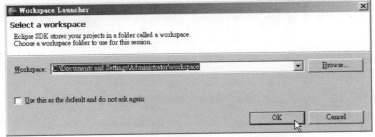

图 2-13

以后如果想要改变 workspace 路径,可以在 Eclipse 开发环境内单击 File | Switch Workspace | Other 来设置新的路径，如图 2-14 所示。

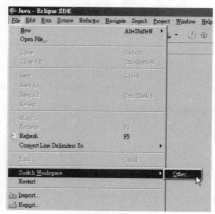

图 2-14

2.4 ADT 与 Android SDK 安装

如果开发者是第一次安装 Android 相关开发工具，请参见"2.4.1 第一次安装"，如果已经安装过较旧版本 Android 开发工具，只想要更新，请参见"2.4.2 旧版升级成新版"。

2.4.1 第一次安装

ADT 是专为 Eclipse 设计的插件。安装 ADT 的方式有两种：在线安装与离线安装。在线安装是通过网络连接服务器来安装 ADT，不需要预先下载 ADT 文件；离线安装则是利用已下载的 ADT 文件来安装，可至 Android SDK 官方网站 http://developer.android.com/sdk/installing/installing-adt.html 下载，如图 2-15 所示。官方网站建议除非无法在线安装再使用离线安装。

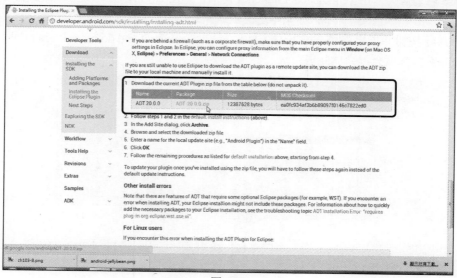

图 2-15

开发工具下载与安装　02

要安装 ADT 之前，必须先启动 Eclipse；接下来请按照下列步骤安装：

STEP 1　在 Eclipse 开发环境内单击功能菜单 Help | Install New Software，如图 2-16 所示。

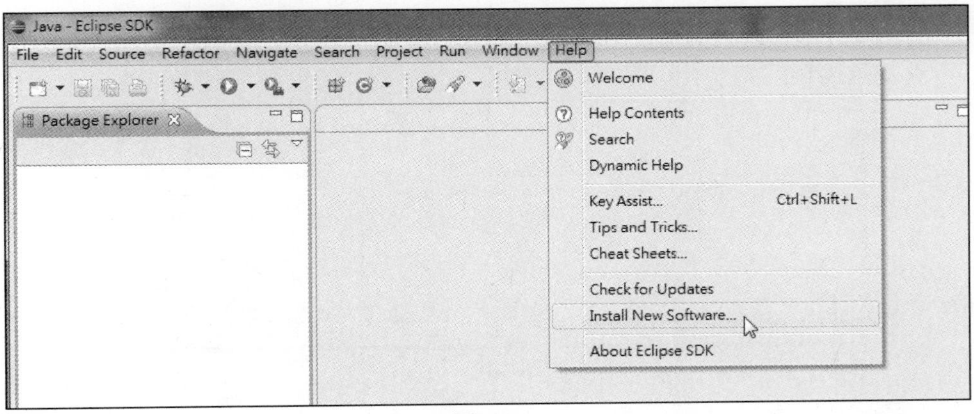

图 2-16

STEP 2　按下 Add 按钮以新建 ADT 插件，如图 2-17 所示。

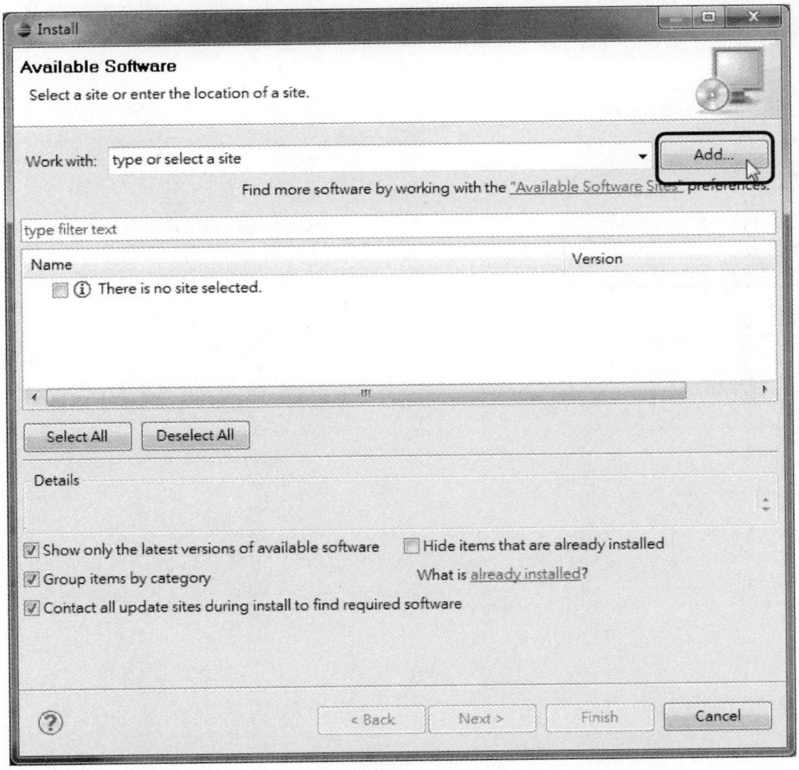

图 2-17

STEP 3　Name 字段输入插件名称(例如：ADT)；Location 字段，如果是在线安装必须输入 https://dl-ssl.google.com/android/eclipse/[①]网址，如图 2-18 所示。如果是离线安装，请按右边 Archive 按钮并指定已下载的 ADT 压缩文件路径(ADT 下载参见图 2-15)，之后按下 OK 按钮。

图 2-18

STEP 4　勾选 Developer Tools(下面所有项目会自动勾选)后按下 Next 按钮，如图 2-19 所示，此时计算机需要花几分钟检查安装项目。

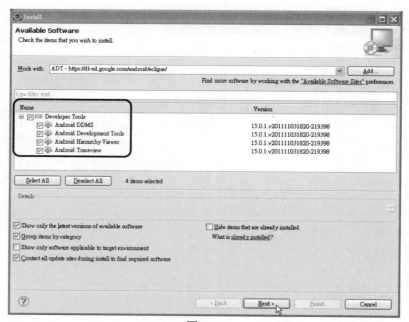

图 2-19

STEP 5　图 2-20 的窗口只是再次确认欲安装的项目，确认后按下 Next 按钮。

---

① 如果输入该网址无法取得 ADT，请将 https 改成 http，然后再试一次。

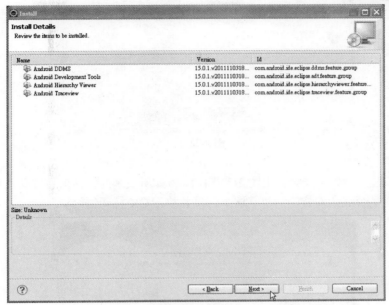

图 2-20

STEP 6　可以单击各个授权条款以浏览其内容，确认后请按 Finish 按钮，如图 2-21 所示。

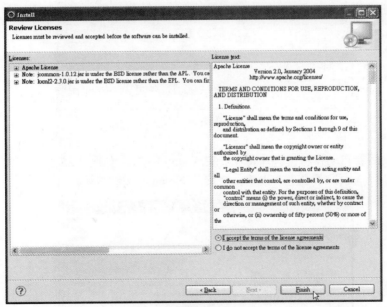

图 2-21

STEP 7　图 2-22 窗口会显示现在的安装进度，Run in Background 按钮可以让安装程序在后台运行。

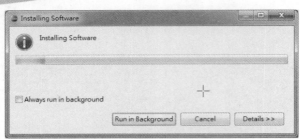

图 2-22

STEP 8 在安装过程中可能会弹出图 2-23 窗口，告知所安装的软件未通过操作系统验证，并询问是否继续安装，请忽略这个信息并按下 OK 按钮继续。

图 2-23

STEP 9 弹出图 2-24 的窗口代表安装完成，它会建议重新启动 Eclipse，请按下 Restart Now 按钮重新启动 Eclipse。

图 2-24

STEP 10 重启后会自动弹出窗口并要求安装 Android SDK，如图 2-25 所示，默认会勾选安装最新版选项，不必更改设置，直接按下 Finish 按钮继续。

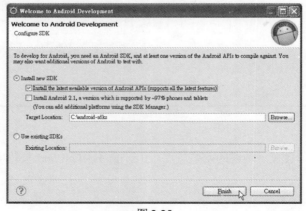

图 2-25

开发工具下载与安装   02

STEP 11 图 2-26 显示包的详细信息与授权条款，单击右下方 Accept All 选项以确定安装全部包，之后按下 Install 按钮开始安装。

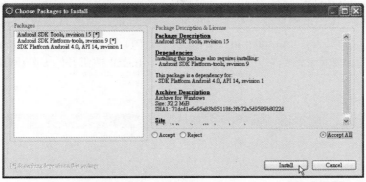

图 2-26

STEP 12 安装完毕会要求重启 ADB(Android Debug Bridge)，如图 2-27 所示，请按下 Yes 按钮重新启动 ADB。

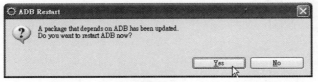

图 2-27

STEP 13 安装完毕后最好检查一下 Android SDK 的路径是否正确。单击 Eclipse 的主菜单 Window｜Preferences 会弹出首选设置窗口，如图 2-28 所示，单击左边导航栏的 Android 项目，并检查右边窗格 SDK Location 字段是否有路径，而且是否正确指向 Android SDK 目录。

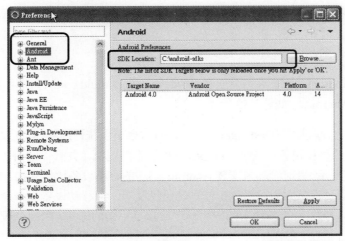

图 2-28

27

STEP 14　安装并设置好 Android SDK 后,即可按下 Eclipse 工具栏上的 " " 按钮或是单击菜单 Window│Android SDK Manager 以启动 Android SDK 管理界面,如图 2-29 所示。Android SDK 管理界面的主要功能是让开发者可以新建/删除 Android SDK 各个版本相关包,虽然之前已经安装好 Android SDK,但是 Documentation for Android SDK(说明文件)、Samples for SDK(范例程序)、Google APIs by Google Inc.(Google 专属的函数库)①都尚未安装,请勾选后按下 Install 按钮开始安装(网络必须保持连接状态)。

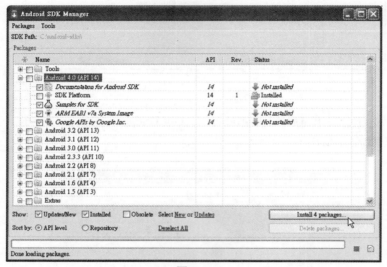

图 2-29

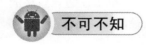

Android SDK 有许多版本(Android 4.0, 3.2, 3.0, 2.3.3, ……),到底要开发哪一个版本的应用程序比较好?如果有商业考虑,就必须知道哪一个版本的移动设备市场占有率比较高。Android 官方网站提供统计数据,进入该网页的步骤如下:

STEP 1　进入 Android 官方网站首页(http://developer.android.com),向下卷动单击左下方 About Android,如图 2-30 所示。

STEP 2　单击左边导航栏的 Dashboards 项目,右边会呈现统计最新 14 天内在线至 Android Market 的移动设备所安装 Android 各种版本的比例,并且以圆饼图形方式呈现,如图 2-31 所示。目前看起来安装 Android 2.3.3 的移动设备比例最高,尚未看到 Android 4.0 移动设备是因为虽然该操作系统已经发布,但移动设备尚未销售。若从商业角度考虑,开发 Android 2.1 的应用程序应该是不错的选择,因为一般而言 Android 2.1

---

① 例如 Google Map API,专门提供开发者开发 Google Map 应用程序的所需功能。

的应用程序可以在 Android 2.1 之后的版本执行,但不选择 Android 2.1 之前版本是因为太旧,功能支持度较差。

图 2-30

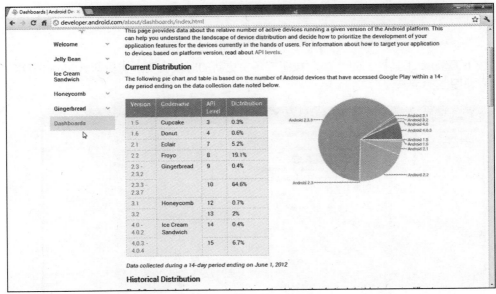

图 2-31

2.4.2 旧版升级成新版

如果开发者使用较旧版本的 Android SDK 工具，必须先更新 ADT 至最新版，如图 2-32 所示。

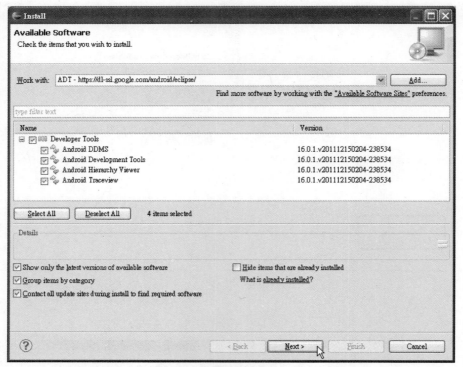

图 2-32

当按下 Eclipse 工具栏上的""按钮准备要更新的时候会要求先更新 Android SDK 的相关工具，如图 2-33 所示。

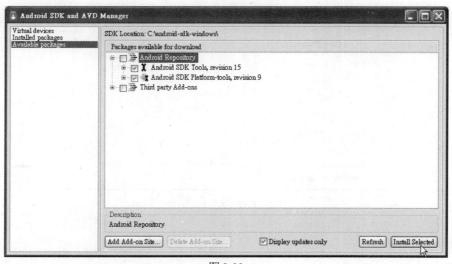

图 2-33

按照指示更新完成后才会弹出如图 2-29 的窗口让开发者安装新版的 Android SDK。

2.5 Eclipse 编码设置成 UTF-8

Windows 版本的 Eclipse 默认编码为 MS950(其就是 Big5)，但是 MS950 并非每个操作系统或开发工具都支持，所以建议改成支持度最好的 UTF-8 编码格式，这样一来编写完的文件即使在其他操作系统或是使用其他开发工具打开时才不会出现乱码情况，即使会出现，也可以很快地排除。为了达到最佳兼容性，本书所有范例都是 UTF-8 编码格式，如果读者未将 Eclipse 编码格式同样设置成 UTF-8，导入或打开本书范例时会出现乱码情况，如图 2-34 所示的程序代码被框起来的部分，这样很可能会导致编译失败。

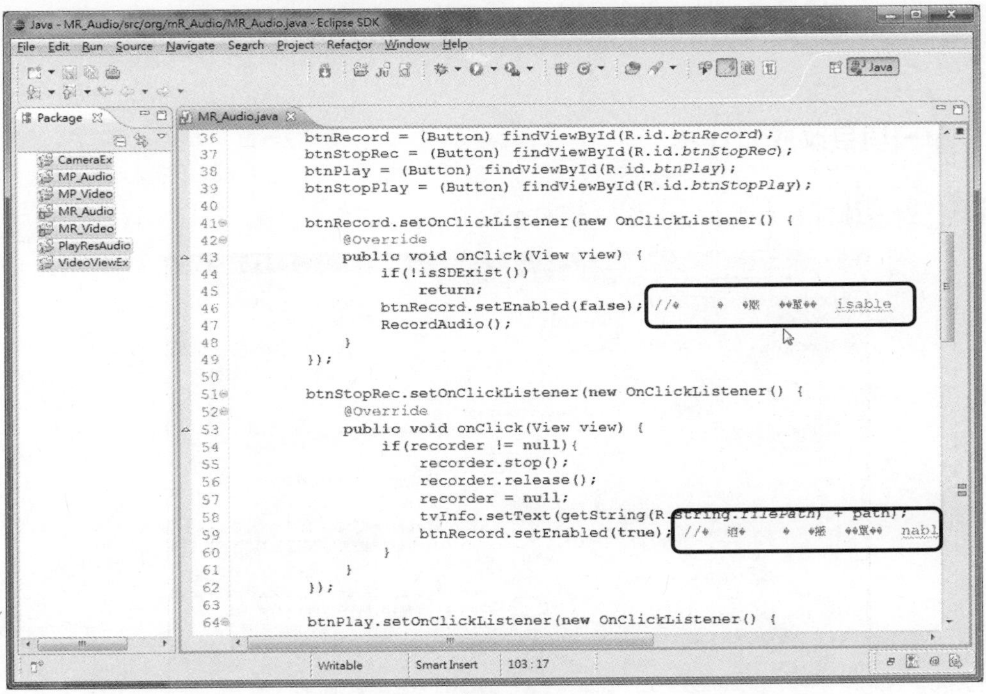

图 2-34

建议读者将整个 workspace 编码方式改成 UTF-8，当然也可以只将出现乱码的项目改变编码方式，分别说明如下。

2.5.1 整个 workspace 都改成 UTF-8

单击 Eclipse 功能菜单的 Window｜Preferences 会弹出如图 2-35 所示的 Preferences 窗口，在左边窗格单击 General｜Workspace，并在对应的右边窗格单击 Other 并选择 UTF-8 之后按下 OK 按钮。

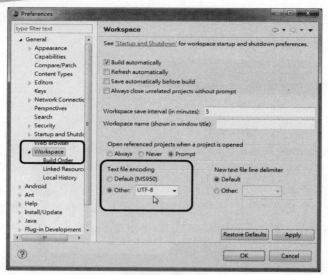

图 2-35

2.5.2 单一项目改成 UTF-8

STEP 1 对着项目右击选 Properties，如图 2-36 所示。

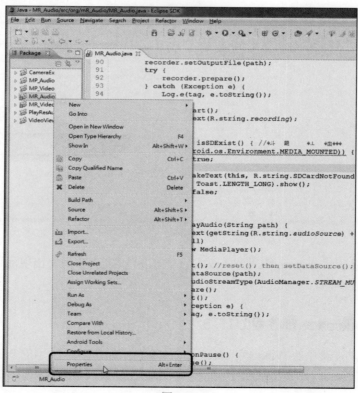

图 2-36

开发工具下载与安装

STEP 2 单击图 2-37 的左边窗格 Resource，在对应的右边窗格单击 Other 并选择 UTF-8 之后按下 OK 按钮。

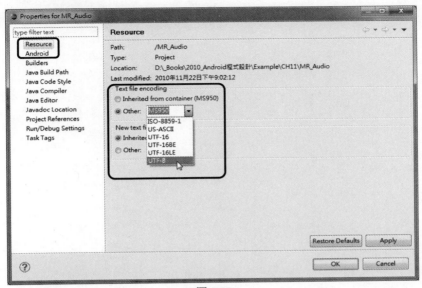

图 2-37

第3章

Android项目与系统架构

本章学习目标：

- 设置 Android 模拟器
- 创建与管理 Android 项目
- DDMS 的使用
- Android 系统架构介绍
- 项目目录架构与 Manifest 文件介绍
- 应用程序本地化

3.1 设置 Android 模拟器

在编写 Android 应用程序过程中，需要不断测试程序结果，为了方便开发人员测试程序，Android SDK 提供 Android 模拟器(Android Virtual Device，简称 AVD；或称 emulator)；所以在编写第一个 Android 应用程序前，应该先创建 Android 模拟器。创建 Android 模拟器的步骤如下。

STEP 1 在 Eclipse 开发环境内，按工具栏上的"📱"按钮，或单击菜单 Window | AVD Manager 打开模拟器管理界面，如图 3-1 所示，按下右边 New 按钮创建模拟器。

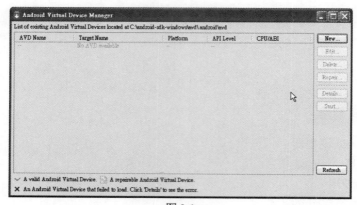

图 3-1

STEP 2 图 3-2 是创建 Android 模拟器的对话窗口，各字段说明如下，输入完毕后按下 Create AVD 按钮继续。

图 3-2

(1) Name：Android 模拟器的名称，必须由英文字母或数字组成，不可以有空格符或特殊符号。

(2) Target：运行平台的版本。

(3) SD Card。
- Size：SD 记忆卡容量大小。
- File：SD 记忆卡图像文件的路径。

(4) Snapshot：勾选后每次关闭模拟器都会自动存储该模拟器的快照集，用来缩短打开模拟器所需的时间。

(5) Skin。
- Built-in：内置模拟器屏幕尺寸[1]。

    QVGA(240x320, low density, small screen)
    WQVGA400(240x400, low density, normal screen)
    WQVGA432(240x432, low density, normal screen)
    HVGA(320x480, medium density, normal screen)
    WVGA800(480x800, high density, normal screen)
    WVGA854(480x854 high density, normal screen)
    WXGA720(1280x720, extra-high density, normal screen)
    WSVGA(1024x600, medium density, large screen)
    WXGA(1280x800, medium density, xlarge screen)

- Resolution：自定义模拟器分辨率，也就是自定义模拟器屏幕大小。

(6) Hardware：按右边 New 按钮可以修改模拟器的硬件配置[2]。

STEP 3　创建完毕后模拟器会在 AVD Manager 窗口中出现，如图 3-3 所示。

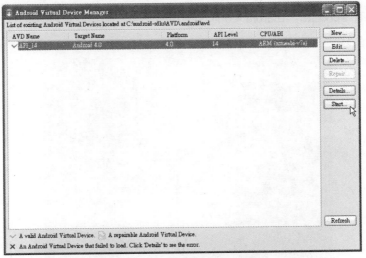

图 3-3

---

[1] 请参见 http://developer.android.com/tools/revisions/platforms.html。
[2] 详细设置说明请参见 http://developer.android.com/guide/developing/devices/managing-avds.html#hardwareopts。

按下 Start 按钮会弹出如图 3-4 的窗口，让开发者可以在打开模拟器前设置，说明如下。

- Scale display to real size：设置模拟器屏幕尺寸，Screen Size 字段的值表示移动设备的真实尺寸，例如，6 表示屏幕尺寸为 6 英寸。
- Wipe user data：清除用户在模拟器上的数据。
- Launch from/ Save to snapshot：当初创建模拟器时有勾选 Snapshot 即可使用此功能达到缩短打开模拟器所需的时间。

按下 Launch 按钮即可启动该模拟器，启动 Android 模拟器需要花一些时间，运行成功后会呈现如图 3-5 所示的结果。因为启动模拟器要花不短的时间，所以确定不需要开发应用程序时再关闭模拟器。

图 3-4

图 3-5

如果计算机有 webcam，Android 4.0 以上的模拟器会在启动时试图连接它，并用以拍照，如图 3-6 所示。

图 3-6

 不可不知

Android 模拟器创建完成后会自动产生目录(以模拟器名称为目录名称)存放模拟器的配置文件与其他相关的图像文件(例如 SD 记忆卡的图像文件)。在不同的操作系统，模拟器目录的所在路径也会不同，说明如下[①]。
- Windows Vista / Windows 7：C:\Users\<user>\.android\avd。
- Windows XP：模拟器目录在 C:\Documents and Settings\<user>\.android\avd。
- Linux/ Mac：模拟器目录在~/.android/avd/。

一般 Windows 操作系统的用户可能会在安装操作系统时将用户名称设成中文名称，而模拟器目录的默认路径又放在用户名称的目录内，这样一来模拟器目录的路径会有中文字符而导致无法打开，如图 3-7 所示。

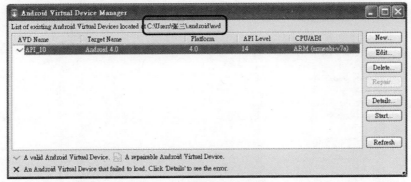

图 3-7

因为 AVD Manager 窗口无法直接更改模拟器目录的路径，必须采取下列步骤方能更改：

STEP 1　创建环境变量[②]ANDROID_SDK_HOME，值为想要指定的路径，如图 3-8 所示，切记路径不可有中文字符。

图 3-8

STEP 2　环境变量创建完毕后，要重新启动 Eclipse(请单击 File｜Exit 结束后再启动；而不是 File｜Restart)。再打开 AVD Manager 窗口，就会发现 AVD 的路径已经更改。

---

① 请参阅 http://developer.android.com/tools/devices/emulator.html#diskimages。
② Windows XP："开始"｜"设置"｜"控制面板"｜"系统"｜"高级"｜"环境变量"｜"系统变量"。
　Windows 7："开始"｜"控制面板"｜"系统"｜"高级系统设置"｜"环境变量"｜"系统变量"。

3.2 创建与管理 Android 项目

3.2.1 创建 Android 项目

创建 Android 项目的步骤如下。

STEP 1 在 Eclipse 开发环境内，按下工具栏上 " " 按钮或单击菜单 File │ New │ Android Project 来创建 Android 项目。

STEP 2 填写项目名称等相关字段，如图 3-9 所示。

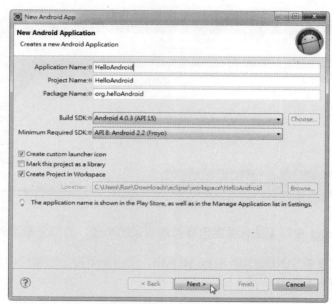

图 3-9

- Application Name：应用程序名称。
- Project Name：项目名称，此名称会成为项目目录的名称。
- Package Name：Java 包名称，必须由两个部分组成(例如：org.helloAndroid)。
- Build SDK：此项目所使用的 API 层级(API level)。
- Minimum Required SDK：要运行此项目，移动设备至少要具备的 API 层级[①]。
- Create custom launcher icon：自定义应用程序图标。如有勾选，下一步就会弹出自定义图标窗口。
- Mark this project as a library：将此项目设成函数库，不需要勾选。
- Create Project in Workspace：将创建的新项目放在默认的 workspace 目录。

---

① API 层级与 Android 平台版本号的对照与说明请自行参见第 1 章。

STEP 3 自定义应用程序图标，如图 3-10 所示。

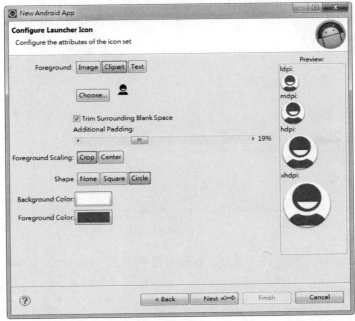

图 3-10

(1) Foreground。
- Image：选择图形文件当作前景图。
- Clipart：可以按 Choose 按钮选择适合的图标。
- Text：输入文字。
- Trim Surrounding Blank Space：勾选则去除空白区域，让选定的图案贴齐边线。
- Additional Padding：设置图案与边线间的填充大小。

(2) Foreground Scaling。
- Crop：修剪。
- Center：居中对齐。

(3) Shape。
- None：只有图案，没有外框。
- Square：方形外框。
- Circle：圆形外框。

(4) Background Color：设置背景色。
(5) Foreground Color：设置前景色。

STEP 4 创建 Activity，如图 3-11 所示。

- BlankActivity：空白 Activity。
- MasterDetailFlow：主从式 Activity。Android API 等级至少要 11 才能使用。

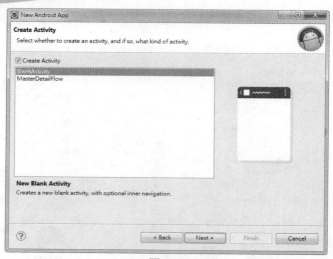

图 3-11

STEP 5　填写 Activity 名称等相关字段，如图 3-12 所示。

- Activity Name：Activity 名称。
- Layout Name：Activity 所使用的 layout 名称。
- Navigation Type：Activity 浏览方式。
- Hierarchical Parent：设置按下 UP 键(就是 Activity 标题的<键)会到的 Activity。
- Title：Activity 标题。

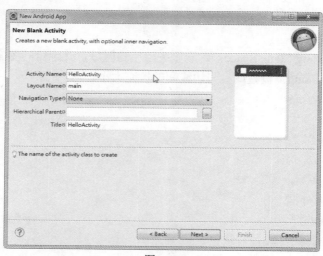

图 3-12

STEP 6　安装 Android Support library，如图 3-13 所示。因为没有安装 Android Support library，所以会弹出要求安装的画面，可以按下 Install/Upgrade 按钮开始安装。安装完毕后即可按下 Finish 按钮完成 Android 项目创建程序。

Android 项目与系统架构

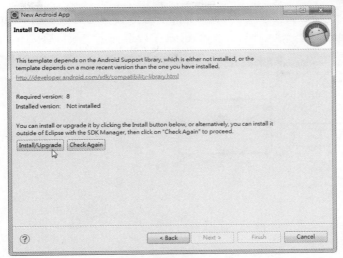

图 3-13

3.2.2 运行 Android 项目

1. 在模拟器上运行项目

在 Android 模拟器上运行项目的步骤如下：

STEP 1 第一次运行项目时，对着项目(例如：HelloAndroid)右击选择 Run As | Run Configurations 运行设置。之后若再次运行同样项目，直接对着项目右击选择 Run As | Android Application 即可运行，无须再运行设置，除非想要更改运行设置。如图 3-14 所示。

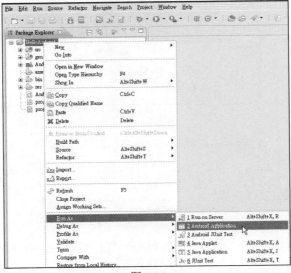

图 3-14

43

STEP 2 在图 3-15 左边窗格双击 Android Application 会自动产生 New_configuration，在右边窗格 Name 字段输入项目名称，Project 字段选择指定项目后按下右下角 Apply 按钮。

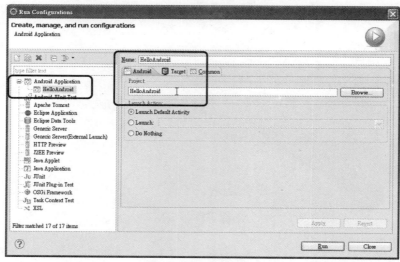

图 3-15

右边窗格单击标签 Target，如图 3-16 所示，最常用到下列两个选项。
- Always prompt to pick device：运行项目时会询问欲启动的 Android 装置。
- Automatically pick compatible device：在其正下方勾选欲启动的 Android 模拟器，以后会直接以选定的模拟器运行项目，不再询问。

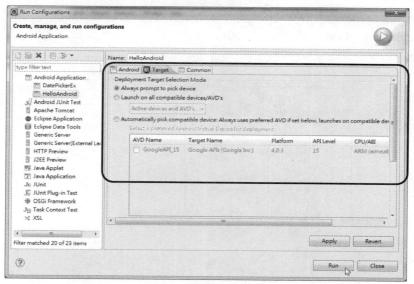

图 3-16

STEP 3 在前一步骤时选择了 Manual，所以会弹出窗口要求选择欲启动的 Android 模拟器，请依照项目设置的 Android 版本选择适当的模拟器并按 OK 按钮，如图 3-17 所示。

Android 项目与系统架构

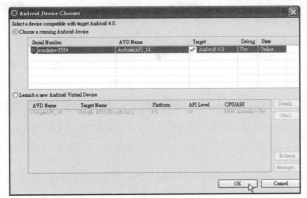

图 3-17

表 3-1 列出模拟器与计算机键盘的对应表。

表 3-1[①]

模拟器键盘	计算机键盘
Home	HOME
Menu (left softkey)	F2 或 Page-up
Star (right softkey)	Shift-F2 或 Page Down
Back	ESC
Call/dial button	F3
Hangup/end call button	F4
Search	F5
Power button	F7
Audio volume up button	KEYPAD_PLUS, Ctrl-5
Audio volume down button	KEYPAD_MINUS, Ctrl-F6
Camera button	Ctrl-KEYPAD_5, Ctrl-F3
Switch to previous layout orientation (例如，portrait, landscape)	KEYPAD_7, Ctrl-F11
Switch to next layout orientation (例如，portrait, landscape)	KEYPAD_9, Ctrl-F12
Toggle cell networking on/off	F8
Toggle code profiling	F9 (only with -trace startup option)
Toggle fullscreen mode	Alt-Enter
Toggle trackball mode	F6
Enter trackball mode temporarily (while key is pressed)	Delete
DPad left/up/right/down	KEYPAD_4/8/6/2
DPad center click	KEYPAD_5
Onion alpha increase/decrease	KEYPAD_MULTIPLY(*)/ KEYPAD_DIVIDE(/)

① 参见 http://developer.android.com/tools/help/emulator.html。

2. 在移动设备上运行项目

在移动设备(实机)上运行项目的步骤如下：

STEP 1 移动设备以 USB 线连接到计算机。

STEP 2 移动设备打开 USB Debugging(USB 调试)：按移动设备上的 Home 按键(⌂)回到主画面，再按 Menu(MENU)按键，选择 System Settings | Developer options | 勾选 USB debugging。

STEP 3 使用 Eclipse 运行项目时，在 Android 设备列表中会显示移动设备；选择该移动设备后按 OK 按钮即可将项目安装至移动设备上并运行[①]。

3.2.3 移除 Android 项目

1. 移除 Eclipse 的 Android 项目

对着项目右击选择 Delete 会弹出确认窗口，如图 3-18 所示，不勾选 Delete project contents on disk(cannot be undone)表示只会将该项目自 Eclipse 项目列表上移除，但不会删除所属文件；勾选则表示不仅会将该项目自 Eclipse 移除，也会直接将所属文件全部删除，而且删除的文件无法还原。

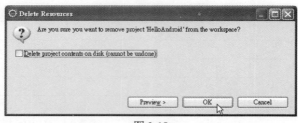

图 3-18

2. 停止或移除模拟器/移动设备上的应用程序

可以直接在模拟器/移动设备上停止或移除应用程序，在主画面(按设备上的 Home 按钮可以回到主画面)按设备上的 Menu 按键并单击 Manage apps(管理应用程序)会列出所有安装的应用程序，单击欲移除的应用程序后会弹出如图 3-19 的画面。

- Force stop(强制停止)按钮：强制停止应用程序。
- Uninstall(卸载)按钮：移除应用程序。
- Clear data(清除数据)按钮：将应用程序存储的用户数据清除。
- Move to SD card(移至 SD 卡)按钮：将应用程序移至 SD 卡上。

[①] 运行应用程序会历经下列步骤：(1) 将 Android 项目以 ZIP 格式压缩成 apk 文件；(2) 将 apk 文件安装在模拟器或移动设备上；(3)运行 apk 文件。apk 文件一般被称做 Android App，其实就是 Android 项目编译过的压缩文件(只有 dex 文件而没有 java 或 class 文件)。

Android 项目与系统架构　03

图 3-19

3.2.4　导入 Android 项目

如果已经有 Android 项目(例如本书的范例程序)，可以使用 Eclipse 的导入功能。

导入 Android 项目的步骤如下：

STEP 1　单击菜单 File | Import，如图 3-20 所示。

STEP 2　单击 General | Existing Projects into Workspace，如图 3-21 所示。

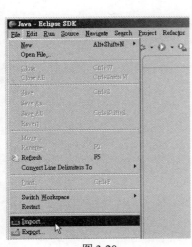

图 3-20

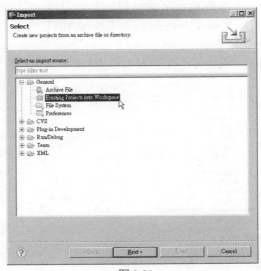

图 3-21

STEP 3　按 Browse 按钮以指定 Android 项目路径，然后按 Finish 按钮，如图 3-22 所示。

47

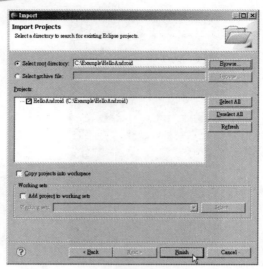

图 3-22

STEP 4 导入项目成功后，即会在 Eclipse 的 Package Explorer 窗格内显示该项目内容。

3.2.5 打开范例程序

如果安装 Android SDK 时又安装了范例程序(Samples for SDK，参见第 2 章图 2-29)就可以打开范例程序作为参考，其中又以 ApiDemos 范例程序对初学者最具参考价值。打开范例程序的步骤如下：

STEP 1 单击 Eclipse 菜单 File | New | Other 会弹出如图 3-23 的窗口。单击 Android | Android Sample Project 后按下 Next 按钮。

STEP 2 勾选范例程序的 Android 平台版本后按下 Next 按钮，如图 3-24 所示。

图 3-23

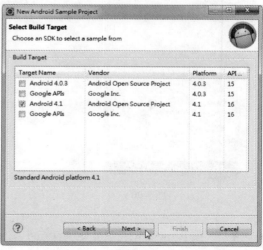

图 3-24

Android 项目与系统架构　03

STEP 3　选择所需的范例程序(例如：ApiDemos)[①]后按下 Finish 按钮，如图 3-25 所示。

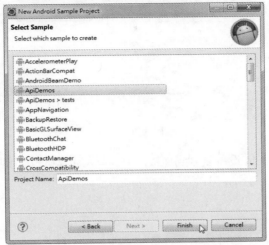

图 3-25

3.3 DDMS 的使用

DDMS(Dalvik Debug Monitor Service)内有许多重要的功能可以协助开发者，单击 Eclipse 菜单 Window | Open Perspective | DDMS 即可打开。DDMS 提供以下重要功能。

1. 文件管理

文件管理功能在模拟器、移动设备皆适用。如欲浏览 Android 设备的文件或目录，可单击 File Explorer 标签。如果要将文件复制到 Android 设备上[②]，可以使用 push 功能，步骤如下(请参见图 3-26)：

STEP 1　单击模拟器(例如：emulator-5554)。

STEP 2　单击 File Explorer 即可显示该设备的文件夹与文件。

STEP 3　按下 "　" (Push a file onto the device)按钮会打开窗口让操作者指定欲复制的来源文件。

STEP 4　来源文件指定完毕后会开始复制到 Android 设备上，复制完毕可以从文件列表上看到该文件。

[①] 模拟器已经安装了 ApiDemos 范例，所以无须安装即可在模拟器上运行该范例，使用 Eclipse 打开该范例程序的目的是为了看源代码学习。

[②] 若欲复制文件到模拟器的 SD card 上，创建模拟器时必须同时设置 SD card 大小。

49

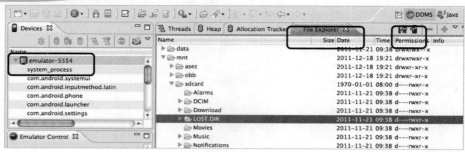

图 3-26

如果要将文件从 Android 设备复制到连接的 PC(个人计算机)上,则可使用 pull 功能,步骤如下(请参见图 3-26):

STEP 1 单击模拟器。

STEP 2 单击 File Explorer 即可显示该设备的文件夹与文件。

STEP 3 单击欲复制到 PC 的来源文件。

STEP 4 按下"🗂"(Pull a file from the device)按钮会打开窗口让操作者指定欲存放的目的路径。

STEP 5 复制完毕后,PC 上即有该文件。

2. 获取屏幕画面

可按 DDMS 工具栏上的 Screen Capture(屏幕获取)按钮即可捕获设备的屏幕画面,如图 3-27 所示。

图 3-27

3. 模拟来电、来短信

DDMS 的 Emulator Control 窗格>Telephony Actions，输入 Incoming number(来电号码)并选择 Voice 或 SMS，可以模拟来电、来短信等情况。仅模拟器适用此功能。

4. 模拟自己的位置

DDMS 的 Emulator Control 窗格>Location Controls，输入 Longitude(经度)与 Latitude(纬度)可以模拟自己的位置。仅模拟器适用此功能。

3.4 Android 系统架构介绍①

Android 是一种以 Linux 为核心的移动平台，可以安装在智能手机与平版计算机等移动设备上。Android 整个系统架构如图 3-28 所示，并说明如后。

1. Linux Kernel(Linux 核心)

Android 以 Linux 2.6 版作为整个操作系统的核心，Linux 提供 Android 主要的系统服务如：安全性管理(Security)、内存管理(Memory Management)、进程管理(Process Management)、网络栈(Network Stack)、驱动模型(Driver Model)、电源管理(Power Management)等。

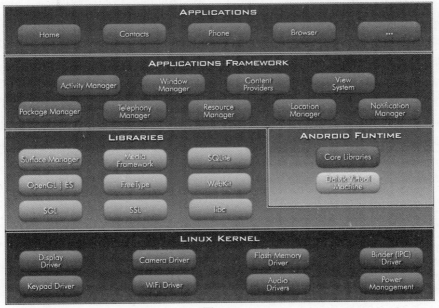

图 3-28

---

① 请参见 http://developer.android.com/guide/basics/what-is-android.html。

2. Libraries(函数库)

Android 有一个内部函数库，此函数库主要以 C/C++编写而成。Android 应用程序开发人员并非直接使用此函数库，而是通过更上层的应用程序框架(Application Framework)来使用此函数库功能，所以有人称此类函数库为原生函数库(Native Libraries)。此函数库依照功能又可细分成各种类型的函数库，以下列出比较重要的函数库。

- Media Framework(媒体函数库)：此函数库让 Android 可以播放与录制许多常见的音频与视频文件，支持的文件类型包括 MPEG4、H.264、MP3、AAC、AMR、JPG 与 PNG 等。
- Surface Manager(外观管理函数库)：管理图形界面的操作与 2D、3D 图层的显示。
- WebKit[①]：Android 内置的浏览器，其引擎就是 WebKit，与 Google 的 Chrome[②]、Apple 的 Safari[③] 浏览器引擎相同。
- SGL：专门处理 Android 的 2D 图形。
- OpenGL|ES[④]：适合嵌入式系统使用的 3D 图形函数库，此函数库实现 OpenGL ES 1.0 版以上的功能。如果 Android 手机本身有 3D 硬件加速器，程序会直接使用该硬件加速器，否则会使用软件加速功能。
- SQLite[⑤]：属轻量级但功能齐全的关系数据库引擎，方便让 Android 所有的应用程序访问数据。

3. Android Runtime(Android 运行环境)

Android Runtime 可分成 Android Core Libraries(Android 核心函数库)与 Dalvik Virtual Machine(Dalvik VM，Dalvik 虚拟机)。

- Android Core Libraries：Android 核心函数库所提供的功能，大部分与 Sun 的 Java 核心函数库相同[⑥]。
- Dalvik Virtual Machine：一般编写好的 Java 程序编译后会产生 class 文件(或称 Bytecode)，而且由 JVM(Java Virtual Machine)运行；但是 Android 不使用 JVM，而改用 Google 自行研发的 Dalvik VM，所运行的文件则是 dex 文件(Dalvik Executable)，而非 class 文件。在 Dalvik VM 运行 dex 文件之前，必须使用 Android 开发工具(Android SDK) 内的 dx 工具将 class 文件转成 dex 文件[⑦]，然后交给 Dalvik VM 运行，如图 3-29 所示。dex 文件比 class 文件更精简、运行性能更佳，而且更省电，可以说是为了移动设备量身打造的。由前述可知，开发者仍需要以 Java 程序语言编写 Android 应用程序，而最后 dx 工具会将 java 文件产生的 class 文件转成 dex 文件。

---

① http://www.webkit.org。
② http://en.wikipedia.org/wiki/Google_Chrome。
③ http://www.apple.com/safari/what-is.html。
④ http://www.khronos.org/opengles/。
⑤ http://www.sqlite.org/。
⑥ Android 核心函数库是由 Google 所开发，Java 核心函数库则属 Oracle Sun，两者并不相同，只不过 Google 刻意让 Android 核心函数库尽量支持 Java 核心函数库的功能 其目的是为了方便Java程序设计师能够快速转移至 Android 平台上开发应用程序,但 Android 核心函数库并不完全支持 Java 核心函数库，Java Swing、AWT 就不支持。
⑦ Eclipse 会自动执行此步骤(只要 Android 相关包安装好)，开发者不需另外下命令。

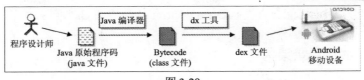

图 3-29

4. Application Framework(应用程序框架)

Application Framework 其实就是 Android 的 API(Application Programming Interface)，开发者只要善用此 API 即可快速开发出 Android 应用程序。重要部分说明如下。

- View System(用户界面)：Android 提供多样化的 UI(User Interface，用户界面)组件，例如：按钮、文本框、列表选项等。
- Activity Manager(活动管理器)：管理 Activity 的生命期,并提供浏览回溯堆栈(Navigation Backstack)，让用户可以通过按下退回键，返回上一页内容。
- Content Providers(内容提供器)：通过此项功能，可以让各个应用程序间分享彼此数据。
- Resource Manager(资源管理器)：用来访问非程序资源，例如：字符串、图形以及页面信息等。
- Notification Manager(信息管理器)：在状态栏显示指定信息，以通知或提醒用户。

5. Applications(应用程序)

用户就是通过编写好的应用程序(通常简称为"程序")与 Android 手机交互，或者可以说用户就是通过这些程序来操控 Android 设备。这些应用程序在设备上都是以一个小图标来表示，用户通过单击图标来运行程序。Android 系统一般内置有 Email、短信收发程序、浏览器、联系人等功能的应用程序。除了内置的应用程序外，开发者可以编写更多的应用程序，让用户可以使用更多便利的功能。

3.5 项目目录架构与 Manifest 文件介绍

为了实现 MVC(Model-View-Controller)架构，Android 应用程序的开发分成 UI 设计与程序设计两大部分，前者是 MVC 的 View 部分；后者包含 Controller 与 Model 部分。UI 设计特别从程序内容抽离出来，而改用 XML 文件来设置，以方便 UI 设计师专心于美学设计，不需要接触程序代码(MVC 架构是假设 UI 设计师不喜欢看程序代码)；程序设计师则专心于程序设计而不需要处理 UI 组件与页面配置等视觉设计的问题(一般而言，程序设计师都不太具有美学概念)。这种 UI 与程序分离的概念很类似网页设计，网页 UI 由 HTML 文件控制，并交由网页 UI 设计师负责，而网页流程与商业逻辑部分则由程序代码处理，交由程序设计师负责；由此可知 Android 手机 UI 的 XML 配置文件(一般放在 res/layout 目录内)类似网页 UI 的 HTML 文件。而且手机 UI 设计部分，未来可能会有类似 Dreamweaver 软件协助 UI 设计师快速完成页面配置[①]。

3.5.1 Android 项目目录架构

当使用 Eclipse 创建 Android 项目，会自动产生 3 个主要目录：src、res、gen。
- src 目录：存放着该项目运行时所需的 Java 程序文件。
- res 目录：项目所需的资源(例如：文字、图形与声音文件等)以及与 UI 设置有关的 layout 文件皆存放于 res 目录的对应子目录中。要特别注意的是在此目录内的文件名只能为小写字母、数字、_(下划线)、.(点)，以正确表达式来表示即为：[a-z0-9_.]。
- gen 目录：存放着 R.java 文件，该文件是根据 res 目录内容而自动产生；开发者就是通过此文件取得 res 目录内的相关信息。不需要修改 R.java 文件，因为 res 目录内的文件内容一有改动，就会重新产生该文件。

```
AndroidProject/              • Android 项目目录
    src/                     • src 目录内存放着 Java 原始程序文件
        HelloAndroid.java
    gen/                     • gen 目录内的 R.java 文件主要是根据 res 目录内容而自动产生
        R.java
    res/                     • res 是资源文件的根目录，提供项目所需资源
        drawable/            • drawable 目录提供图形相关资源，例如图形文件(支持 PNG、JPG、GIF 文件)[②]
            background.png
        layout/              • layout 目录专门存放与 UI 设计有关的 layout 文件(默认为 main.xml)
            main.xml
        values/              • values 目录存储 UI 所需用到的文字[③](默认为 strings.xml)
            strings.xml
```

 范例 AndroidBasic

范例(如图 3-30 所示)说明：
- 显示文字在画面上。
- 显示图形在画面上。

[①] 现在已有简易型的 UI 设计软件，名为 DroidDraw，虽然尚未成熟，但已可节省不少 UI 设计时间，请参见 http://www.droiddraw.org/。
[②] drawable 目录可以按照屏幕分辨率高低而分成 drawable-hdpi(高分辨率)、drawable-mdpi(中分辨率)、drawable-ldpi(低分辨率)3 个目录。一般建议相同图形按照分辨率不同而制作成 3 份，分别存放在前述 3 个对应目录内，当移动设备不支持高分辨率时，会自动取得较低分辨率图形，以呈现最适当的图形给用户观看。分辨率对照信息请参见本书 39 页注①。
[③] 文本文件专门存放 Android 应用程序所需用到的文字，虽然文字可以直接写在 layout 文件内，但不建议这样做，原因是翻译人员只要直接将文本文件翻译即可，而无须在 layout 文件内寻找要翻译的文字，方便应用程序本地化。

Android 项目与系统架构   03

图 3-30

AndroidBasic/res/layout/main.xml

```
1.  <?xml version="1.0" encoding="utf-8"?>
2.  <LinearLayout xmlns:android="http://schemas.android.com/apk/res/android"
3.      android:orientation="vertical"
4.      android:background="@drawable/background"
5.      android:layout_width="match_parent"
6.      android:layout_height="match_parent" >
7.      <TextView
8.          android:layout_width="match_parent"
9.          android:layout_height="wrap_content"
10.         android:textSize="20sp"
11.         android:textColor="#FFFF00"
12.         android:background="#666666"
13.         android:text="@string/str" />
14. </LinearLayout>
```

2 行：LinearLayout 是 layout 组件的一种，layout 组件会在第 4 章作详细说明。xmlns 定义此文件的名称空间，必须指定为 "http://schemas.android.com/apk/res/android"。

3 行：设置组件是垂直走向。

4 行：以项目的 res/drawable 目录内的 background 图形文件(虽然文件名为 background.jpg，但是不用标示扩展名)当作背景。@ 与 at 意思相同，代表"在"的意思。

5 行：layout_width 代表组件宽度，属于 ViewGroup.LayoutParams 类的 XML 属性，值可以为 fill_parent、match_parent、wrap_content 或尺寸(dimension)。

- fill_parent：代表填满父组件，也就是与父组件一样宽。但从 API Level 8 开始列为 deprecated(未来可能不支持)，建议改用 match_parent。
- match_parent：API Level 8 开始支持，意思与 fill_parent 同。
- wrap_content：组件的宽度只要能包含其内容(例如：文字)即可。
- 尺寸：直接以数字加上单位指定组件的宽度，例如 "12dp"。

6 行：layout_height 代表组件高度，设置方式与 layout_width 相同，不再赘述。

7 行：TextView 组件就是文本框，专门用来显示文字。

10 行：设置文字大小[①]。

11 行：设置文字的颜色(以 RGB 三原色指定)。

12 行：设置组件的背景颜色(也是以 RGB 三原色指定)。

13 行：组件上的文字存储在文本文件内(在此文本文件为 strings.xml)，str 为其标识符。

AndroidBasic/res/values/strings.xml

```
1.  <?xml version="1.0" encoding="utf-8"?>
2.  <resources>
3.      <string name="app_name">Android Basic Example</string>
4.      <string name="str">滚滚长江东逝水，浪花淘尽英雄。</string>
5.  </resources>
```

文本文件存储的是应用程序使用到的所有文字。

3 行：文字的标识符为 app_name，方便程序代码或 layout 文件访问，值为"Android Basic Example"。

AndroidBasic/gen/org/androidbasic/R.java

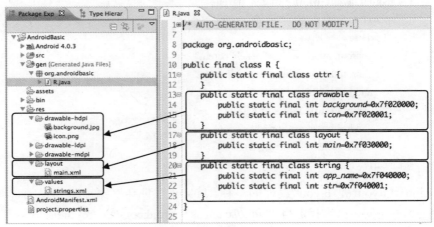

图 3-31

---

① sp(scaled-pixels)、dp (density-independent pixels)、px (pixels)、pt (points)、in (inches)、mm (millimeters)都属于尺寸单位，一般建议使用 sp 与 dp，因为该尺寸单位可以按照不同的屏幕自动调整。sp 是用在文字的尺寸，而 dp 是用在一般 UI 组件的尺寸(例如按钮的宽度)。详细信息请参见 http://developer.android.com/guide/topics/resources/more-resources.html#Dimension。

R.java 文件会帮每个资源作分类，参见图 3-31，并给予每个资源一个整数 ID，以方便开发者访问该资源。

- drawable 类：res/drawable 目录内的资源都归属于此类；background 与 icon 都是该目录内的图形文件。
- layout 类：res/layout 目录内的资源都归属于此类；main 就是 main.xml 文件。
- string 类：res/values 目录内的文本文件都归属于此类。但 string 类不是指 strings.xml 文件，而是指文本文件内 XML 标签为 string 的文字资源。

AndroidBasic/src/org/androidbasic/AndroidBasic.java

```
1.    package org.androidbasic;
2.    import android.app.Activity;
3.    import android.os.Bundle;
4.    public class AndroidBasic extends Activity {
5.        @Override
6.        public void onCreate(Bundle savedInstanceState) {
7.            super.onCreate(savedInstanceState);
8.            setContentView(R.layout.main);
9.        }
10.   }
```

4 行：创建 Android 项目时必须填写欲创建的 Activity 类名称(参见前图 3-13 的 Create Activity 字段)，在此为 AndroidBasic 就代表当初填写的 Activity 类名称即为 AndroidBasic。

6 行：Activity 一启动时就会调用 onCreate()方法。Activity 有其生命周期，必须靠改写 Activity 类对应的方法(例如 onCreate())来控制流程，详细说明请参见第 6 章。

8 行：R.layout.main 会取得项目内 res/layout/main.xml 文件的内容。SetContentView(R.layout.main)就代表将 main.xml 所设置的 UI 画面加载。

3.5.2 Manifest 文件

每一个 Android 应用程序都需要 manifest 文件(例如置于项目根目录内的 AndroidManifest.xml 文件)，此文件存储该应用程序的重要信息。在 Android 系统运行应用程序之前，会先获取该应用程序的 manifest 文件内容，若找不到该文件或该文件有错误，Android 系统将无法运行对应的应用程序。

AndroidBasic/AndroidManifest.xml

```
1.    <?xml version="1.0" encoding="utf-8"?>
2.    <manifest xmlns:android="http://schemas.android.com/apk/res/android"
3.        package="org.androidbasic"
4.        android:versionCode="1"
5.        android:versionName="1.0">
6.        <application
7.            android:icon="@drawable/icon"
```

```
 8.         android:label="@string/app_name">
 9.     <activity
10.         android:name=".AndroidBasic"
11.         android:label="@string/app_name">
12.         <intent-filter>
13.             <action android:name="android.intent.action.MAIN" />
14.             <category android:name="android.intent.category.LAUNCHER" />
15.         </intent-filter>
16.     </activity>
17.     </application>
18.     <uses-sdk android:minSdkVersion="8" />
19. </manifest>
```

3 行：Android 应用程序的完整包(Package)名称。

4 行：Android Market 管理与控制的版本号码，必须为整数值，而非字符串。用户不会看到版本号码。

5 行：对外发布的版本名称，值为字符串，与上述版本号码不同。

7 行：Android 应用程序在手机上的代表图标，在此引用到 drawable 目录内的图形文件。

8 行：Android 应用程序在手机上的代表名称，在此引用到文本文件内的字符串。

10 行：Android 应用程序中继承 Activity 类的子类名称。本来应该要使用完整合法的类名称(fully qualified class name)，但这里使用省略完整包名称的简略写法，表示完整包名称与第 3 行相同。

11 行：Activity 的名称，会在画面的标题栏上显示。

13 行：代表指定的 Activity 为主要启动点(在此为 10 行指定的 AndroidBasic)。

14 行：指定的 Activity 将被添加到 Launcher 的应用程序行程表上，表示在应用程序安装完毕后会自动启动[①]。

18 行：minSdkVersion 代表要运行该应用程序最低需要的 API 层级(API level)，而非 Android 系统平台的版本号码(Android platform version)。这个设置相当重要，因为用户的移动设备会依照这个号码来决定是否安装该应用程序[②]。

3.6 应用程序本地化

Android 移动设备早已销售遍及全球，所以在开发 Android 应用程序的时候，必须注意各

[①] 如果未加上这行，在 Eclipse 运行 Android 应用程序时，只会更新模拟器上对应的应用程序，而无法自动启动该应用程序，此时必须手动单击模拟器上对应图标以启动应用程序。

[②] 如果 Android 应用程序的 minSdkVersion 设置为 15，而用户移动设备的系统版本为 8。代表该移动设备系统不支持该应用程序，便会拒绝安装该应用程序。如果应用程序没有设置 minSdkVersion，就会采用默认值 1。当用户移动设备联机到 Android Market 时，只会看到移动设备操作系统支持的应用程序。

地区语言的问题，才不会让用户看到不熟悉的语言文字，这就是本地化(Localization，简称L10N——开头为L，结尾为N，中间有10个英文字母)的话题。所谓本地化就是让用户通过地区(locale)或语言的设置来切换画面显示的语言(例如：用户将手机切换成简体中文，则应用程序就显示简体中文)。简单地说，就是支持多国语言(Internationalization，简称I18N)。为了达到这个目的，Android特别将应用程序所使用到的文字内容独立出来成为一个文本文件(例如strings.xml)，这样一来，只要将该文本文件内容翻译成其他语言(例如将英文翻译成中文)并放在项目内，应用程序即可显示其他语言。

3.6.1 Android 支持的地区与语言

要设计具有本地化的应用程序，首先必须了解Android支持哪些地区、语言。截至目前为止，Android支持的地区、语言如表3-2所示[①]。

表 3-2

地 区	语 言
Arabic, Egypt (ar_EG)	Croatian, Croatia (hr_HR)
Arabic, Israel (ar_IL)	Hungarian, Hungary (hu_HU)
Bulgarian, Bulgaria (bg_BG)	Indonesian, Indonesia (id_ID)
Catalan, Spain (ca_ES)	Italian, Switzerland (it_CH)
Czech, Czech Republic (cs_CZ)	Italian, Italy (it_IT)
Danish, Denmark(da_DK)	Japanese (ja_JP)
German, Austria (de_AT)	Korean (ko_KR)
German, Switzerland (de_CH)	Lithuanian, Lithuania (lt_LT)
German, Germany (de_DE)	Latvian, Latvia (lv_LV)
German, Liechtenstein (de_LI)	Norwegian bokmål, Norway (nb_NO)
Greek, Greece (el_GR)	Dutch, Belgium (nl_BE)
English, Australia (en_AU)	Dutch, Netherlands (nl_NL)
English, Canada (en_CA)	Polish (pl_PL)
English, Britain (en_GB)	Portuguese, Brazil (pt_BR)
English, Ireland (en_IE)	Portuguese, Portugal (pt_PT)
English, India (en_IN)	Romanian, Romania (ro_RO)
English, New Zealand (en_NZ)	Russian (ru_RU)
English, Singapore(en_SG)	Slovak, Slovakia (sk_SK)
English, US (en_US)	Slovenian, Slovenia (sl_SI)
English, Zimbabwe (en_ZA)	Serbian (sr_RS)
Spanish (es_ES)	Swedish, Sweden (sv_SE)
Spanish, US (es_US)	Thai, Thailand (th_TH)
Finnish, Finland (fi_FI)	Tagalog, Philippines (tl_PH)
French, Belgium (fr_BE)	Turkish, Turkey (tr_TR)
French, Canada (fr_CA)	Ukrainian, Ukraine (uk_UA)
French, Switzerland (fr_CH)	Vietnamese, Vietnam (vi_VN)
French, France (fr_FR)	Chinese, PRC (zh_CN)
Hebrew, Israel (he_IL)	
Hindi, India (hi_IN)	

[①] 地区参见 http://en.wikipedia.org/wiki/ISO_3166-2；语言参见 http://en.wikipedia.org/wiki/List_of_ISO_639-1_codes。

3.6.2 创建支持多国语言的应用程序

要支持多国语言，可以增加多个 res/values 目录并在其内存放对应的文本文件。res/values 会被当作默认语言的目录，当 Android 系统找不到对应语言时，就会使用默认语言目录内的文本文件[①]。新增其他语言目录的步骤如下(以前述范例 HelloAndroid 为例)：

STEP 1 选择指定项目后，按下"🗋"(Opens a wizard to help create a new Android XML file) 按钮，如图 3-32 所示。

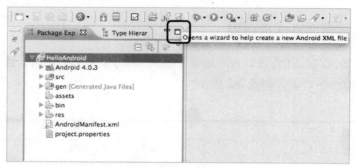

图 3-32

STEP 2 在图 3-33 填写数据后按下 Next 按钮。

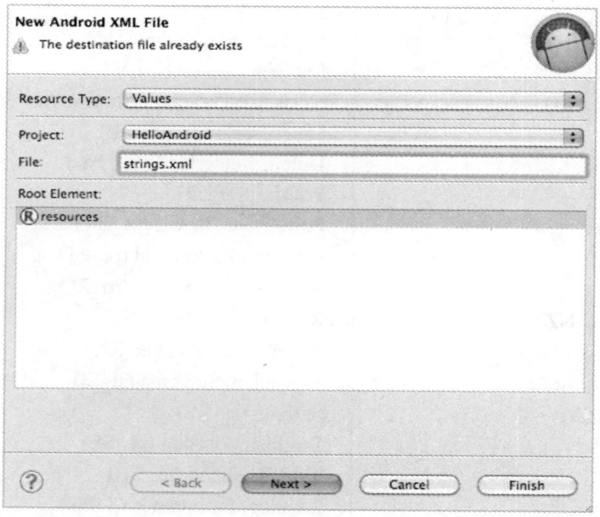

图 3-33

- Resource type：选择 Values 资源类型。
- Project：指定项目。
- File：欲创建的文本文件名称。

[①] 考虑要将应用程序销售遍及全球，建议把默认语言设置为英文，也就是 res/values 内的文本文件内容最好为英文。如果应用程序默认语言为简体中文，即便有支持英文(但非默认语言)，德国用户仍会看到中文而非英文，除非该用户将移动设备的语言改成英文。简单来说，建议默认语言为英文是因为非华人的用户可能看得懂英文，却看不懂简体中文。

STEP 3 在图 3-34 填写数据后按下 Next 按钮。

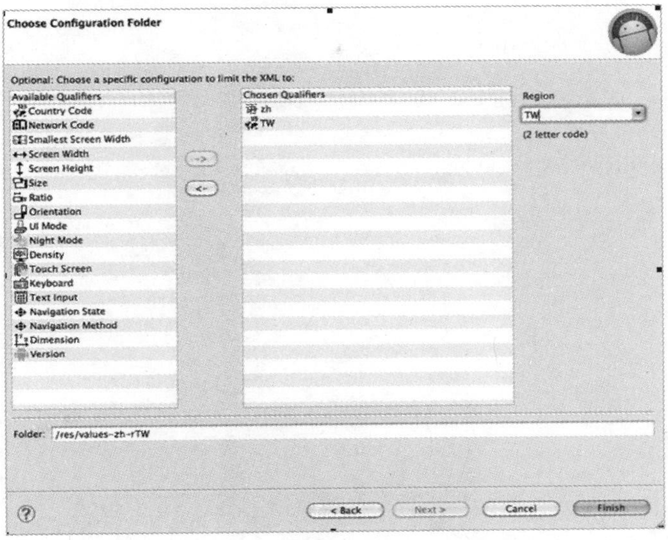

图 3-34

- 将左边窗格的 Language、Region 选项新增至右边窗格并分别输入 zh、TW。
- Folder：设置完成后会自动产生对应的语言目录，不需要自行输入。
- 完成后按 Finish 按钮结束。

STEP 4 将原来 res/values 目录内的 strings.xml 复制一份并覆盖 res/values-zh-rTW 内的 strings.xml，然后将英文字改成对应的简体中文字即可，如图 3-35 所示。

运行应用程序后更改语言设置[1]，如果成功即可看到如图 3-36 所示的画面[2]。

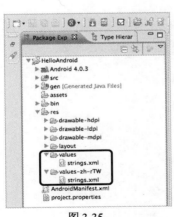

图 3-35

图 3-36

---

[1] 更改 Android 移动设备的地区语言设置的步骤为：按下 Home 按键>Menu 按键>System settings(设置)>Language & input(语言与输入)>选择 language(语言)。

[2] 模拟器可能发生已更改语言但应用程序语言仍未改变的情况；这是因为 Android 系统没有加载新的文本文件。强制停止该应用程序后再打开，即可解决此问题。

第 4 章

UI设计基本概念

本章学习目标：

- Android UI 设计基本观念
- Widget 组件介绍
- 页面配置与 layout 组件介绍
- 设置 UI 样式——使用 style 与 theme
- 触控与手势

4.1 Android UI 设计基本观念[1]

用户与计算机交互的画面被称作 UI(User Interface，用户界面)，而 UI 上面放满了各种组件，其中最容易被了解的组件就是可以让用户输入的组件，例如：按钮、文字输入方块、下拉菜单等，这些组件被称为 UI 组件(UI elements)。

一个 Android 应用程序的 UI 组件可分成 widget 与 layout 组件，而它们的根类(root class)都是 View 类。UI 组件的相关类大部分都放在 android.widget 包内。

- widget 组件：已经是 UI 的最基本单位，换句话说，不能在这类组件内再放入其他的组件，这与 layout 组件内还可以放入其他组件的性质不同。最具代表性的 widget 组件为：Button(按钮)、EditText(文字输入方块)、CheckBox(复选框)等。
- layout 组件：像一个容器般还可以再置入其他 widget 组件或 layout 组件。最具代表性的 layout 组件为 LinearLayout、RelativeLayout、TableLayout；而 ViewGroup 类是 layout 组件的根类。ViewGroup 类定义了 ViewGroup.LayoutParams 内部类，该内部类用来定义子组件如何在 layout 父组件上配置。

Android 的 UI 组件大多可以使用 XML 文件来创建并作页面配置，该 XML 文件一般称为 layout 文件，默认名称为 main.xml。虽然开发者也可以不通过 XML 文件而直接使用程序代码来动态创建 UI 组件并同时完成页面配置，不过除非是特殊情况，否则还是建议将 UI 创建与页面配置从程序代码中抽离出来，单独放在 layout 文件内，以符合 MVC 架构，方便以后维护。而 UI 上面的文字也属于应用程序文字的一部分，建议从 layout 文件抽离出来放在文本文件内(文本文件默认名称为 strings.xml)，以方便之后多国语言版本的制作。

为了简化 UI 设计，Android 开发工具提供了 Android Layout Editor(可视化设计工具)，让开发者可以通过拖曳 UI 组件与属性设置方式来完成 UI 画面的设计。只要双击 layout 文件(例如 main.xml)即可自动打开 Android Layout Editor，如图 4-1 所示。如果要看源代码，可以单击下方 main.xml 的标签即可。

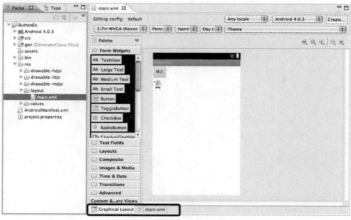

图 4-1

---

[1] 参见 http://developer.android.com/guide/topics/ui/overview.html。

UI 设计基本概念 04

如果不能打开可视化设计工具，可以对着 layout 文件右击选择 Open With│Android Layout Editor，如图 4-2 所示，然后重启 Eclipse 即可。

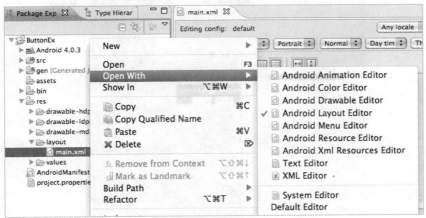

图 4-2

下面将说明如何使用 Android Layout Editor 来设计 UI 画面。

- 项目默认采用 LinearLayout，可以对空白处右击选择 Change Layout 来改变页面配置，参见图 4-3。关于各种 layout 的说明请参见 4.3 节。

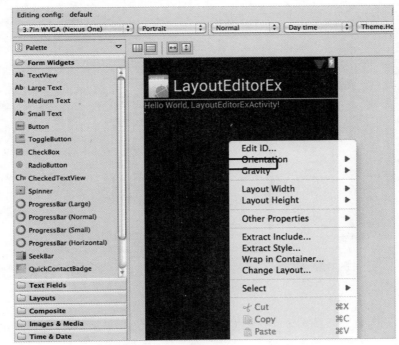

图 4-3

- 要增加 UI 组件，只要从左边 Palette 窗格拖曳指定组件至画面区即可，参见图 4-4。

图 4-4

- 要改变组件属性，只要对着该组件右击选择 Show In | Properties，即可弹出属性列表，可以按照需要修改属性设置，如图 4-5 所示。

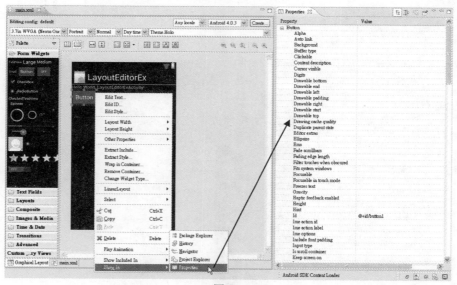

图 4-5

4.2 Widget 组件介绍

widget 组件的主要功能就是可以让用户与 Android 系统交互。Android 提供大量 widget 组件的相关类与方法以方便开发者设计各种丰富且实用的 UI。本章将会介绍简单且常用的 widget 组件，下一章则会介绍比较高级的 widget 组件。

在学习 widget 组件时，最重要的就是事件处理(event handling)机制，所以在介绍各个 widget 组件时，会将该 widget 组件所触发的事件与处理方式加以详细说明。而所有 widget 组件事件处理机制当中，最简单的莫过于 Button 事件处理，而且这种处理模式可以应用在其他的 widget 组件上，所以下面会先介绍 Button 事件处理机制。

4.2.1 与用户间的交互——以 Button 事件处理为例

一般用户最常使用到的 widget 组件就是按钮，在 Android 系统中就是 Button 组件。按钮可以在 layout 文件内使用<Button>来定义。每当按钮被按下，会触发 Button 事件(event)，而系统有对应的监听器(listener)可以获取到该事件；如果不要求监听器监听，系统将会忽略此事件，则按钮被按下后，不会有任何结果。如果希望当用户按下按钮时，可以有对应的响应，就必须要求监听器监听该按钮，并且将对应的处理结果编写好让监听器自动调用，等到按钮真的被按下，编写好的程序内容就会被运行，而用户就会得到相对响应，这就是所谓的事件处理机制(event handling)。由上述可知，事件处理机制是一种委托机制(delegation)，也就是一种自动机制。而要让此机制运作，必须按照下列步骤(请参见图 4-6)。

STEP 1 Button 对象向监听器注册：让监听器去监听指定的按钮是否被按下，Button 对象必须先调用 setOnClickListener() 向 OnClickListener 监听器[①]注册，这样该监听器才会开始监听。

STEP 2 实现 OnClickListener 的 onClick()：当按钮被按下时，OnClickListener 监听器会知道，而且它会自动调用 onClick()以响应用户按下按钮的操作[②]。所以必须实现 OnClickListener.onClick()方法，将该方法内容写好，才能让监听器顺利运行。

图 4-6

与 Button 事件处理机制有关的方法说明如表 4-1 所示。

---

① OnClickListener 是专门监听 UI 组件是否被按下的监听器。
② 如果开发者想要自行触发按钮事件,不等用户按下按钮,可以直接调用 View 类的 performClick()方法,例如 button.performClick()。

表 4-1

View 类

public void setOnClickListener (View.OnClickListener listener)

向 OnClickListener 监听器注册，要求该监听器监听指定的组件是否被按下

- listener: 实现 OnClickListener 接口的对象。要求实现 OnClickListener 接口的目的就是要求要实现 onClick() 方法，之后才能顺利执行 onClick() 内容

OnClickListener 界面

public abstract void onClick (View v)

当 UI 组件被按下时会自动调用此方法

- v: 发生按下事件的 UI 组件。例如被按下的是按钮 B，则按钮 B 的对象会传递给参数 v

Toast 是一个非常简易的消息框，因为之后将会经常使用到，所以在此先作说明。Toast 类的重要的参数与方法说明如表 4-2 所示。

表 4-2

public static final int LENGTH_LONG
Toast 方块显示信息的时间为长

public static final int LENGTH_SHORT
Toast 方块显示信息的时间为短

public static Toast makeText (Context context, CharSequence text, int duration)
出现一个含有消息正文的 Toast 方块
- context: 通常是现行 Activity 对象
- text: 欲显示的消息正文。
- duration: 信息会停留多久才消失。有 LENGTH_SHORT (短时间)、 LENGTH_LONG(长时间)两种可以选择

public static Toast makeText (Context context, int resId, int duration)
出现一个含有消息正文的 Toast 方块。跟前一个 makeText() 方法几乎相同，差别在于前一个方法是直接提供 CharSequence 文字来当作消息正文，而此方法则是提供文本文件对应的文字 Id 来当作消息正文
- context: 通常是现行 Activity 对象
- resId: 欲显示的消息正文在文本文件内对应的 Id
- duration: 信息会停留多久才消失

public void show()
显示 Toast 消息框

04 UI 设计基本概念

 范例 ButtonEx

范例(如图 4-7 所示)说明：
- 按下确定按钮(一般按钮)会弹出信息。
- 图形按钮会按照按下、聚焦(就是选择状态)①或正常状态显示对应图标。
- 按下图形按钮，也会弹出信息，并且改变按钮图标。

图 4-7

按钮可分为一般按钮与图形按钮。
- 一般按钮：系统默认按钮，外观固定不变。
- 图形按钮(image button)：以图形代表文字，并且可以按照按钮的状态，例如，正常、按下(pressed)、聚焦(focused)，来显示对应的图形文件，又称作自定义按钮②。

ButtonEx/res/layout/main.xml

```
8.      <Button android:id="@+id/btnSubmit"
9.          android:layout_width="wrap_content"
10.         android:layout_height="wrap_content"
11.         android:text="@string/btnSubmit" />
12.
13.     <Button android:id="@+id/cusbutton"
14.         android:layout_width="wrap_content"
15.         android:layout_height="wrap_content"
16.         android:background="@drawable/stateimages" />
```

---
① 聚焦状态是指使用轨迹球或触控板将焦点移到按钮上的状态。
② 参见 http://developer.android.com/resources/tutorials/views/hello-formstuff.html#CustomButton。

8~11 行：定义一般按钮。

8 行："@+id"表示要求替按钮产生新的 id 作为识别用。如果是"@id"则表示直接取用组件而非产生新的 id。

13~16 行：定义图形按钮。

16 行：指定图形按钮在各个状态的背景图形。因为图形不只一张，所以 stateimages 并非一张图，而是 XML 文件，内容就是定义在何种状态应用哪一个图形文件。

ButtonEx/res/drawable/stateimages.xml

```xml
1.  <?xml version="1.0" encoding="utf-8"?>
2.  <selector xmlns:android="http://schemas.android.com/apk/res/android">
3.      <item android:state_pressed="true"
4.          android:drawable="@drawable/android_pressed" />
5.      <item android:state_focused="true"
6.          android:drawable="@drawable/android_focused"/>
7.      <item android:drawable="@drawable/android_normal" />
8.  </selector>
```

2 行：以<selector>来定义图形按钮有哪些状态选项，每一个<item>子标签都代表一个选项。

3~4 行：state_pressed="true"代表按钮被按下，此时应用 android_pressed 图形文件。

5~6 行：state_focused="true"代表按钮被聚焦，此时应用 android_focused 图形文件。

7 行：Android 系统会从第一个 item 一路比较下来，当 3、5 行的 item 未比较成功，才会来到 7 行，这也是为什么将正常状态选项放在最后的原因。

ButtonEx/src/org/buttonEx/ButtonEx.java

```java
20.     private void findViews() {
21.         btnSubmit = (Button)findViewById(R.id.btnSubmit);
22.         btnSubmit.setOnClickListener(new OnClickListener() {
23.             @Override
24.             public void onClick(View v) {
25.                 CharSequence text = ((Button)v).getText();
26.                 Toast.makeText(
27.                     ButtonEx.this,
28.                     text,
29.                     Toast.LENGTH_SHORT)
30.                     .show();
31.             }
32.         });
33.
34.         cusbutton = (Button)findViewById(R.id.cusbutton);
35.         cusbutton.setOnClickListener(new OnClickListener() {
36.             @Override
37.             public void onClick(View v) {
```

```
38.                    Toast.makeText(
39.                            ButtonEx.this,
40.                            R.string.imagebutton_pressed,
41.                            Toast.LENGTH_SHORT)
42.                            .show();
43.                }
44.            });
45.    }
```

21 行：取得定义在 layout 文件的 btnSubmit 组件，因为该组件属于 Button 组件，所以可以显式强制转换(explicit casting)成 Button 对象。

22~25 行：Button 组件调用 setOnClickListener() 向 OnClickListener 监听器注册，接下来利用匿名内部类(anonymous inner class)[①]实现 OnClickListener 的 onClick()。当按钮被按下时，onClick() 会自动被调用并执行。在此范例，参数 v 是指发生按下事件的 Button 组件，也就是 btnSubmit 所代表的按钮，但数据类型是 View，所以 25 行必须执行强制转换成 Button 对象，方可调用 getText() 取得按钮上的文字。

26 行：调用 Toast 的 makeText() 可以创建 Toast 对象。

28 行：text 定义在 25 行，代表的是按钮上面的文字。

30 行：单纯调用 26 行的 makeText()只是创建 Toast 对象，必须调用 show()方可弹出 Toast 消息框。

40 行：R.string.imagebutton_pressed 代表取得文本文件内 imagebutton_pressed 所代表的文字。

4.2.2 TextView 与 EditText

TextView 组件专门用来呈现文字，无论提示文字或是消息正文，都可以使用 TextView 组件来呈现。EditText 组件是文字输入方块，操作移动设备时常常需要输入，例如输入联系人姓名、Email、电话、地址等，开发者可以在操作画面上放置 EditText 组件以方便用户输入。EditText 输入完毕后按键盘上的按键(包含实体或虚拟键盘)会触发按键事件(key event)，处理步骤如下。

STEP 1 EditText 组件调用 setOnKeyListener()向 OnKeyListener 注册：OnKeyListener 专门监听是否有键盘上的按键被按下。

STEP 2 实现 OnKeyListener 的 onKey()：当键盘上的按键被按下时，onKey()会自动被调用并执行。

与 EditText 组件的按键事件处理机制有关的方法说明如表 4-3 所示。

[①] 匿名内部类说明可参见"SCJP 6.0 认证教战手册"的"8-5 内部类"。

表 4-3

View 类
public void setOnKeyListener (View.OnKeyListener listener) 向 OnKeyListener 监听器注册，要求该监听器监听键盘上的按键是否被按下 ● listener：实现 OnKeyListener 界面的对象。要求实现 OnKeyListener 界面的目的就是要求要实现 onKey() 方法，之后才能顺利执行 onKey()内容

OnKeyListener 界面
public abstract boolean onKey (View v, int keyCode, KeyEvent event) 当按键被按下时会自动调用此方法 ● v：发生按键按下事件的 UI 组件，这里是指 EditText 组件 ● keyCode：被按下的按键对应的代码 ● event：KeyEvent 对象，存储着事件的相关信息

 范例 EditTextEx

范例(如图 4-8 所示)说明：
- 单击 EditText 会弹出虚拟键盘。
- 输入完毕后按虚拟键盘上的 Enter 按键(确定按键)即可将结果以 Toast 消息框呈现出来。

图 4-8

EditTextEx/res/layout/main.xml

```
7.      <TextView
8.          android:layout_width="match_parent"
9.          android:layout_height="wrap_content"
10.         android:textSize="18sp"
11.         android:text="@string/tvInput"/>
12.
```

UI 设计基本概念

```
13.    <EditText android:id="@+id/etInput"
14.        android:layout_width="match_parent"
15.        android:layout_height="wrap_content" />
```

7~11 行：定义 TextView 组件。

13~15 行：定义 EditText 组件。

EditTextEx/src/org/editTextEx/EditTextEx.java

```
20.    private void findViews() {
21.        etInput = (EditText)findViewById(R.id.etInput);
22.        etInput.setOnKeyListener(new OnKeyListener() {
23.            @Override
24.            public boolean onKey(View v, int keyCode, KeyEvent event) {
25.                if(event.getAction() == KeyEvent.ACTION_DOWN &&
26.                    keyCode == KeyEvent.KEYCODE_ENTER){
27.                    Toast.makeText(
28.                            EditTextEx.this,
29.                            etInput.getText(),
30.                            Toast.LENGTH_SHORT)
31.                        .show();
32.                    return true;
33.                }
34.                return false;
35.            }
36.        });
37.    }
```

21 行：取得定义在 layout 文件的 EditText 组件。

22~24 行：EditText 组件调用 setOnKeyListener() 向 OnKeyListener 注册，OnKeyListener 专门监听是否有键盘上的按键被按下。接下来利用匿名内部类实现 OnKeyListener 的 onKey()。当键盘上的按键被按下时，onKey() 会自动被调用并执行。

25 行：调用 getAction() 会取得 key 事件的操作描述，有 3 种可能情况：ACTION_DOWN(按键被按下)、ACTION_UP(放开按键)或 ACTION_MULTIPLE(按键多次被按下)。在此检查是否发生 ACTION_DOWN 的情况。

26 行：检查被按下的按键是否为 Enter 按键。

32、34 行：返回 true 代表事件到此为止，不要再往下一个组件传送，返回 false 就代表事件还会往下传送[1]。

EditText 组件的重要 XML 属性(XML attribute)[2]请参见表 4-4。

---

[1] 事件往往是一种连锁反应，UI 组件如果处理事件，通常会消耗掉(consume)该事件，如果不处理，该事件会传至下一个组件(通常是父组件)；如果事件一路传下去，都没有任何组件处理，则该事件就会被舍弃。例如一个 TextView 组件放在 LinearLayout 上，单击 TextView 组件，如果没有处理单击事件(click event)，事件就会传至 LinearLayout；如果没有任何组件处理，单击事件就会被舍弃。

[2] 一般而言，要查询特定组件 XML 属性的说明，可以查询该组件所属类或父类的 API 说明文件，说明文件会详细说明每个 XML 属性，并列出对应功能的方法。以 EditText 组件为例，参见 http://developer.android.com/reference/android/widget/EditText.html。

表 4-4

属 性 名 称	说　　　明	属　性　值
android:password	是否定义成密码输入类型，如果是，则输入的文字改以"·"(点)来呈现，避免旁人窥视	布尔值，"true" 或 "false"
android:hint	提示文字。当 EditText 组件没有输入任何文字时，会显示提示文字	文字值
android:maxLength	限制可以输入的字数长度	数字值，"20" 代表可以输入 20 个字
android:phoneNumber	限制只能输入电话号码。输入时会弹出只有数字按键的模拟键盘	布尔值，"true" 或 "false"
android:numeric	限制只能输入数字	只可输入下列 3 种值： ● integer–只能输入整数(不能有负号) ● signed–只能输入整数(可以有负号) ● decimal–可以输入整数或小数点(不能有负号)
android:digits	限制能输入的数字	例如："android:digits="123""，代表只能输入 1、2 或 3 数字，而无法输入其他数字(例如：4、5 等)

 范例 InputEx

范例(如图 4-9 所示)说明：
- 身份证字号字段定义成密码输入类型。
- 手机号码字段限制用户仅能输入电话号码而不能输入文字或小数点。
- 按下"确定"按钮后会将输入的数据以 Toast 消息框呈现。

图 4-9

InputEx/res/layout/main.xml

```
13.     <EditText android:id="@+id/etId"
14.         android:layout_width="match_parent"
15.         android:layout_height="wrap_content"
16.         android:maxLength="10"
17.         android:password="true"
18.         android:hint="@string/id" />
    . . .
37.     <EditText android:id="@+id/etMobile"
38.         android:layout_width="match_parent"
39.         android:layout_height="wrap_content"
40.         android:maxLength="10"
41.         android:phoneNumber="true"
42.         android:hint="@string/mobile" />
```

17 行：定义成密码输入类型，输入的文字会以"•"来呈现。
18 行：EditText 组件没有输入任何文字时，会显示提示文字。
40 行：限制用户最多可以输入 10 个字符。
41 行：限制用户只能输入电话号码。输入时会弹出只有数字按键的虚拟键盘。

InputEx/src/org/inputEx/InputEx.java

```
22.     private void findViews() {
23.         etId = (EditText)findViewById(R.id.etId);
24.         etName = (EditText)findViewById(R.id.etName);
25.         etMobile = (EditText)findViewById(R.id.etMobile);
26.         btnSubmit = (Button)findViewById(R.id.btnSubmit);
27.         btnSubmit.setOnClickListener(new OnClickListener() {
28.             @Override
29.             public void onClick(View v) {
30.                 String id = etId.getText().toString();
31.                 String name = etName.getText().toString();
32.                 String mobile = etMobile.getText().toString();
33.                 String msg =
34.                     getString(R.string.yourInfo) + id + "\n" + name + "\n" + mobile;
35.                 Toast.makeText(
36.                     InputEx.this,
37.                     msg,
38.                     Toast.LENGTH_LONG)
39.                     .show();
40.             }
41.         });
42.     }
```

23~26 行:取得定义在 layout 文件的对应组件。

29~40 行:当按钮被按下时,onClick()会自动被调用。30~32 行先获取各个 EditText 组件内已经输入的文字,之后将输入的文字以 Toast 消息框呈现。

4.2.3 CheckBox、RadioButton 与 ToggleButton

在 Android API,CheckBox、RadioButton 与 ToggleButton 类都是 Button 类的子类,所以这 3 个组件可以说都是按钮,只不过是属于单选按钮,可以按照用户的选择,执行对应的结果。CheckBox 组件适用于多重选择型架构,也就是同一时间可以有多个选项被选择;RadioButton 组件则属于单选型架构,也就是同一时间只有一个选项可以被选择;ToggleButton 组件的外型类似电源开关,打开代表选择,关闭代表取消选择。

范例 SelectButtons

范例(如图 4-10 所示)说明:

当 CheckBox、RadioButton 与 ToggleButton 等不同单选按钮的选择状态改变时,会取得该单选按钮上面的文字,并以 Toast 消息框呈现。

图 4-10

SelectButtons/res/layout/main.xml

```
7.      <CheckBox android:id="@+id/cbPlace"
8.          android:layout_width="wrap_content"
9.          android:layout_height="wrap_content"
```

```
10.         android:text="@string/place" />
11.
12.     <RadioGroup
13.         android:layout_width="match_parent"
14.         android:layout_height="wrap_content"
15.         android:orientation="vertical">
16.         <RadioButton android:id="@+id/rbFemale"
17.             android:layout_width="wrap_content"
18.             android:layout_height="wrap_content"
19.             android:checked="true"
20.             android:text="@string/female" />
21.         <RadioButton android:id="@+id/rbMale"
22.             android:layout_width="wrap_content"
23.             android:layout_height="wrap_content"
24.             android:text="@string/male" />
25.     </RadioGroup>
26.
27.     <ToggleButton android:id="@+id/tbVibrate"
28.         android:layout_width="wrap_content"
29.         android:layout_height="wrap_content"
30.         android:textOn="@string/vibrateOn"
31.         android:textOff="@string/vibrateOff" />
```

7~10 行：定义 CheckBox 组件。

12 行：同一个群组的 RadioButton 组件必须放在同一对<RadioGroup></RadioGroup>内，这样同一时间只有一个 RadioButton 组件可以被选择，也就是所谓的单选型架构。

15 行：以垂直方式排列 RadioButton 组件。

16~20 行：定义第 1 个 RadioButton 组件。

19 行：代表此 RadioButton 组件一开始即为选择状态。

21~24 行：定义第 2 个 RadioButton 组件。

27~31 行：定义 ToggleButton 组件。

30 行：当按钮被选择时(代表开关被打开)，指定按钮上显示的文字。

31 行：当按钮被取消选择时(代表开关被关闭)，指定按钮上显示的文字。

SelectButtons/src/org/selectButtons/SelectButtons.java

```
23.     private void findViews() {
24.         cbPlace = (CheckBox)findViewById(R.id.cbPlace);
25.         cbPlace.setOnClickListener(new OnClickListener() {
26.             @Override
27.             public void onClick(View v) {
28.                 String msg = cbPlace.getText().toString();
29.                 if(((CheckBox)v).isChecked())
```

```
30.              msg = getString(R.string.checked) + " " + msg;
31.           else{
32.              msg = getString(R.string.unchecked) + " " + msg;
33.           }
34.           Toast.makeText(
35.                  SelectButtons.this,
36.                  msg,
37.                  Toast.LENGTH_SHORT)
38.                  .show();
39.       }
40.    });
41.
42.    rbFemale = (RadioButton)findViewById(R.id.rbFemale);
43.    rbMale = (RadioButton)findViewById(R.id.rbMale);
44.    OnClickListener rbListener = new OnClickListener() {
45.       @Override
46.       public void onClick(View v) {
47.           RadioButton rb = (RadioButton)v;
48.           Toast.makeText(
49.                  SelectButtons.this,
50.                  rb.getText(),
51.                  Toast.LENGTH_SHORT)
52.                  .show();
53.       }
54.    };
55.    rbFemale.setOnClickListener(rbListener);
56.    rbMale.setOnClickListener(rbListener);
57.
58.    tbVibrate = (ToggleButton)findViewById(R.id.tbVibrate);
59.    tbVibrate.setOnClickListener(new OnClickListener() {
60.       @Override
61.       public void onClick(View v) {
62.           ToggleButton tb = (ToggleButton)v;
63.           Toast.makeText(
64.                  SelectButtons.this,
65.                  tb.getText(),
66.                  Toast.LENGTH_SHORT)
67.                  .show();
68.       }
69.    });
70. }
```

24 行：取得定义在 layout 文件的 CheckBox 组件。

25~27 行：CheckBox 组件调用 setOnClickListener()向 OnClickListener 注册，并利用匿名内部类实现 OnClickListener 的 onClick()。

29~38 行：参数 v 是指发生按下事件的 UI 组件，这里是指 CheckBox 组件，但数据类型是 View，所以 29 行必须进行强制转换，方可调用 isChecked()检查该 CheckBox 组件是否被选择。无论选择或取消选择都以 Toast 消息框呈现对应信息。

42~43 行：取得定义在 layout 文件的 RadioButton 组件。

44 行：创建 OnClickListener 对象 rbListener，以方便 55~56 行 RadioButton 组件注册。

47 行：在此参数 v 是指 RadioButton 组件，所以强制转换成 RadioButton 组件。

50 行：取得该组件上面的文字。

58 行：取得定义在 layout 文件的 ToggleButton 组件。

59~69 行：与前面 CheckBox、RadioButton 组件观念相同，不再赘述。

4.2.4　RatingBar

如图 4-11 所示，在 Android Market 每个应用软件都有评分，让用户可以了解何种应用程序评价较高，供用户参考。用户当然也可以添加评分行列，如果想设计这种让用户以单击几颗星的方式来对指定项目评分，可以利用 RatingBar 组件。

图 4-11

当 RatingBar 组件评分状态改变时会触发事件，处理步骤如下：

STEP 1　RatingBar 组件调用 setOnRatingBarChangeListener()向 OnRatingBarChangeListener 注册：OnRatingBarChangeListener 专门监听 RatingBar 组件的评分状态。

STEP 2　实现 OnRatingBarChangeListener 的 onRatingChanged()：当 RatingBar 组件评分状态改变时(增加或减少星星数)，onRatingChanged()会自动被调用并执行。

与 RatingBar 组件评分状态改变事件处理机制有关的方法说明如表 4-5 所示。

表 4-5
RatingBar 类

public void setOnRatingBarChangeListener (RatingBar.OnRatingBarChangeListener listener)
向 OnKeyListener 监听器注册，要求该监听器监听键盘上的按键是否被按下
- listener：实现 OnRatingBarChangeListener 接口的对象

OnRatingBarChangeListener 界面

public abstract void onRatingChanged (RatingBar ratingBar, float rating, boolean fromUser)
当 RatingBar 组件评分状态改变时会自动调用此方法
- ratingBar：发生评分状态改变的 RatingBar 组件
- rating：目前的评分
- fromUser：如果用户以触控或轨迹球方式改变评分状态，则会返回 true

 范例 RatingBarEx

范例(如图 4-12 所示)说明：
- 创建 RatingBar 组件，让用户以单击几颗星的方式来评分。
- 会以 Toast 消息框呈现选择几个星。

图 4-12

RatingBarEx/res/layout/main.xml

```
7.     <RatingBar android:id="@+id/rtbBook"
8.         android:layout_width="wrap_content"
9.         android:layout_height="wrap_content"
10.        android:numStars="5"
11.        android:stepSize="1.0" />
```

10 行：总共有几颗星可以让用户选择。

11 行：代表用户可选择的值距，例如 stepSize="0.5"代表用户一次可以增加或减少 0.5，也就是可以选择半颗星。

RatingBarEx/src/org/ratingBarEx/RatingBarEx.java

```
17.     private void findViews() {
18.         rtbBook = (RatingBar)findViewById(R.id.rtbBook);
19.         rtbBook.setOnRatingBarChangeListener(new OnRatingBarChangeListener() {
20.             @Override
21.             public void onRatingChanged(RatingBar ratingBar,
22.                     float rating, boolean fromUser) {
23.                 String msg = getString(R.string.start_number) + rating;
24.                 Toast.makeText(RatingBarEx.this, msg, Toast.LENGTH_SHORT).show();
25.             }
26.         });
27.     }
```

18 行：取得定义在 layout 文件的 RatingBar 组件。

19~22 行：RatingBar 组件调用 setOnRatingBarChangeListener()向 OnRatingBarChange-Listener 注册，OnRatingBarChangeListener 专门监听 RatingBar 组件评分状态。接下来利用匿名内部类实现 OnRatingBarChangeListener.onRatingChanged()。当 RatingBar 组件评分状态改变时，onRatingChanged() 会自动被调用并执行。

4.2.5 SeekBar

SeekBar 组件其实就是一种滚动条式的调整组件，通过拖曳方式达到改变数值大小的目的，适合用来调整屏幕的明亮度、声音的大小等。当 SeekBar 组件所代表的值改变时会触发事件，处理步骤如下。

STEP 1 SeekBar 组件调用 setOnSeekBarChangeListener() 向 OnSeekBarChangeListener 注册；OnSeekBarChangeListener 专门监听 SeekBar 组件所代表的值是否有改变。

STEP 2 实现 OnSeekBarChangeListener 的 3 个方法如下。

- onProgressChanged()：当 SeekBar 组件所代表的值正在改变时(无论增加或减少)，此方法会自动被调用并执行。

- onStopTrackingTouch()：当 SeekBar 组件所代表的值改变结束时，此方法会自动被调用并执行。

- onStartTrackingTouch()：当 SeekBar 组件所代表的值改变结束时，此方法会自动被调用并执行。

与 SeekBar 组件事件处理机制有关的方法说明如表 4-6 所示。

表 4-6

SeekBar 类
public void setOnSeekBarChangeListener (SeekBar.OnSeekBarChangeListener listener) 向 OnSeekBarChangeListener 监听器注册，要求该监听器监听 SeekBar 组件所代表的值是否有改变 ● listener：实现 OnSeekBarChangeListener 接口的对象
OnSeekBarChangeListener 界面
public abstract void onProgressChanged (SeekBar seekBar, int progress, boolean fromUser) 当 SeekBar 组件所代表的值正在改变时，此方法会自动被调用并执行 ● seekBar：值发生改变的 SeekBar 组件 ● progress：目前 SeekBar 组件所代表的值 ● fromUser：如果值的改变是由用户造成，则返回 true
OnSeekBarChangeListener 界面
public abstract void onStopTrackingTouch (SeekBar seekBar) 当 SeekBar 组件所代表的值改变结束时，此方法会自动被调用并执行 ● seekBar：值发生改变的 SeekBar 组件
public abstract void onStartTrackingTouch (SeekBar seekBar) 当 SeekBar 组件所代表的值开始改变时，此方法会自动被调用并执行 ● seekBar：值发生改变的 SeekBar 组件

 范例 SeekBarEx

范例(如图 4-13 所示)说明：
- 创建 SeekBar 组件，让用户以拖曳方式来设置屏幕背景的明亮度。
- 拖曳 SeekBar 组件会改变屏幕背景的明亮度，拖曳停止时会以 Toast 消息框呈现红、绿、蓝色的值。

图 4-13

SeekBarEx/res/layout/main.xml

```
13.      <SeekBar android:id="@+id/sbColor"
14.              android:layout_width="match_parent"
15.              android:layout_height="wrap_content"
16.              android:max="255"
17.              android:progress="0" />
```

16 行：SeeBar 组件的最大值设为 255[①]。

17 行：SeeBar 组件的默认值设为 0。

SeekBarEx/src/org/seekBarEx/SeekBarEx.java

```
20.      private void findViews() {
21.          linear = (LinearLayout)findViewById(R.id.linear);
22.          sbColor = (SeekBar)findViewById(R.id.sbColor);
23.          sbColor.setOnSeekBarChangeListener(new OnSeekBarChangeListener() {
24.              @Override
25.              public void onProgressChanged(SeekBar seekBar, int progress,
26.                      boolean fromUser) {
27.                  linear.setBackgroundColor(
28.                          Color.rgb(progress, progress, progress));
29.              }
30.
31.              @Override
32.              public void onStopTrackingTouch(SeekBar seekBar) {
33.                  String msg = getString(R.string.color_value) + seekBar.getProgress();
34.                  Toast.makeText(
35.                          SeekBarEx.this,
36.                          msg,
37.                          Toast.LENGTH_SHORT)
38.                          .show();
39.              }
40.
41.              @Override
42.              public void onStartTrackingTouch(SeekBar seekBar) {
43.                  //Notification that the user has started a touch gesture.
44.              }
45.          });
46.      }
```

21 行：取得定义在 layout 文件的 LinearLayout 组件。

22 行：取得定义在 layout 文件的 SeekBar 组件。

---

① 数字影像 RGB(红绿蓝)3 原色的值域皆为：0~255。

23 行：SeekBar 组件调用 setOnSeekBarChangeListener() 向 OnSeekBarChangeListener 注册，OnSeekBarChangeListener 专门监听 SeekBar 组件所代表的值是否有改变。接下来利用匿名内部类实现 OnSeekBarChangeListener 的 3 个方法，onProgressChanged()、onStopTrackingTouch()、onStartTrackingTouch()，详细说明请参见前表 4-6，不再赘述。

27~28 行：当 SeekBar 组件所代表的值改变时，LinearLayout 组件调用 setBackgroundColor()，并按照 progress(SeekBar 组件所代表的值)来改变其背景色。

33~38 行：当 SeekBar 组件所代表的值改变结束时，调用 getProgress() 以取得目前 SeekBar 组件所代表的值，并利用 Toast 消息框呈现该值。

4.2.6 ImageView

如果要呈现单一图片，ImageView 组件是一个不错的选择，可以用 layout 文件设置图形文件来源，也可直接以程序代码配置文件来源。

  范例 ImageViewEx

范例(如图 4-14 所示)说明：
- 创建 TextView 组件呈现文字。
- 创建 ImageView 组件呈现图片。

图 4-14

ImageViewEx/res/layout/main.xml

```
14.    <ImageView android:id="@+id/ivPhoto"
15.        android:src="@drawable/photo"
16.        android:contentDescription="@string/desc"
17.        android:layout_width="wrap_content"
18.        android:layout_height="wrap_content"
```

```
19.         android:layout_margin="5dp"
20.         android:layout_marginTop="10dp" />
```

15 行：指定图形文件来源。

16 行：设置图形文件的文字描述。

ImageViewEx/src/org/imageViewEx/ImageViewExActivity.java

```
8.      public class ImageViewExActivity extends Activity {
9.          ImageView iv_photo;
10.         TextView tv_desc;
11.
12.         @Override
13.         public void onCreate(Bundle savedInstanceState) {
14.             super.onCreate(savedInstanceState);
15.             setContentView(R.layout.main);
16.             findViews();
17.         }
18.
19.         private void findViews() {
20.             iv_photo = (ImageView)findViewById(R.id.ivPhoto);
21.             // set image resource by id
22.             // iv_photo.setImageResource(R.drawable.photo);
23.             tv_desc = (TextView)findViewById(R.id.tv_desc);
24.             tv_desc.setText(iv_photo.getContentDescription());
25.         }
26.     }
```

22 行：根据 Resource ID 指定 ImageView 组件的图形文件来源为项目 res/drawable 目录内的文件。

4.2.7 WebView

想要让用户在 Android 手机上浏览网页，必须使用 WebView 组件。WebView 也是 UI 组件的一种，所以可以与其他 UI 组件一样配置在 layout 文件内。创建 WebView 组件的步骤如下。

STEP 1 在 layout 文件内新建下列设置：

```
<WebView android:id="@+id/wvBrowser"
    android:layout_width="match_parent"
    android:layout_height="match_parent" />
```

STEP 2 在 Java 文件内新建下列程序代码：

```
//从 layout 文件取得 WebView 组件的设置并初始化
```

```
WebView wvBrowser = (WebView) findViewById(R.id.wvBrowser);
//让 JavaScript 语法可以在 WebView 组件上运行
wvBrowser.getSettings().setJavaScriptEnabled(true);
// WebView 组件一开始欲加载的 URL
wvBrowser.loadUrl("http://www.google.com");
```

STEP 3 因为要浏览网页必须使用到 Internet,所以必须在 manifest 文件内添加 INTERNET permission:

```
<uses-permission android:name="android.permission.INTERNET" />
```

STEP 4 如果要让用户有更大的浏览空间,可以应用 "NoTitleBar" theme 移除标题栏。

```
<application android:icon="@drawable/icon"
    android:label="@string/app_name"
    android:theme="@android:style/Theme.NoTitleBar" >
```

STEP 5 现在用户已经有一个简易型的网页浏览器,不过用户点击任何超链接都会自动打开 Android 默认的浏览器,并将内容显示在该浏览器上,而非显示在原先 WebView 组件上。如果想要使用自行设计的 WebView 组件来处理 URL 请求,必须自定义类继承 WebViewClient 类并改写 shouldOverrideUrlLoading():

```
private class MyWebViewClient extends WebViewClient {
    @Override
    public boolean shouldOverrideUrlLoading(WebView view, String url) {
        view.loadUrl(url);
        return true;
    }
}
```

STEP 6 将 WebView 组件的 WebViewClient 设成自定义的 WebViewClient:

```
wvBrowser.setWebViewClient(new MyWebViewClient());
```

STEP 7 按手机上的实体返回键无法回到前面的网页,而是返回前一个应用程序,所以最好对返回键的按下事件加以处理:

```
@Override
public boolean onKeyDown(int keyCode, KeyEvent event) {
    // KEYCODE_BACK 代表返回键;canGoBack()代表是否可以返回前一页
    if ((keyCode == KeyEvent.KEYCODE_BACK) && wvBrowser.canGoBack()) {
        wvBrowser.goBack(); //返回前一页
```

```
                return true;
        }
        return super.onKeyDown(keyCode, event);
    }
```

使用到的 WebView 类相关方法说明如表 4-7 所示。

表 4-7

public void loadUrl (String url)   加载指定的 URL
public WebSettings getSettings ()   取得 WebSettings 对象方便控制 WebView 组件
public void goBack ()   返回前一网页
public boolean canGoBack ()   如果有可以返回的浏览记录，则返回 true，否则返回 false

范例 WebViewEx

范例(如图 4-15 所示)说明：
- 默认网址为 http://www.google.com，并自动选择 www.google.com 部分方便让用户修改网址。
- 输入完网址后按下"确定"按钮会前往指定的网页。

图 4-15

WebViewEx/src/org/webViewEx/WebViewEx.java

```java
13.   public class WebViewEx extends Activity {
14.       private EditText etUrl;
15.       private Button btnSubmit;
16.       private WebView wvBrowser;
17.
18.       @Override
19.       public void onCreate(Bundle savedInstanceState) {
20.           super.onCreate(savedInstanceState);
21.           setContentView(R.layout.main);
22.           findViews();
23.       }
24.       private void findViews() {
25.           etUrl = (EditText)findViewById(R.id.etUrl);
26.           int start = "http://".length();
27.           int stop = etUrl.getText().toString().length();
28.           etUrl.setSelection(start, stop);
29.           btnSubmit = (Button)findViewById(R.id.btnSubmit);
30.           btnSubmit.setOnClickListener(new OnClickListener() {
31.               @Override
32.               public void onClick(View v) {
33.                   String url = etUrl.getText().toString();
34.                   wvBrowser.loadUrl(url);
35.               }
36.           });
37.           wvBrowser = (WebView) findViewById(R.id.wvBrowser);
38.           wvBrowser.getSettings().setJavaScriptEnabled(true);
39.           wvBrowser.loadUrl(getString(R.string.googleUrl));
40.           wvBrowser.setWebViewClient(new MyWebViewClient());
41.       }
42.
43.       private class MyWebViewClient extends WebViewClient {
44.           @Override
45.           public boolean shouldOverrideUrlLoading(WebView view, String url) {
46.               view.loadUrl(url);
47.               return true;
48.           }
49.       }
50.
51.       @Override
52.       public boolean onKeyDown(int keyCode, KeyEvent event) {
53.           if ((keyCode == KeyEvent.KEYCODE_BACK) && wvBrowser.canGoBack()) {
54.               wvBrowser.goBack();
55.               return true;
```

UI 设计基本概念

```
        56.        }
        57.        return super.onKeyDown(keyCode, event);
        58.    }
        59. }
```

26~28 行：取得 "http://" 长度当作文字选择的起始 index(最后一个 "/" 的下一个字开始选择)；取得 EditText 文字的长度当作文字选择的结束 index。

38 行：让 JavaScript 语法可以在 WebView 组件上运行。

39 行：WebView 组件一开始会加载指定的网页。

40 行：使用 MyWebViewClient 来处理 WebView 组件的 URL 请求。

43~49 行：自定义 MyWebViewClient 类继承 WebViewClient 类并改写 shouldOverrideUrl-Loading() 以处理 URL 请求。

52~56 行：检查按下的是否为返回键，同时检查是否有可返回的网页，如果符合，则返回到前一页。

页面配置与 layout 组件介绍

如何将 UI 组件放在指定的位置上？大部分人都会觉得应该使用 x、y 坐标定位方式来解决这个问题。使用 x、y 坐标来指定组件的位置虽然十分准确、易懂，但是程序设计者必须花很大的精力在坐标的计算上，否则一疏忽就很有可能造成两个组件重叠。为了减少程序设计者花费在 UI 页面设计上的时间，Android 提供不同的 layout 组件以达到页面配置功能。ViewGroup 类是 layout 组件的根类，换句话说，layout 组件属于 ViewGroup 组件，而 ViewGroup 组件是一种可以装载其他组件的容器。以下介绍常用到的 layout 组件。

4.3.1 LinearLayout

LinearLayout 以线性方式呈现 UI 组件，所谓线性方式就是将 UI 组件以垂直或水平排列，页面看起来就像是由一条条直线或横线组成。如果一个页面既需要垂直又需要水平排列组件，不妨使用嵌套式来设置页面。LinearLayout 可以在 layout 文件内使用<LinearLayout>来定义。放在 LinearLayout 内的子组件可以使用 LinearLayout.LayoutParams 内部类所定义的属性来说明自己如何在 LinearLayout 上配置，表 4-8 列出常使用到的 XML 属性：如何在 layout 父组件上配置。

表 4-8

属 性 名 称	说　明
android:layout_gravity	将组件放在父组件的指定位置
android:layout_weight	组件在父组件所占的比重，比重值越大，代表该组件扩展到剩余空间[1]的比例会越大

---

[1] 每个组件都会有基本所需大小，一个页面扣除所有组件所需的基本空间后剩下来的空间即称为剩余空间。

范例 LinearLayoutEx

范例(如图 4-16 所示)说明：
- 使用<LinearLayout>创建 LinearLayout。
- 利用嵌套方式在第 1 个 LinearLayout 内创建两个子 LinearLayout。
- 第 1 个子 LinearLayout 为水平排列，并新建 4 个 TextView 组件；第 2 个子 LinearLayout 为垂直排列，亦新建 4 个 TextView 组件。

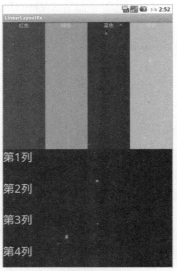

图 4-16

LinearLayoutEx/res/layout/main.xml

```
1.   <?xml version="1.0" encoding="utf-8"?>
2.   <LinearLayout xmlns:android="http://schemas.android.com/apk/res/android"
3.       android:orientation="vertical"
4.       android:layout_width="match_parent"
5.       android:layout_height="match_parent">
6.   
7.       <LinearLayout
8.           android:orientation="horizontal"
9.           android:layout_width="match_parent"
10.          android:layout_height="match_parent"
11.          android:layout_weight="1">
12.          <TextView
13.              android:text="@string/red"
14.              android:gravity="center_horizontal"
15.              android:background="#AA0000"
```

```
16.            android:layout_width="wrap_content"
17.            android:layout_height="match_parent"
18.            android:layout_weight="1"/>
       …
42.    <LinearLayout
43.        android:orientation="vertical"
44.        android:layout_width="match_parent"
45.        android:layout_height="match_parent"
46.        android:layout_weight="1">
47.            android:layout_weight="1"/>
```

2 行：<LinearLayout>标签内有 7 行、42 行等两个嵌套式的子标签。

3 行：将最外层的 LinearLayout 设置成垂直式；8 行将内层的第一个 LinearLayout 设置成水平式排列；43 行将内层的第二个 LinearLayout 设置成垂直式排列。所以两个内层的 LinearLayout 组件会由上至下垂直排列。

14 行：android:gravity="center_horizontal"代表 TextView 组件放在 LinearLayout 水平中间的地方。

18 行：android:layout_weight="1"代表 TextView 组件在 LinearLayout 所占比重为 1。

4.3.2　RelativeLayout

一般情况下，大多会使用 LinearLayout 来设置页面，不过如果页面过于复杂，需要大量使用嵌套式 LinearLayout，那么不妨改用 RelativeLayout，可以减少嵌套方式设置页面。RelativeLayout 以相对位置来呈现 UI 组件，也就是描述一个 UI 组件在其他附近组件的对应位置(例如 B 按钮在 A 按钮的右边位置)。为了能够比较精确地确定一个组件位置，最好提供该组件与其他至少两个以上组件间的相对关系。RelativeLayout 可以在 layout 文件内使用<RelativeLayout>来定义。与 RelativeLayout 有关的 XML 属性都定义在 RelativeLayout.LayoutParams 中，表 4-9 列出常使用到的 XML 属性。

表 4-9

属 性 名 称	说　　明
android:layout_above	将组件放在指定组件的上方
android:layout_alignBottom	将组件下边对齐指定组件的下边
android:layout_alignLeft	将组件左边对齐指定组件的左边
android:layout_alignParentBottom	true 代表将组件下边对齐父组件的下边
android:layout_alignParentLeft	true 代表将组件左边对齐父组件的左边
android:layout_alignParentRight	true 代表将组件右边对齐父组件的右边
android:layout_alignParentTop	true 代表将组件上边对齐父组件的上边
android:layout_alignRight	将组件右边对齐指定组件的右边
android:layout_alignTop	将组件上边对齐指定组件的上边
android:layout_below	将组件放在指定组件的下方

(续表)

属 性 名 称	说　　明
android:layout_centerHorizontal	true 代表将组件放在父组件的水平中央位置
android:layout_centerInParent	true 代表将组件放在父组件的水平及垂直中央位置
android:layout_centerVertical	true 代表将组件放在父组件的垂直中央位置
android:layout_toLeftOf	将组件放在指定组件的左方
android:layout_toRightOf	将组件放在指定组件的右方

 范例 RelativeLayoutEx

范例(如图 4-17 所示)说明：

- 使用 <RelativeLayout> 以创建 RelativeLayout。
- 由上至下，先新建 TextView 组件。
- EditText 组件放在 TextView 组件的下方。
- "取消"按钮放在 EditText 组件的下方，并对齐父组件(此为 RelativeLayout)的右边。
- "确定"按钮放在"取消"按钮的左方，而且"确定"按钮上边对齐"取消"按钮的上边。

图 4-17

RelativeLayoutEx/res/layout/main.xml

```
1.    <?xml version="1.0" encoding="utf-8"?>
2.    <RelativeLayout xmlns:android="http://schemas.android.com/apk/res/android"
3.        android:layout_width="match_parent"
4.        android:layout_height="match_parent">
5.
```

```
6.      <TextView android:id="@+id/tvPrompt"
7.          android:layout_width="match_parent"
8.          android:layout_height="wrap_content"
9.          android:text="@string/prompt"/>
10.
11.     <EditText android:id="@+id/etUserName"
12.         android:layout_width="match_parent"
13.         android:layout_height="wrap_content"
14.         android:layout_below="@id/tvPrompt"/>
15.
16.     <Button android:id="@+id/btnCancel"
17.         android:layout_width="wrap_content"
18.         android:layout_height="wrap_content"
19.         android:layout_below="@id/etUserName"
20.         android:layout_alignParentRight="true"
21.         android:layout_marginLeft="10dp"
22.         android:text="@string/cancel" />
23.
24.     <Button
25.         android:layout_width="wrap_content"
26.         android:layout_height="wrap_content"
27.         android:layout_toLeftOf="@id/btnCancel"
28.         android:layout_alignTop="@id/btnCancel"
29.         android:text="@string/submit" />
30. </RelativeLayout>
```

2 行：定义 RelativeLayout 组件。

14 行：将 EditText 组件放在 id 为 tvPrompt 的 TextView 组件下方。tvPrompt 组件就是 6 行定义的 TextView 组件。

20 行：将 Button 组件右边对齐父组件的右边。这里的父组件就是 2 行定义的 RelativeLayout 组件。

21 行：Button 组件与其左边组件的距离设置为 10dp[①]。

27 行：将 Button 组件放在 id 为 btnCancel 的 Button 组件左方。

28 行：将 Button 组件上边对齐 id 为 btnCancel 的 Button 组件上边。

4.3.3 TableLayout

TableLayout 是以行和列的方式来作页面配置，简单地说就是将画面分成数行与数列，设置方式非常类似于 HTML 的 TABLE(表格)。可以在 layout 文件内使用表 4-10 所列的 XML 标签来设置 TableLayout。

---

① 相对像素，160dp 等于 1 inch。参见 http://developer.android.com/guide/topics/resources/more-resources.html#Dimension 的 dp 说明。

表 4-10

标 签 名 称	说　　明
<TableLayout>	使用 TableLayout 页面配置[①]
<TableRow>	定义一行[②]

与 TableLayout 有关且常用的 XML 属性列于表 4-11。

表 4-11

属 性 名 称	说　　明
android:stretchColumns	加长指定字段(以 index 指定，第 1 栏 index 为 0)。例如：android:stretchColumns="0,1"代表要将 index 为 0、1 的字段空间加长
android:shrinkColumns	缩短指定字段，用法与 android:stretchColumns 相同

与 TableRow.LayoutParams 有关且常用的 XML 属性列于表 4-12。

表 4-12

属 性 名 称	说　　明
android:layout_column	将组件放在指定 index 的字段，如果 TableRow 内的每个字段都按照顺序放好，可以省略此属性
android:layout_span	指定组件可以横跨的字段数[③]

 范例 TableLayoutEx

范例(如图 4-18 所示)说明：

- 使用<TableLayout>以创建 TableLayout，并加长第 1、2 字段的长度。
- 新建 3 组<TableRow>以创建 3 行。
- 让最后一行的文字输入方块横跨两个字段。

图 4-18

[①] 类似 HTML 标签的<TABLE>。
[②] 类似 HTML 标签的<TR>。
[③] 类似 HTML 表格字段的 COLSPAN 属性。

TableLayoutEx/res/layout/main.xml

```xml
1.  <?xml version="1.0" encoding="utf-8"?>
2.  <TableLayout xmlns:android="http://schemas.android.com/apk/res/android"
3.      android:layout_width="match_parent"
4.      android:layout_height="match_parent"
5.      android:stretchColumns="0,1" >
6.
7.      <TableRow>
8.          <EditText android:layout_column="0"
9.              android:padding="3dp"
10.             android:hint="@string/editText" />
11.         <Button
12.             android:padding="3dp"
13.             android:textSize="16sp"
14.             android:text="@string/submit" />
15.         <Button
16.             android:padding="3dp"
17.             android:textSize="16sp"
18.             android:text="@string/cancel" />
19.     </TableRow>
20.
21.     <TableRow>
22.         <EditText
23.             android:padding="3dp"
24.             android:hint="@string/editText" />
25.         <Button
26.             android:padding="3dp"
27.             android:textSize="16sp"
28.             android:text="@string/submit" />
29.         <Button
30.             android:padding="3dp"
31.             android:textSize="16sp"
32.             android:text="@string/cancel" />
33.     </TableRow>
34.
35.     <TableRow>
36.         <EditText
37.             android:layout_span="2"
38.             android:padding="3dp"
```

```
39.                    android:hint="@string/editText" />
40.              <Button
41.                    android:padding="3dp"
42.                    android:textSize="16sp"
43.                    android:text="@string/submit" />
44.          </TableRow>
45.      </TableLayout>
```

2 行：定义 TableLayout 组件。

5 行：将 index 为 0、1 的字段加长。

7 行：新建 1 行。

8 行：将此组件放在 index 为 0 的字段。因为此 EditText 放在第一个，就是在 index 为 0 的字段上，所以即使不设置放在 index 为 0 的字段也没关系，因此可以省略 android:layout_column="0"。

9 行：设置此组件距离边界 3dp。

37 行：指定此组件可以横跨两个字段。

4.3.4 ScrollView

ScrollView 是 FrameLayout 的一种(ScrollView 是 FrameLayout 的 subclass)，所以它也是 layout 组件。它所显示的范围可以超过移动设备实际屏幕的大小，超过部分用户可以用翻阅方式来浏览。通常会在 ScrollView 内置入其他 layout 组件，最常见的为垂直排列的 LinearLayout，让用户可以上下翻阅画面以浏览内容。常用的 XML 属性说明如表 4-13 所示。

表 4-13

属 性 名 称	说　　明
android:scrollbars	当画面翻阅时是否要显示滚动条，可以设置的值如下 ● none：不显示滚动条 ● horizontal：仅显示水平滚动条 ● vertical：仅显示垂直滚动条 如果有两个以上的值要使用 | 符号分隔，例如：android:scrollbars="vertical | horizontal"

范例 ScrollViewEx

范例(如图 4-19 所示)说明：
- 使用<ScrollView>以创建 ScrollView，并在其内创建垂直排列的 LinearLayout。
- 当用户翻阅画面时会显示滚动条(方框部分)。
- 动态产生 20 个文字输入方块与按钮，长度超过屏幕大小，所以用户可以上下翻阅画面。

UI 设计基本概念

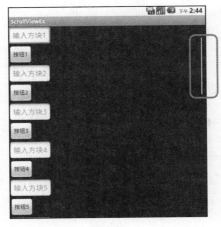

图 4-19

ScrollViewEx/res/layout/main.xml

```
1.   <?xml version="1.0" encoding="utf-8"?>
2.   <ScrollView xmlns:android="http://schemas.android.com/apk/res/android"
3.       android:layout_width="match_parent"
4.       android:layout_height="match_parent"
5.       android:scrollbars="vertical|horizontal">
6.
7.       <LinearLayout android:id="@+id/linear"
8.           android:orientation="vertical"
9.           android:layout_width="match_parent"
10.          android:layout_height="match_parent" />
11.  </ScrollView>
```

2 行：定义 ScrollView 组件。

5 行：按照用户可以翻阅方向，显示对应的垂直或水平滚动条。如图 4-19 所示的方框部分。

ScrollViewEx/src/org/ScrollViewEx/ScrollViewEx.java

```
9.   public class ScrollViewEx extends Activity {
10.
11.      @Override
12.      protected void onCreate(Bundle savedInstanceState) {
13.          super.onCreate(savedInstanceState);
14.          setContentView(R.layout.main);
15.
16.          LinearLayout linear = (LinearLayout)findViewById(R.id.linear);
17.          for (int i=1; i<=20; i++) {
18.              EditText editText = new EditText(this);
19.              editText.setHint(getString(R.string.editText) + i);
```

```
20.              LinearLayout.LayoutParams params =
21.                  new LinearLayout.LayoutParams(
22.                      LinearLayout.LayoutParams.WRAP_CONTENT,
23.                      LinearLayout.LayoutParams.WRAP_CONTENT);
24.              linear.addView(editText, params);
25.
26.              Button button = new Button(this);
27.              button.setText(getString(R.string.button) + i);
28.              linear.addView(button, params);
29.          }
30.      }
31. }
```

16 行：取得定义在 layout 文件的 LinearLayout 组件。

18 行：以程序代码动态产生 EditText 组件，而非使用 layout 文件来设置该组件。

19 行：将 strings.xml 文件内的文字取出后设置成 EditText 组件的提示文字，若以 layout 文件设置则为 android:hint="@string/editText"。

21~24 行：设置 linear(LinearLayout)的 width 与 height 值都为 WRAP_CONTENT，若以 layout 文件设置则为 android:layout_width="wrap_content"、android:layout_height="wrap_content"。

26~28 行：以程序代码动态产生 Button 组件，其设置方式与上述 EditText 组件几乎相同，不再赘述。

4.4 设置 UI 样式——使用 style 与 theme

style(样式)就是把与 UI 外观有关系的样式属性从 layout 文件独立出来并放在一个<style>标签内，以方便各个 UI 组件应用。这些样式属性包含：宽(width)、高(height)、文字颜色(font color)、文字大小(font size)、背景颜色(background color)等。换句话说，就是将各个 UI 组件的样式属性群组化成一个 style，需要使用到该样式的组件时再应用该 style。这种观念非常类似网页设计的样式表单(Cascading Stylesheets，CSS)功能。这些 style 设置会放在一个 XML 文件内，而这个文件要放在 res/values 目录内。

style 的设计是方便个别 UI 组件的样式应用，如果要将 Activity 或整个应用程序应用到指定样式，就要考虑到该范围内的所有 UI 组件(例如 Button、TextView、CheckBox 等)分别要应用何种样式，也就是 theme(主题)的观念。不过 theme 也就是 style，只不过要设置的样式多且复杂。

4.4.1 定义 style

以下说明的对应范例为 StyleRes，请自行打开项目内容参考。

layout 文件(参见 StyleRes/res/layout/main.xml)内容如下：

```xml
<TextView
    android:layout_width="match_parent"
    android:layout_height="wrap_content"
    android:textSize="16sp"
    android:textColor="#00FFFF"
    android:background="#666622"
    android:text="@string/text" />
```

可以把与 UI 样式属性独立出来，成为下面的内容，并放在一个独立的 XML 文件(请参见 StyleRes/res/values/style.xml)内：

```xml
<resources>
  <style name="style01">
    <item name="android:layout_width">match_parent</item>
    <item name="android:layout_height">wrap_content</item>
    <item name="android:textSize">16sp</item>
    <item name="android:textColor">#00FFFF</item>
    <item name="android:background">#666622</item>
  </style>
</resources>
```

- 样式设置内容必须放在 <resources> 标签内。
- <style>标签的使用目的在于给予同一群组的样式设置一个代号(也就是标识符)，便于之后取用。
- <item>标签用来指定原来样式属性的名称并设置对应的值。

这样一来，就可以将原来 layout 文件内容精简成下面内容：

```xml
<TextView
    style="@style/style01"
    android:text="@string/text" />
```

- style——style.xml 文件。
- style01——指定应用前述 style01 的样式内容。

4.4.2 继承 style

style 也提供类似 Java 继承(inheritance)的功能。<style>标签只要搭配 parent 属性即可继承指定的父 style，而子 style 可新建自己的样式属性或改变原来在父 style 样式属性的值。

以下说明的对应范例仍为 StyleRes，请自行打开项目内容参见。

一个已经定义好的 style(参见 StyleRes/res/values/style.xml)，内容如下：

```xml
<style name="style01">
    <item name="android:layout_width">match_parent</item>
```

```xml
        <item name="android:layout_height">wrap_content</item>
        <item name="android:textSize">16sp</item>
        <item name="android:textColor">#00FFFF</item>
        <item name="android:background">#666622</item>
    </style>
```

一个 style 可以利用 parent 属性继承上述 style，也可以新建自己的样式属性，或改变原来 style 的设置值。

```xml
    <style name="style02" parent="@style/style01">
        <item name="android:textColor">#FFFF00</item>
        <item name="android:background">#666666</item>
        <item name="android:layout_marginTop">5dp</item>
    </style>
```

- style02 利用 parent 属性继承 style01 的样式设置。
- style02 改变了 style01 的 android:textColor、android:background 设置值。
- style02 新建 android:layout_marginTop 样式属性，而 style01 没有设置该属性。

4.4.3 应用 theme

如前所述，要定义一个完整的 theme 相当复杂，必须考虑到各个 UI 组件的样式设置。所以一般都是应用 Android 系统默认的 theme[①]，之前范例虽然都没有作 theme 的设置，其实已经自动应用默认的 theme。Android 系统有其他 theme 可供选择，参见 R.style 类内开头为 Theme 的参数(例如：Theme_Wallpaper)。

要将 Activity 或应用程序应用到指定的 theme，必须打开 manifest 文件，并编辑 <activity> 或<application>内容。以下说明的对应范例为 ThemeEx，请自行打开项目内容参考。

1. Activity 应用指定的 theme

先打开项目的 manifest 文件(参见 ThemeEx/AndroidManifest.xml)。一个应用程序可能有多个 Activity，想要应用 theme 的 Activity，就必须在<Activity>标签内添加 android:theme 属性，如果不加上该属性，就代表使用默认的 theme，也就是 R.style.Theme。

```xml
    <activity android:name=".ThemeEx" android:label="@string/app_name"
        android:theme="@android:style/Theme.Wallpaper">
        <intent-filter>
            <action android:name="android.intent.action.MAIN" />
            <category android:name="android.intent.category.LAUNCHER" />
        </intent-filter>
    </activity>
```

如果要应用 Android 系统内置的 theme，开头都会有 android:，而且 Theme_Wallpaper(定义

[①] Android 2.X 版本默认 theme 的名称就是 Theme；Android 3.X 默认的 theme 则是 Theme.Holo。

在 R.style 的参数)必须改为 Theme.Wallpaper。

2. 应用程序应用指定的 theme

若应用程序应用到指定的 theme，该应用程序内的所有 Activity 都会直接应用到该 theme。应用方式与 Activity 相同，只不过改在 <application> 标签内添加 android:theme 属性。

```
<application android:theme="@android:style/Theme.Wallpaper"
    android:icon="@drawable/icon" android:label="@string/app_name">
    <activity android:name=".ThemeEx" android:label="@string/app_name" >
        <intent-filter>
            <action android:name="android.intent.action.MAIN" />
            <category android:name="android.intent.category.LAUNCHER" />
        </intent-filter>
    </activity>
</application>
```

4.4.4 继承 theme

前述继承 style 的观念也可应用在继承 theme。<style> 标签只要搭配 parent 属性即可继承指定的 theme。

```
<resources>
    <style name="MyTheme"   parent="@android:style/Theme.Wallpaper">
        <item name="android:background">#666622</item>
    </style>
</resources>
```

- 自定义 MyTheme 继承 Android 系统内置的 theme。
- Theme.Wallpaper 会应用现有的背景图当作背景，可以利用 android:background 将背景改成其他颜色。

若想应用自定义的 theme，必须修改 manifest 文件内 android:theme 的设置。

```
<application android:theme="@style/MyTheme"
    android:icon="@drawable/icon" android:label="@string/app_name">
    <activity android:name=".ThemeEx" android:label="@string/app_name" >
        <intent-filter>
            <action android:name="android.intent.action.MAIN" />
            <category android:name="android.intent.category.LAUNCHER" />
        </intent-filter>
    </activity>
</application>
```

4.5 触控与手势

4.5.1 触控事件处理

触控屏幕让用户与移动设备之间的交互更丰富，也提供更具亲和力的操作方式。下面范例将说明如何监听触控事件并取得用户触摸点的坐标。

 范例 TouchEx

范例(如图 4-20 所示)说明：
用户触摸屏幕后会显示触摸状态、手指数与触摸点的坐标。

图 4-20

范例创建步骤如下：

STEP 1 在 res/layout/main.xml 文件内新建 LinearLayout，之后会处理用户触摸 LinearLayout 的情况。

```
<LinearLayout xmlns:android="http://schemas.android.com/apk/res/android"
    android:id="@+id/linearLayout"
    android:layout_width="match_parent"
    android:layout_height="match_parent"
    android:orientation="vertical" >
</LinearLayout>
```

STEP 2 在 TouchExActivity.java 文件内添加下列程序代码。

```java
public class TouchExActivity extends Activity {
    private LinearLayout linearLayout;
    private TextView textView;

    @Override
    public void onCreate(Bundle savedInstanceState) {
        super.onCreate(savedInstanceState);
        setContentView(R.layout.main);
        findViews();
    }

    private void findViews() {
        linearLayout = (LinearLayout) findViewById(R.id.linearLayout);
        /*调用 setOnTouchListener()方法向 MyTouchListener 注册，MyTouchListener
        会监听用户是否触摸该组件*/
        linearLayout.setOnTouchListener(new MyTouchListener());
        textView = (TextView) findViewById(R.id.textView);
    }
    /* MyTouchListener 实现 OnTouchListener 的 onTouch()方法，当用户触摸该组件，
    系统会自动调用 onTouch()方法*/
    class MyTouchListener implements OnTouchListener {

        @Override
        public boolean onTouch(View v, MotionEvent event) {
            StringBuilder sb = new StringBuilder();
            /*如果用户仅用一根手指触摸，可以直接调用 getX()、getY()方法取得触摸点的
            坐标*/
            sb.append(String.format("current pointer: (%.1f,%.1f) %n",
                    event.getX(), event.getY()));

            sb.append("touch state: ");
            /*调用 getAction()方法可以取得用户触摸方式，ACTION_DOWN 代表触摸*/
            if (event.getAction() == MotionEvent.ACTION_DOWN) {
                sb.append("touch down!\n");
            }
            /*ACTION_MOVE 代表移动，也就是持续改变触摸点位置*/
            else if (event.getAction() == MotionEvent.ACTION_MOVE) {
                sb.append("touch move!\n");
            }
            /*ACTION_UP 代表触摸结束*/
            else if (event.getAction() == MotionEvent.ACTION_UP) {
                sb.append("touch up!\n");
            }
            /*调用 getPointerCount()取得触摸点数，也就是手指数*/
            int pointerCount = event.getPointerCount();
            sb.append(String.format("point count: %d %n", pointerCount));
            /*如果有多点触控，使用循环将各点的 ID、坐标一一取得*/
            for (int i = 0; i < pointerCount; i++) {
```

```
                sb.append(String.format("pointer %d: (%.1f,%.1f) %n",
                        event.getPointerId(i), event.getX(i), event.getY(i)));
            }
            textView.setText(sb);
            return true;
        }
    }
}
```

4.5.2 手势

有了触控屏幕，用户可以使用许多不同触控方式来下命令，这些方式统称为手势(gesture)[①]，常见的手势有：轻拍(tap)、拖曳(drag)、滑过(slide, swipe)等手势，也可以使用自定义手势。如果要使用自定义手势，就必须先创建手势数据库(gestures library)。

在模拟器上有一个名为 Gestures Builder 的应用程序，开发者可以使用该应用程序预先创建各种需要的手势，之后会产生手势数据库文件，可以将该文件复制到指定的应用程序中，方便比较用户的手势是否符合手势数据库内的手势，如果符合再做进一步操作。手势数据库的创建步骤如下：

STEP 1 如前所述，模拟器上有一个名为 Gestures Builder 的应用程序，图标为"📱"，其实它是一个内建的范例程序。如果模拟器没有该应用程序，参见 3.2.5 节，并将其安装至模拟器后运行。

STEP 2 一开始运行 Gestures Builder 应用程序，尚无创建任何手势，所以按下 Add gesture 按钮准备创建第一个手势，如图 4-21 所示。

STEP 3 创建手势时必须替该手势命名，并画出该手势，如图 4-22 所示。完毕后按下 Done 按钮就会创建该手势，并产生手势数据库文件(文件名为 gestures)，存储在 SD 卡内。

图 4-21

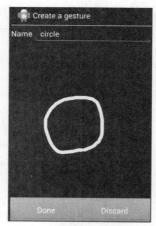

图 4-22

[①] Android 1.6 版开始支持手势。

UI 设计基本概念

STEP 4 使用 DDMS 将手势数据库文件复制出来，如图 4-23 所示，之后只要将该文件复制至指定应用程序的 res/raw 目录内即可用来比较用户的手势操作是否符合要求。

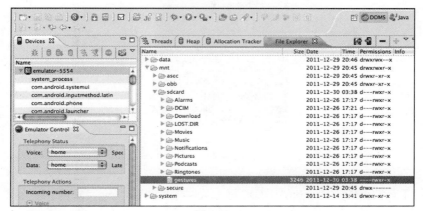

图 4-23

 范例 GestureEx

范例(如图 4-24 所示)说明：

用户的手势会与手势数据库内的手势比较，如果符合，会显示该手势名称与正确程度(会以分数表示)。

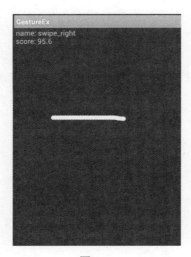

图 4-24

范例创建步骤如下：

STEP 1 使用 Gestures Builder 应用程序创建手势数据库，如图 4-25 所示。

105

图 4-25

STEP 2 将手势数据库文件 gestures 复制至本范例项目的 res/raw 目录内。

STEP 3 在 res/layout/main.xml 文件内新建 GestureOverlayView 组件，因为该组件不属于 android.widget 包，所以必须使用完整名称。之后用户就会在此组件上画出他们的手势。

```
<android.gesture.GestureOverlayView android:id="@+id/gestureOverlay"
    android:layout_width="match_parent"
    android:layout_height="match_parent"
    android:orientation="vertical"/>
<!--orientation 为 vertical(默认值)代表不将垂直手势视为手势，以避免与其他组件的手势混淆，例如 ScrollView-->
```

STEP 4 在 GestureExActivity.java 文件内添加下列程序代码。

```
public class GestureExActivity extends Activity {
    TextView textView;
    GestureOverlayView gestureOverlay;
    GestureLibrary gestureLibrary;
    @Override
    public void onCreate(Bundle savedInstanceState) {
        super.onCreate(savedInstanceState);
        setContentView(R.layout.main);
        /*通过资源 ID 来指定欲加载的手势数据库*/
        gestureLibrary = GestureLibraries.fromRawResource(this, R.raw.gestures);
        /*开始加载手势数据库，成功加载则返回 true，否则返回 false*/
        if (!gestureLibrary.load()) {
            finish();
        }
        findViews();
```

```
}
private void findViews() {
    textView = (TextView)findViewById(R.id.textView);
    /*调用 findViewById()方法取得定义在 main.xml 的 GestureOverlayView 组件*/
    gestureOverlay = (GestureOverlayView)findViewById(R.id.gestureOverlay);
    /*调用 addOnGesturePerformedListener()方法注册 OnGesturePerformedListener，并实现其
    onGesturePerformed()方法，当用户画出手势时系统会自动调用 onGesturePerformed()方法*/
    gestureOverlay.addOnGesturePerformedListener(new OnGesturePerformedListener() {

        @Override
        /*传入 Gesture 对象代表用户的手势*/
        public void onGesturePerformed(GestureOverlayView overlay, Gesture gesture) {
            /*将用户手势与手势数据库比较，会返回比较结果*/
            ArrayList<Prediction> predictions = gestureLibrary.recognize(gesture);
            /*比较结果可能符合数据库内的多个手势，会按照分数从高至低排序，分数越高代
            表相似度越高，所以取第一个(索引为 0)即取得相似度最高的手势，并限制分数至少大
            于 1 分*/
            if (predictions.size() > 0 && predictions.get(0).score > 1.0) {
                StringBuilder sb = new StringBuilder();
                /*取得手势名称*/
                String gestureName = predictions.get(0).name;
                /*取得手势分数*/
                double gestureScore = predictions.get(0).score;
                sb.append(String.format(" name: %s %n score: %.1f", gestureName, gestureScore));
                textView.setText(sb);
            }
        }
    });
}
}
```

第 5 章

UI 高级设计

本章学习目标:

- Menus
- 对话窗口与日期选择器
- Spinner 与 AutoCompleteTextView
- Gallery 与 GridView
- ListView
- 自定义 View 组件与 2D 绘图
- 补间动画
- Drawable 动画

5.1 Menus

大部分 Android 设备上都有一个 Menu 实体按钮,方便用户与设备交互并作对应设置。如果在设备的首页(Home)按下 Menu 按钮,会弹出如图 5-1 所示的功能菜单,这也是一般人最常用到的 Menu(菜单)功能。

图 5-1

实际上 Android 提供了 3 种 Menu 的功能。

(1) Options Menu(功能菜单):图 5-1 即属于 Options Menu。

(2) Context Menu(上下文菜单):当用户在指定组件上久按不放(long-press),即会弹出 Context Menu。

(3) Submenu(子菜单):当用户单击 Options Menu 或 Context Menu 上面的选项时可以显示第二层选项,第二层选项即属于 Submenu。

5.1.1 Options Menu

虽然可以在程序代码内直接利用 Menu 类来创建 Options Menu,但是除非特殊情况,否则一般建议通过 XML 文件来定义 Options Menu。要产生一个用户可以操作的 Options Menu,需要下列两个步骤:

STEP 1 使用 XML 文件创建 Menu 组件:如上所述,一般会使用 XML 文件来创建 Menu 组件相关内容,所以下面范例将会说明如何通过 XML 文件来创建 Menu 组件。

STEP 2 改写与 Options Menu 有关的方法:虽然已经使用 XML 文件创建 Menu 组件,但画面仍不会显示该 Menu 组件,必须改写与 Options Menu 有关的方法才能真正在画面上呈现 Menu。

1. 使用 XML 文件创建 Menu 组件

将定义 Menu 组件的 XML 文件放在 res/menu 目录内，在该文件内会使用下列两个重要标签。

(1) <menu>标签：创建 Menu 组件，该组件专门用来存放 MenuItem 组件(选项)，图 5-1 中弹出来的功能菜单就是利用<menu>标签创建的。

(2) <item>标签：创建 MenuItem 组件，也就是创建 Menu 的选项；图 5-1 中的"新建"、"壁纸"等选项就是利用<item>标签创建的选项。而<item>标签则有 3 个重要属性：

- android:id——MenuItem 组件的资源 ID，方便在程序代码中访问指定的 MenuItem 组件。
- android:icon——选项上呈现的图标。
- android:title——选项上呈现的文字。

 范例 OptionsMenuEx

范例(如图 5-2 所示)说明。

- 画面显示：请按 Menu 按钮。
- 按下实体 Menu 按钮后会弹出 5 个选项，其中"阳明山公园"、"我的位置"、"结束"等 3 个选项有图标。
- 单击"结束"选项会结束该程序；单击其余 4 个选项会以 Toast 信息方式呈现选项上面的文字。

图 5-2

OptionsMenuEx/res/menu/mymenu.xml

```
1.    <?xml version="1.0" encoding="utf-8"?>
2.    <menu xmlns:android="http://schemas.android.com/apk/res/android">
3.        <item android:id="@+id/yangmingshan" android:title="@string/yangmingshan"
```

```
4.            android:icon="@drawable/yangmingshan" />
5.        <item android:id="@+id/yushan" android:title="@string/yushan" />
6.        <item android:id="@+id/taroko" android:title="@string/taroko" />
7.        <item android:id="@+id/myloc" android:title="@string/myloc"
8.            android:icon="@drawable/ic_menu_mylocation" />
9.        <item android:id="@+id/exit" android:title="@string/exit"
10.           android:icon="@drawable/ic_menu_close_clear_cancel" />
11.    </menu>
```

mymenu.xml 文件主要设置 Menu 的 MenuItem 组件(选项)，3、5、6、7、9 行共有 5 个 <item> 标签，代表 Menu 组件有 5 个 MenuItem 组件。也可以在程序代码内调用 Menu 的 add() 来动态新建 MenuItem 组件。

3~4 行如下。
- android:id：值为"@+id/yangmingshan"，代表该选项的资源 ID 为 yangmingshan，方便在程序代码中通过 id 访问此组件。
- android:title：值为"@string/yangmingshan"，代表取得放在文本文件内的"阳明山公园"文字，所以图 5-2 的第一个选项会呈现该文字。
- android:icon：值为"@drawable/yangmingshan"，代表取得放在 res/drawable 目录内主文件名为 yangmingshan 的图形文件，所以图 5-2 的第一个选项会呈现该图形。

 不可不知

选项为 6 个以下(含 6 个)，会全部显示出来。选项超过 6 个，则第 6 个选项会显示"更多"，如图 5-3 所示。当点击"更多"选项，才会将真正的第 6 个选项以及剩下的其他选项显示出来。

图 5-3

2. 改写与 Options Menu 有关的方法

要在画面上呈现可以让用户操作的 Options Menu，还需要改写下列 Activity 类的两个方法。

public boolean onCreateOptionsMenu(Menu menu)
当 Options Menu 要呈现时，会调用此方法。此方法目的在于初始化 Options Menu，所以必须加载定义 Menu 组件的 XML 文件。此方法只会被调用一次(意味着产生 Option Menu 后，就无法再修改其选项)，并产生一个真正会在画面上呈现的 Options Menu。

- menu：会显示在画面上的 Menu 组件，所以必须将选项放置在此组件上。
- 返回值：返回 true，Options Menu 才能显示在画面上；返回 false 则无法显示 Options Menu。

public boolean onOptionsItemSelected(MenuItem item)
Options Menu 上的选项被单击时，此方法会被调用。

- item：被选择的 MenuItem 组件会被传递进来。
- 返回值：返回 true 代表开发者已经对选项被选择的情况作了对应的处理，所以不需要系统进行后续处理。false 代表允许系统进行后续处理。

OptionsMenuEx/src/org/optionsMenuEx/OptionsMenuEx.java

```
18.     @Override
19.     public boolean onCreateOptionsMenu(Menu menu) {
20.         MenuInflater inflater = getMenuInflater();
21.         inflater.inflate(R.menu.mymenu, menu);
22.         return true;
23.     }
24.
25.     @Override
26.     public boolean onOptionsItemSelected(MenuItem item) {
27.         String msg = "";
28.         switch (item.getItemId()) {
29.         case R.id.yangmingshan:
30.             msg = getString(R.string.yangmingshan);
31.             break;
32.         case R.id.yushan:
33.             msg = getString(R.string.yushan);
34.             break;
35.         case R.id.taroko:
36.             msg = getString(R.string.taroko);
37.             break;
38.         case R.id.myloc:
39.             msg = getString(R.string.myloc);
40.             break;
41.         case R.id.exit:
```

```
42.                this.finish();
43.            default:
44.                return super.onOptionsItemSelected(item);
45.            }
46.            Toast.makeText(this, msg, Toast.LENGTH_SHORT).show();
47.            return true;
48.        }
```

20~21 行：调用 getMenuInflater() 取得 MenuInflater 对象，有了该对象才能调用 inflate() 载入 mymenu.xml 资源文件，并将文件内容转化成显示在画面上的 Menu 组件。

26 行：当用户选择选项时，会将被选择的 MenuItem 组件传递给 item 参数。

28~47 行：调用 getItemId() 可以取得选项所对应的资源 ID。用户选择选项后，使用 switch-case 判断被选择的是哪个选项，并使用 Toast 消息框呈现该选项的文字。因为都有对应的处理(43~44 行的情况除外)，所以 47 行可以直接返回 true，代表不需要系统进行后续处理。

43~44 行：一般建议如果没有对应的处理方式，应该调用父类(Activity)的 onOptionsItem-Selected()，由该方法返回 false；而非直接返回 false[①]。

5.1.2 Context Menu

Context Menu 与 Options Menu 的产生方式十分类似，要产生一个用户可以操作的 Context Menu，需要下列 3 个步骤。

STEP 1 使用 XML 文件创建 Menu 组件：这点与 Options Menu 相同，不再赘述。另外，Context Menu 可以使用<group>标签将数个选项组合起来设置成复选框(checkbox)与单选按钮(radio button)。

STEP 2 改写与 Context Menu 有关的方法：虽然改写的方法名称与 Options Menu 不同，但功能大致相同。

STEP 3 注册指定 UI 组件：必须指定在何种 UI 组件上久按才会弹出 Context Menu。Options Menu 没有此项设置。

使用 XML 文件创建 Menu 组件

 范例 ContextMenuEx

范例(如图 5-4 所示)说明。

- 画面显示：请久按本画面。
- 久按画面后会弹出 5 个选项，其中"阳明山公园"、"玉山公园"、"太鲁阁公园"选项为同一选项按钮组。

---

① 请参见 http://developer.android.com/guide/topics/ui/menus.html 的 Creating an Options Menu 部分，有详细的说明。

- 单击"结束"选项会结束该程序；单击其余 4 个选项会以 Toast 消息框呈现选项上面的文字。

图 5-4

ContextMenuEx/res/menu/mymenu.xml

```
1.  <?xml version="1.0" encoding="utf-8"?>
2.  <menu xmlns:android="http://schemas.android.com/apk/res/android">
3.      <group android:checkableBehavior="single">
4.          <item android:id="@+id/yangmingshan" android:title="@string/yangmingshan" />
5.          <item android:id="@+id/yushan" android:title="@string/yushan" />
6.          <item android:id="@+id/taroko" android:title="@string/taroko" />
7.      </group>
8.      <item android:id="@+id/myloc" android:title="@string/myloc" />
9.      <item android:id="@+id/exit" android:title="@string/exit" />
10. </menu>
```

3~7 行：使用<group>标签将 4、5、6 行的选项组合起来并设置成单选按钮。3 行的 android:checkableBehavior 属性有 3 个值可选择。

- single：设置成单选按钮。
- all：设置成复选框。
- none：设置成普通选项。

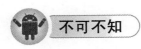

Context Menu 不支持图标选项(icon menu)，但支持复选框与单选按钮；而 Options Menu 则正好相反，支持图标选项，但不支持复选框与单选按钮。

3. 改写与 Context Menu 组件有关的方法

要在画面上呈现可以让用户操作的 Context Menu，还需要改写下列 Activity 类的两个方法。

public void onCreateContextMenu (ContextMenu menu, View v, ContextMenu.ContextMenuInfo menuInfo)
当 Context Menu 要呈现时，会调用此方法。此方法目的在于初始化 Context Menu，所以必须加载定义 Menu 组件的 XML 文件。每次要呈现 Context Menu 时，此方法就会被调用一次；意味着产生 Context Menu 后，还可以修改其选项，这点与 Options Menu 不同。
- menu：须将选项放置在此 Menu 组件上。
- v：被久按的 UI 组件。
- menuInfo：根据 v 所属的 View 类不同会显示不同的额外信息。

public boolean onContextItemSelected (MenuItem item)
Context Menu 上的选项被单击时，此方法会被调用。
- item：被选择的 MenuItem 组件会被传递进来。
- 返回值：返回 true 代表开发者已经对选项被选择的情况进行了对应的处理，所以不需要系统进行后续处理。false 代表允许系统进行后续处理。

ContextMenuEx/src/org/contextMenuEx/ContextMenuEx.java

```java
15.     @Override
16.     public void onCreate(Bundle savedInstanceState) {
17.         super.onCreate(savedInstanceState);
18.         setContentView(R.layout.main);
19.         LinearLayout linear = (LinearLayout)findViewById(R.id.linear);
20.         registerForContextMenu(linear);
21.     }
22.
23.     @Override
24.     public void onCreateContextMenu(ContextMenu menu, View v,
25.                                     ContextMenuInfo menuInfo) {
26.         MenuInflater inflater = getMenuInflater();
27.         inflater.inflate(R.menu.mymenu, menu);
28.     }
29.
30.     @Override
31.     public boolean onContextItemSelected(MenuItem item){
32.         String msg = "";
33.         switch (item.getItemId()) {
34.         case R.id.yangmingshan:
35.             msg = getString(R.string.yangmingshan);
36.             break;
37.         case R.id.yushan:
38.             msg = getString(R.string.yushan);
39.             break;
40.         case R.id.taroko:
41.             msg = getString(R.string.taroko);
42.             break;
43.         case R.id.myloc:
```

```
44.            msg = getString(R.string.myloc);
45.            break;
46.        case R.id.exit:
47.            this.finish();
48.        default:
49.            return super.onContextItemSelected(item);
50.        }
51.        Toast.makeText(this, msg, Toast.LENGTH_SHORT).show();
52.        return true;
53.    }
```

20 行：调用 registerForContextMenu()并指定在 LinearLayout 组件上久按时会弹出 Context Menu。

26~27 行：调用 getMenuInflater()取得 MenuInflater 对象，有了该对象才能调用 inflate()载入 mymenu.xml 资源文件，并将文件内容转化成显示在画面上的 Menu 组件。

33~52 行：调用 getItemId()可以取得选项所对应的资源 ID，通过 switch-case 判断被选择的选项，并使用 Toast 消息框呈现该选项的文字。因为都有对应的处理(48~49 行除外)，所以 52 行可以直接返回 true，代表不需要系统进行后续处理。

48~49 行：如果没有对应处理方式，应该调用父类的 onContextItemSelected()。

5.1.3 Submenu

单击 Options Menu 与 Context Menu 的选项后还可以再产生子菜单，也就是所谓的 Submenu，可以提供更多的选项让用户单击；不过在 Submenu 内不可再创建 Submenu。要产生一个用户可以操作的 Submenu，需要下列两个步骤。

STEP 1 使用 XML 文件创建 Submenu 组件：前述的<item>标签内再以嵌套方式置入<menu>与对应的<item>标签即可。

STEP 2 改写与 Menu 有关的方法：根据 Submenu 依附的对象(Options Menu 或 Context Menu)来改写对应的方法。

1. 使用 XML 文件创建 Submenu 组件

范例(如图 5-5、图 5-6 所示)说明：

- 画面显示：请按 Menu 按钮，按下实体 Menu 按钮后会弹出 5 个选项，如图 5-5 所示。
- 单击"阳明山公园"会弹出子菜单，其上有"园区导航"、"交通位置"两个子选项，如图 5-6 所示。
- 单击任何一个子选项都会以 Toast 消息框呈现父选项与子选项上面的文字。

图 5-5

图 5-6

SubmenuEx/res/menu/mymenu.xml

```
1.    <?xml version="1.0" encoding="utf-8"?>
2.    <menu xmlns:android="http://schemas.android.com/apk/res/android">
3.        <item android:id="@+id/yangmingshan" android:title="@string/yangmingshan" >
4.            <menu>
5.                <item android:id="@+id/guide" android:title="@string/guide" />
6.                <item android:id="@+id/traffic" android:title="@string/traffic" />
7.            </menu>
8.        </item>
9.        <item android:id="@+id/yushan" android:title="@string/yushan" />
10.       <item android:id="@+id/taroko" android:title="@string/taroko" />
11.       <item android:id="@+id/myloc" android:title="@string/myloc" />
12.       <item android:id="@+id/exit" android:title="@string/exit" />
13.   </menu>
```

3~7 行：为 3 行的选项创建子菜单，并在 5、6 行定义子选项。也可以在程序代码内调用 Menu 的 addSubMenu()动态新建 SubMenu(Menu 的子界面)组件，然后通过 SubMenu 组件调用 add() 动态新建子选项。

2. 改写与 Menu 有关的方法

如果 Submenu 依附的对象是 Options Menu，应改写 onCreateOptionsMenu()、onOptionsItemSelected()；如果依附的对象是 Context Menu，则应改写 onCreateContextMenu()、onContextItemSelected()。

SubmenuEx/src/org/submenuEx/SubmenuEx.java

```
25.       @Override
26.       public boolean onOptionsItemSelected(MenuItem item) {
27.           String msg = "";
28.           switch (item.getItemId()) {
29.           case R.id.yangmingshan:
30.               msg = getString(R.string.yangmingshan);
31.               break;
32.           case R.id.guide:
33.               msg = getString(R.string.yangmingshan) + " > " +
34.                   getString(R.string.guide);
```

```
35.         break;
36.     case R.id.traffic:
37.         msg = getString(R.string.yangmingshan) + " > " +
38.                 getString(R.string.traffic);
39.         break;
40.
41.     case R.id.yushan:
42.         msg = getString(R.string.yushan);
43.         break;
44.     case R.id.taroko:
45.         msg = getString(R.string.taroko);
46.         break;
47.     case R.id.myloc:
48.         msg = getString(R.string.myloc);
49.         break;
50.     case R.id.exit:
51.         this.finish();
52.     default:
53.         return super.onOptionsItemSelected(item);
54.     }
55.     Toast.makeText(this, msg, Toast.LENGTH_SHORT).show();
56.     return true;
57. }
```

26 行：无论单击主菜单(Options Menu)选项或是子菜单(Submenu)选项都会调用此方法。

32~39 行：通过 switch-case 判断被选择的选项，如果是子菜单的选项被选择，则使用 Toast 消息框呈现被选择的子选项与所属父选项的文字。

5.2 对话窗口与日期选择器

当用户要删除一联系人时，为了避免删错，通常会弹出一个对话窗口再次询问"删除此联系人？"，让用户有反悔的机会，如图 5-7 所示。

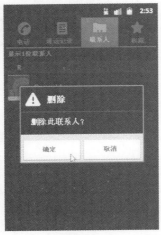

图 5-7

以上所述的对话窗口属于 AlertDialog(警示对话窗口)，除此之外 Android 还提供 DatePickerDialog(日期选择器)与 TimePickerDialog(时间选择器)等其他功能的对话窗口。

5.2.1 AlertDialog

AlertDialog 是 Dialog 的子类，是最常见的对话窗口。创建 AlertDialog 时必须设置 3 个部分：
- 对话窗口的标题文字。
- 对话窗口的消息正文。
- 对话窗口的按钮(包含 Button 事件处理)及按钮上面的文字。

使用 AlertDialog.Builder 类可以快速创建并设置 AlertDialog，其常用方法如表 5-1 所示。

表 5-1

AlertDialog.Builder 类
构 造 函 数
public AlertDialog.Builder (Context context) 创建 AlertDialog.Builder 对象 ● context：对话窗口所依附的对象，通常是指 Activity 对象
方　　法
public AlertDialog.Builder setTitle (int titleId) 设置对话窗口的标题文字 ● resId：标题文字的资源 ID，通常是指文本文件内对应的文字 ID
public void setIcon (int resId) 设置对话窗口的图标 ● titleId：图标的资源 ID
public AlertDialog.Builder setMessage (int messageId) 设置对话窗口欲显示的消息正文 ● messageId：消息正文的资源 ID，通常是指文本文件内对应的文字 ID
public AlertDialog.Builder setPositiveButton (int textId, DialogInterface.OnClickListener listener) 设置对话窗口的 Positive 按钮[①]及上面的文字 ● textId：按钮文字的资源 ID，通常是指文本文件内对应的文字 ID ● listener：监听 Positive 按钮是否被按下的监听器
AlertDialog.Builder 类
方　　法
public AlertDialog.Builder setNegativeButton (int textId, DialogInterface.OnClickListener listener) 设置对话窗口的 Nagative 按钮及上面的文字 ● textId：按钮文字的资源 ID，通常是指文本文件内对应的文字 ID ● listener：监听 Negative 按钮是否被按下的监听器
public AlertDialog.Builder setCancelable (boolean cancelable) 设置对话窗口是否可以被取消 ● cancelable：true 代表可以取消，例如按下实体的返回按钮可以取消对话窗口；false 代表不可以取消。默认为 true
public AlertDialog show () 按照 AlertDialog.Builder 的设置创建 AlertDialog，并显示在画面上

[①] AlertDialog 一共可设 3 种按钮：PositiveButton、NegativeButton 和 NeutralButton。此种分类与按钮真正的功能不一定有关系，但建议功能与按钮的分类尽量相符，例如：将"确定"按钮设置在 PositiveButton。

UI 高级设计 05

 范例 AlertDialogEx

范例(如图 5-8 所示)说明：
- 按下"结束程序"按钮后会弹出 AlertDialog 并显示对应的标题、图标、信息与按钮。
- 按下"确定"按钮会结束并离开此应用程序。
- 按下"取消"按钮会返回到主窗口。

图 5-8

AlertDialogEx/src/org/alertDialogEx/AlertDialogEx.java

```
19.     private void findViews() {
20.         btnExit = (Button)findViewById(R.id.btnExit);
21.         btnExit.setOnClickListener(new OnClickListener() {
22.             @Override
23.             public void onClick(View v) {
24.                 new AlertDialog.Builder(AlertDialogEx.this)
25.                 .setTitle(R.string.title).setIcon(R.drawable.icon)
26.                 .setMessage(R.string.prompt)
27.                 .setPositiveButton(R.string.submit,
28.                     new DialogInterface.OnClickListener() {
29.                         public void onClick(DialogInterface dialog, int id) {
30.                             AlertDialogEx.this.finish();
31.                         }
32.                     })
33.                 .setNegativeButton(R.string.cancel,
34.                     new DialogInterface.OnClickListener() {
35.                         public void onClick(DialogInterface dialog, int id) {
```

121

```
36.                    dialog.cancel();
37.                  }
38.                })
39.             .setCancelable(false)
40.             .show();
41.       }
42.    });
40. }
```

25 行：设置对话窗口的标题文字与图标。

26 行：设置对话窗口欲显示的消息正文。

27 行：设置对话窗口的 Positive 按钮及上面的文字。

28~32 行：利用匿名内部类实现 DialogInterface.OnClickListener.onClick()，当 Positive 按钮被按下时，onClick() 会自动被调用，执行 AlertDialogEx.this.finish();并关闭 Activity。onClick() 方法签名为 public abstract void onClick (DialogInterface dialog, int id)。

- dialog：按钮所在的对话窗口。
- id：被按下的按钮所对应的 ID。

36 行：调用 cancel() 会取消对话窗口。

39 行：false 代表不可以取消对话窗口。

40 行：将创建好的 AlertDialog 显示在画面上。

5.2.2 DatePickerDialog 与 TimePickerDialog

为了让用户能够更直觉地选择日期/时间，开发者可以使用 DatePickerDialog/TimePickerDialog 可视化选择组件来达到此目的。创建 DatePickerDialog/TimePickerDialog 组件所需使用到的方法说明如表 5-2 所示。

表 5-2

DialogFragment 类
public Dialog onCreateDialog (Bundle savedInstanceState) 改写此方法以创建自定义对话窗口 savedInstanceState：存储 Fragment 最新状态
DatePickerDialog 类
public DatePickerDialog(Context context,DatePickerDialog.OnDateSetListener callBack,int year, int monthOfYear, int dayOfMonth) 创建 DatePickerDialog 对话窗口<ul><li>context：通常是现行 Activity 对象</li><li>callBack：OnDateSetListener 对象，为了调用已实现的 onDateSet()</li><li>year：DatePickerDialog 对话窗口呈现时，预选的年</li><li>monthOfYear：预选的月</li><li>dayOfMonth：预选的日</li></ul>

(续表)

DatePickerDialog.OnDateSetListener 界面
public abstract void onDateSet (DatePicker view, int year, int monthOfYear, int dayOfMonth) 用户选定日期并按下确定按钮后会自动调用此方法 ● view：发生事件的 DatePicker 组件 ● year：选定的年 ● monthOfYear：选定的月，值为 0~11；换句话说，1 月是以 0 代表，这是为了配合 Calendar 类 ● dayOfMonth：选定的日
TimePickerDialog 类
public TimePickerDialog(Context context, TimePickerDialog.OnTimeSetListener callBack, int hourOfDay, int minute, boolean is24HourView) 创建 TimePickerDialog 对话窗口 ● context：通常是现行 Activity 对象 ● callBack：OnTimeSetListener 对象，为了调用已实现的 onTimeSet() ● hourOfDay：TimePickerDialog 对话窗口呈现时，预选的小时 ● minute：预选的分钟 ● is24HourView：是否为 24 小时制
TimePickerDialog.OnTimeSetListener 界面
public abstract void onTimeSet (TimePicker view, int hourOfDay, int minute) 用户选定时间并按下确定按钮后会自动调用此方法 ● view：发生事件的 TimePicker 组件 ● hourOfDay：选定的小时 ● minute：选定的分钟

产生 DatePickerDialog 对话窗口的步骤如下：

STEP 1 API Level 11(Android 3.0)开始支持 DialogFragment，专门提供 DatePickerDialog 的内容；所以第一步先继承 DialogFragment 类并改写 onCreateDialog()方法以提供 Dialog 内容。之后实现 OnDateSetListener.onDateSet()，日期选择器的"设置"按钮被按下会调用此方法。

```java
public static class DatePickerFragment extends DialogFragment implements
    DatePickerDialog.OnDateSetListener {
@Override
// 改写此方法以提供 Dialog 内容
public Dialog onCreateDialog(Bundle savedInstanceState) {
    DatePickerDialog datePickerDialog = new DatePickerDialog(
        getActivity(), this, mYear, mMonth, mDay);
    return datePickerDialog;
}
@Override
// 日期选择完成会调用此方法，并传入选择的年月日
public void onDateSet(DatePicker view, int year, int month, int day){
}
}
```

STEP 2 调用 DialogFragment.show()在画面上呈现对应的 Dialog。

```
DatePickerFragment datePickerFragment = new DatePickerFragment();

FragmentManager fm = getFragmentManager();

datePickerFragment.show(fm, "datePicker");
```

 范例 DatePickerEx

范例(如图 5-9 所示)说明：
- 按下"改变日期"按钮会弹出 DatePickerDialog 让用户设置日期。
- 按下"设置"按钮会将用户选定的日期显示在主窗口的 TextView 组件上。按下"取消"按钮则取消日期设置并返回到主窗口。

图 5-9

范例创建步骤如下：

STEP 1 创建 res/layout/main.xml 作为此范例的 layout 文件，并在其内创建 TextView 与 Button 组件，然后呈现日期以及让用户按下按钮以弹出日期选择器。

```
<LinearLayout xmlns:android="http://schemas.android.com/apk/res/android"
    android:layout_width="match_parent"
    android:layout_height="match_parent"
    android:orientation="vertical" >

    <!-- 显示日期 -->
    <TextView
```

```xml
            android:id="@+id/tvDateDisplay"
            android:layout_width="match_parent"
            android:layout_height="wrap_content"
            android:layout_marginTop="10dp"
            android:textSize="16sp" />

    <!-- "改变日期"按钮 -->
    <Button
            android:id="@+id/btnPickDate"
            android:layout_width="wrap_content"
            android:layout_height="wrap_content"
            android:layout_marginTop="10dp"
            android:text="@string/btnPickDate" />

</LinearLayout>
```

STEP 2 在 DatePickerEx.java 文件内添加下列程序代码。

```java
public class DatePickerEx extends Activity {
    private static TextView tvDateDisplay;
    private Button btnPickDate;
    private static int mYear;
    private static int mMonth;
    private static int mDay;

    @Override
    public void onCreate(Bundle savedInstanceState) {
        super.onCreate(savedInstanceState);
        setContentView(R.layout.main);
        findViews();
        // 取得现在日期
        final Calendar c = Calendar.getInstance();
        mYear = c.get(Calendar.YEAR);
        mMonth = c.get(Calendar.MONTH);
        mDay = c.get(Calendar.DAY_OF_MONTH);
        updateDisplay();
    }

    // 将指定的日期显示在 TextView 上。"mMonth + 1"是因为一月的值是 0 而非 1
    private static void updateDisplay() {
        StringBuilder sb = new StringBuilder().append(mYear).append("-")
            .append(pad(mMonth + 1)).append("-").append(pad(mDay));
        tvDateDisplay.setText(sb);
    }

    // 若数字有十位数,直接显示;若只有个位数则补 0 后再显示。例如 7 会改成 07 后再显示
    private static String pad(int c) {
        if (c >= 10)
            return String.valueOf(c);
```

```java
        else
            return "0" + String.valueOf(c);
    }

    private void findViews() {
        tvDateDisplay = (TextView) findViewById(R.id.tvDateDisplay);
        btnPickDate = (Button) findViewById(R.id.btnPickDate);
        btnPickDate.setOnClickListener(new OnClickListener() {
            @Override
            // 按下"改变日期"按钮会呈现 DatePickerFragment.onCreateDialog()
            // 返回的 DatePickerDialog，也就是日期选择器
            public void onClick(View v) {
                DatePickerFragment datePickerFragment = new DatePickerFragment();
                FragmentManager fm = getFragmentManager();
                datePickerFragment.show(fm, "datePicker");
            }
        });
    }

    // 此 Fragment 为内部类，必须声明为 static
    public static class DatePickerFragment extends DialogFragment implements
            DatePickerDialog.OnDateSetListener {

        @Override
        // 改写此方法以提供 Dialog 内容
        public Dialog onCreateDialog(Bundle savedInstanceState) {
            // 创建 DatePickerDialog 对象
            // this 为 OnDateSetListener 对象
            // mYear、mMonth、mDay 会成为日期选择器预选的年月日
            DatePickerDialog datePickerDialog = new DatePickerDialog(
                    getActivity(), this, mYear, mMonth, mDay);
            return datePickerDialog;
        }

        @Override
        // 日期选择完成会调用此方法，并传入选择的年月日
        public void onDateSet(DatePicker view, int year, int month, int day) {
            mYear = year;
            mMonth = month;
            mDay = day;
            updateDisplay();
        }
    }
}
```

UI 高级设计

产生 TimePickerDialog 对话窗口的步骤与 DatePickerDialog 大致相同，不再赘述。

 范例 TimePickerEx

范例(如图 5-10 所示)说明：
- 按下"改变时间"按钮会弹出 TimePickerDialog 让用户设置时间。
- 按下"设置"按钮会将用户选定的时间显示在主窗口的 TextView 组件上。按下"取消"按钮则取消时间设置并返回到主窗口。

图 5-10

范例创建步骤如下：

 创建 res/layout/main.xml 作为此范例的 layout 文件，并在其内创建 TextView 与 Button 组件，然后呈现时间以及让用户按下按钮以弹出时间选择器。

```xml
<LinearLayout xmlns:android="http://schemas.android.com/apk/res/android"
    android:layout_width="match_parent"
    android:layout_height="match_parent"
    android:orientation="vertical" >

    <!-- 显示时间 -->
    <TextView
        android:id="@+id/tvTimeDisplay"
        android:layout_width="match_parent"
        android:layout_height="wrap_content"
        android:layout_marginTop="10dp"
        android:textSize="16sp" />
```

127

```xml
<!-- "改变时间" 按钮 -->
<Button
    android:id="@+id/btnPickTime"
    android:layout_width="wrap_content"
    android:layout_height="wrap_content"
    android:layout_marginTop="10dp"
    android:text="@string/btnPickTime" />
```
```
</LinearLayout>
```

STEP 2　在 TimePickerEx.java 文件内添加下列程序代码。

```java
public class TimePickerEx extends Activity {
    private static TextView tvTimeDisplay;
    private Button btnPickTime;
    private static int mHour;
    private static int mMinute;

    @Override
    public void onCreate(Bundle savedInstanceState) {
        super.onCreate(savedInstanceState);
        setContentView(R.layout.main);
        findViews();
        // 取得现在时间
        final Calendar c = Calendar.getInstance();
        mHour = c.get(Calendar.HOUR_OF_DAY);
        mMinute = c.get(Calendar.MINUTE);
        updateDisplay();
    }

    // 将指定的时间显示在 TextView 上
    private static void updateDisplay() {
        StringBuilder sb = new StringBuilder().append(pad(mHour)).append(":")
            .append(pad(mMinute));
        tvTimeDisplay.setText(sb);
    }

    // 若数字有十位数，直接显示；若只有个位数则补 0 后再显示。例如 7 会改成 07 后再显示
    private static String pad(int c) {
        if (c >= 10)
            return String.valueOf(c);
        else
            return "0" + String.valueOf(c);
    }
```

```java
private void findViews() {
    tvTimeDisplay = (TextView) findViewById(R.id.tvTimeDisplay);
    btnPickTime = (Button) findViewById(R.id.btnPickTime);
    btnPickTime.setOnClickListener(new OnClickListener() {
        @Override
        // 按下"改变时间"按钮会呈现 TimePickerFragment.onCreateDialog()
        // 返回的 TimePickerDialog,也就是时间选择器
        public void onClick(View v) {
            TimePickerFragment timePickerFragment = new TimePickerFragment();
            FragmentManager fm = getFragmentManager();
            timePickerFragment.show(fm, "timePicker");
        }
    });
}

// 此 Fragment 为内部类,必须声明为 static
public static class TimePickerFragment extends DialogFragment implements
        TimePickerDialog.OnTimeSetListener {

    @Override
    public Dialog onCreateDialog(Bundle savedInstanceState) {
        // 创建 TimePickerDialog 对象
        // this 为 OnTimeSetListener 对象
        // mHour、mMinute 会成为时间选择器预选的时与分
        // false 代表不为 24 时制
        TimePickerDialog timePickerDialog = new TimePickerDialog(
                getActivity(), this, mHour, mMinute, false);
        return timePickerDialog;
    }

    @Override
    // 时间选择完成会调用此方法,并传入选择的时与分
    public void onTimeSet(TimePicker view, int hourOfDay, int minute) {
        mHour = hourOfDay;
        mMinute = minute;
        updateDisplay();
    }
}
```

5.2.3 CalendarView

要让用户选择日期,除了 DatePicker 外,也可以使用 CalendarView (Android 3.0, API Level 11 开始支持)。CalendarView 组件是以日历方式呈现,方便用户选择日期。要使用 CalendarView

组件,最重要的就是处理用户更改日期时所触发的事件,开发者必须实现 CalendarView.OnDateChangeListener 的 onSelectedDayChange()方法,说明如表 5-3 所示。

表 5-3

CalendarView.OnDateChangeListener 界面
public abstract void onSelectedDayChange(CalendarView view, int year, int month, int dayOfMonth) 用户选定日期后会自动调用此方法 • view:发生事件的 CalendarView 组件 • year:选定的年 • month:选定的月,值为 0~11;换句话说,1 月是以 0 代表 • dayOfMonth:选定的日

 范例 CalendarViewEx

范例(如图 5-11 所示)说明:
用户单击日期后,会将选定的日期显示在左上角的 TextView 组件上。

图 5-11

CalendarViewEx/res/layout/main.xml

```
12.     <CalendarView
13.         android:id="@+id/calendarView"
14.         android:layout_width="match_parent"
15.         android:layout_height="match_parent"
16.         android:firstDayOfWeek="2"
17.         android:showWeekNumber="false" />
```

16 行:设置一周的第一天是星期几,1 是星期日,2 是星期一,以此类推。

17 行:是否要在左边显示周次。

CalendarViewEx/src/org/calendarViewEx/CalendarViewExActivity.java

```
11. public class CalendarViewExActivity extends Activity {
12.     private TextView tvDateDisplay;
13.     private CalendarView calendarView;
14.     private int mYear;
15.     private int mMonth;
16.     private int mDay;
17.
18.     /** Called when the activity is first created. */
19.     @Override
20.     public void onCreate(Bundle savedInstanceState) {
21.         super.onCreate(savedInstanceState);
22.         setContentView(R.layout.main);
23.         findViews();
24.         final Calendar c = Calendar.getInstance();
25.         mYear = c.get(Calendar.YEAR);
26.         mMonth = c.get(Calendar.MONTH);
27.         mDay = c.get(Calendar.DAY_OF_MONTH);
28.         updateDisplay();
29.     }
30.
31.     private void updateDisplay() {
32.         tvDateDisplay.setText(
33.             new StringBuilder()
34.                 .append(mYear).append("-")
35.                 .append(mMonth + 1).append("-")
36.                 .append(mDay));
37.     }
38.
39.     private void findViews() {
40.         tvDateDisplay = (TextView)findViewById(R.id.tvDateDisplay);
41.         calendarView = (CalendarView)findViewById(R.id.calendarView);
42.         calendarView.setOnDateChangeListener(new OnDateChangeListener() {
43.             @Override
44.             public void onSelectedDayChange(CalendarView view, int year, int month,
45.                     int dayOfMonth) {
46.                 mYear = year;
47.                 mMonth = month;
48.                 mDay = dayOfMonth;
49.                 updateDisplay();
50.             }
51.         });
```

```
52.     }
53. }
```

31~37 行：将指定的日期显示在 TextView 上。

42 行：CalendarView 组件调用 setOnDateChangeListener() 向 OnDateChangeListener 注册，OnDateChangeListener 专门监听用户是否改变选择的日期。

44~50 行：利用匿名内部类实现 OnDateChangeListener 的 onSelectedDayChange()方法。当用户改变选择的日期时，onSelectedDayChange() 会自动被调用并传递用户选择的年/月/日(year/month/dayOfMonth)，最后在 49 行调用 updateDisplay()方法更新 TextView 组件上显示的日期。

5.3 Spinner 与 AutoCompleteTextView

5.3.1 Spinner

Spinner 是一个非常类似下拉菜单(drop-down list)的 UI 组件，其优点为节省显示空间，因为用户尚未单击时，仅显示一笔数据。如同其他 UI 组件，一般建议通过 layout 文件创建 Spinner 组件。使用 layout 文件(可参见 SpinnerEx/res/layout/main.xml)创建 Spinner 组件以及选项被单击后的事件处理步骤如下。

STEP 1 创建 Spinner 选项：在文本文件内创建字符串数组[①](参见 SpinnerEx/res/values/strings.xml)，而数组内容即为 Spinner 选项。

```xml
<string-array name="food_array">
    <item>蚵仔面线</item>
    <item>臭豆腐</item>
    <item>葱油饼</item>
</string-array>
```

STEP 2 创建 Spinner 组件：使用<Spinner>标签。

STEP 3 设置 Spinner 的提示文字：利用 android:prompt 属性设置提示文字，例如 android:prompt="爱吃什么？"会显示如图 5-12 的提示文字。

STEP 4 设置 Spinner 选项：利用 android:entries 属性设置选项。例如 android:entries="@array/food_array"，而 array/food_array 就是 Step 1 创建好的字符串数组。

---

① 也可以在程序代码内使用数组或 List 集合创建选项，例如：String[] foods = {"蚵仔面线", "臭豆腐", "葱油饼"};，然后使用 ArrayAdapter 来获取内容。

UI 高级设计 05

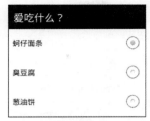

图 5-12

STEP 5 取得 Spinner 组件后调用 setOnItemSelectedListener() 向 OnItemSelectedListener 监听器注册，并实现 onItemSelected() 以响应选项改变的情况。

```
spFood = (Spinner)findViewById(R.id.spFood);
spFood.setOnItemSelectedListener(listener);

Spinner.OnItemSelectedListener listener =
    new Spinner.OnItemSelectedListener(){
    @Override
    public void onItemSelected(AdapterView parent,
        View view, int pos, long id) {
            //ItemSelected
    }

    @Override
    public void onNothingSelected(AdapterView parent) {
        //NothingSelected
    }
};
```

创建 Spinner 组件所需使用到的方法说明如表 5-4 所示。

表 5-4

AdapterView 类[①]
public void setOnItemSelectedListener (AdapterView.OnItemSelectedListener listener)
向 OnItemSelectedListener 监听器注册，要求监听选项是否被选择
listener：实现 OnItemSelectedListener 界面的对象
AdapterView.OnItemSelectedListener 界面
public abstract void onItemSelected (AdapterView<?> parent, View view, int position, long id)
当选项被选择时会自动调用此方法
● parent：发生选项被单击的 UI 组件，这里是指 Spinner 组件
● view：在 AdapterView 中被单击的组件，通常是 TextView 组件
● position：选项的 index
● id：选项的 row ID

---

① Spinner 继承 AbsSpinner；AbsSpinner 继承 AdapterView。

(续表)

public abstract void onNothingSelected (AdapterView<?> parent) 当选项消失时会自动调用此方法 parent：发生选项消失的 UI 组件，这里是指 Spinner 组件

ArrayAdapter 类[①]

构 造 函 数

public ArrayAdapter (Context context, int textViewResourceId, T[] objects) 创建 ArrayAdapter 组件。ArrayAdapter 组件主要用来管理整个选项的内容与样式 ● context：通常是现行 Activity 对象 ● textViewResourceId：选项的样式主要是以 TextView 组件组成，所以可以在 layout 文件内创建 TextView 组件的样式，然后通过资源 ID 应用在选项上。如果不想自行创建选项样式，也可以直接应用 Android 系统内置的样式：android.R.layout.simple_spinner_item ● objects：选项的内容
public void setDropDownViewResource(int resource) 设置整个下拉菜单的样式 esource：通过资源 ID 应用欲呈现的下拉菜单样式。可以直接应用 Android 系统内置的样式：android.R.layout.simple_spinner_dropdown_item

AbsSpinner 类

public void setAdapter (SpinnerAdapter adapter) 将设置好的选项内容与样式应用在 Spinner 组件上 adapter：可以应用上述 ArrayAdapter 对象，也就是选项的内容与样式

 范例 SpinnerEx

范例(如图 5-13 所示)说明

单击任何一个 Spinner 组件都会将被选择的选项文字以 Toast 消息框显示出来。

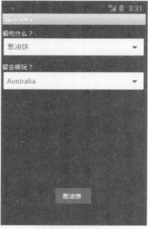

图 5-13

---

① ArrayAdapter 继承 SpinnerAdapter。

UI 高级设计

SpinnerEx/src/org/spinnerEx/SpinnerEx.java

```java
22.    private void findViews() {
23.        spFood = (Spinner)findViewById(R.id.spFood);
24.        spFood.setOnItemSelectedListener(listener);
25.
26.        spPlace = (Spinner)findViewById(R.id.spPlace);
27.        String[] places = {"Australia", "U.K.", "Japan", "Thailand"};
28.        ArrayAdapter<String> adapterPlace =
29.            new ArrayAdapter<String>(this,
30.                    android.R.layout.simple_spinner_item, places);
31.        adapterPlace.setDropDownViewResource(
32.                    android.R.layout.simple_spinner_dropdown_item);
33.        spPlace.setAdapter(adapterPlace);
34.        spPlace.setOnItemSelectedListener(listener);
35.    }
36.
37.    Spinner.OnItemSelectedListener listener =
38.        new Spinner.OnItemSelectedListener(){
39.            @Override
40.            public void onItemSelected(AdapterView parent,
41.                View view, int pos, long id) {
42.                Toast.makeText(parent.getContext(),
43.                        parent.getItemAtPosition(pos).toString(),
44.                        Toast.LENGTH_SHORT).show();
45.            }
46.
47.            @Override
48.            public void onNothingSelected(AdapterView parent) {
49.                Toast.makeText(parent.getContext(),
50.                        "Nothing Selected!",
51.                        Toast.LENGTH_SHORT).show();
52.            }
53.    };
```

24、34、37~53 行：Spinner 组件调用 setOnItemSelectedListener() 向 OnItemSelectedListener 注册，OnItemSelectedListener 专门监听选项是否被选择。接下来 37 行利用匿名内部类实现 OnItemSelectedListener 的 onItemSelected()、onNothingSelected()等两个方法。当选项被选择时，onItemSelected() 会自动被调用，而被选择的选项文字将以 Toast 消息框显示出来。

27~30 行：调用 ArrayAdapter 构造函数以创建选项的内容与样式。选项的内容来自于 27 行的 places 字符串数组；样式则应用系统内置：android.R.layout.simple_spinner_item。

31~32 行：调用 setDropDownViewResource()应用系统内置的下拉菜单样式：android.R.layout.simple_spinner_dropdown_item。

33 行：Spinner 组件调用 setAdapter() 应用指定的 ArrayAdapter 以加载对应的选项内容与样式。

5.3.2 AutoCompleteTextView

AutoCompleteTextView非常类似EditText，都是方便用户输入的组件。不过AutoComplete-TextView组件另外提供提示文字列表，当用户输入的局部文字符合提示文字时，应用程序就会自动列出符合的提示文字列表，让用户可以直接选择欲输入的文字而不必将全部文字输入完毕，是一种体贴用户输入的设计。

AutoCompleteTextView组件的提示列表与Spinner组件的选项创建方式相同，都是在文本文件内创建字符串数组来存储欲提示的文字。

范例 AutoCompleteEx

范例(如图5-14所示)说明：

输入a，应用程序会作比较，并自动将符合的提示文字以列表方式呈现，方便用户以选择方式输入。

图 5-14

AutoCompleteEx/res/layout/main.xml

```
13.     <AutoCompleteTextView android:id="@+id/acCountry"
14.         android:layout_width="match_parent"
15.         android:layout_height="wrap_content"
16.         android:layout_marginLeft="5dp"
17.         android:completionThreshold="1" />
```

13行：定义AutoCompleteTextView组件。

17行：设置至少要输入1个字符才会显示提示文字。如果未设置，则默认为两个字符。

AutoCompleteEx/src/org/autoCompleteEx/AutoCompleteEx.java

```
18.     private void findViews() {
19.         acCountry = (AutoCompleteTextView)findViewById(R.id.acCountry);
20.         String[] countries =
```

```
21.         getResources().getStringArray(R.array.countries_array);
22.     ArrayAdapter<String> adapterCountry =
23.         new ArrayAdapter<String>(this, R.layout.list_item, countries);
24.     acCountry.setAdapter(adapterCountry);
25. }
```

21 行：调用 Context 的 getResources() 会取得 Resources 对象，再调用 getStringArray() 会取得字符串数组，其内容将用来作为 AutoCompleteTextView 组件的提示文字。R.array.countries_array 代表定义在文本文件内，名为 countries_array 的字符串数组。

23 行：R.layout.list_item 代表提示文字将应用"list_item" layout 文件所定义的样式。ArrayAdapter 已在 Spinner 一节说明，不再赘述。

5.4 Gallery 与 GridView

5.4.1 Gallery

Gallery 组件属于 ViewGroup，是以水平方式显示图片，而且可以让用户以滑动方式浏览，被选择的图片会放在 Gallery 组件的正中间。

 范例 GalleryEx

范例(如图 5-15 所示)说明：

单击 Gallery 的图片后会将该图片显示在中间 ImageView 上，并以 Toast 显示该图片的索引。

图 5-15

范例创建步骤如下：

STEP 1 准备图形文件并放在项目的 res/drawable 目录内。

STEP 2 在 res/layout/main.xml 文件内新建 Gallery 组件：

```xml
<Gallery
    android:id="@+id/gallery"
    android:layout_width="match_parent"
    android:layout_height="wrap_content" />
```

STEP 3 创建一个 XML 文件 attrs.xml 并放在 res/values 目录内，内容如下：

```xml
<!-- galleryItemBackground 定义 Gallery 图片的背景样式，将会在下列程序代码中被应用-->
<resources>
    <declare-styleable name="GalleryEx">
        <attr name="android:galleryItemBackground" />
    </declare-styleable>
</resources>
```

STEP 4 GalleryExActivity.java 文件内插入下列程序代码。

```java
public class GalleryExActivity extends Activity {
    ImageView imageView;
    Gallery gallery;

    /*声明 images 数组存储 res/drawable 所有图形文件的资源 ID */
    private Integer[] images = {
            R.drawable.p01, R.drawable.p02, R.drawable.p03, R.drawable.p04,
            R.drawable.p05, R.drawable.p06, R.drawable.p07
    };

    /*继承 BaseAdapter 类并改写 getCount()、getItem()、getItemId()、getView()方法，系统会自动调
      用这些方法*/
    public class ImageAdapter extends BaseAdapter {
        int galleryItemBackground;
        private Context context;

        public ImageAdapter(Context c) {
            context = c;
            /*Context 对象调用 obtainStyledAttributes()并搭配
              R.styleable.GalleryEx 参数来取得存储在 attrs.xml 文件内名称同样为
              GalleryEx 的所有属性*/
            TypedArray attrs = context.obtainStyledAttributes(R.styleable.GalleryEx);
            /* attrs 调用 getResourceId()方法并搭配第一个参数可以取得名称为
```

android:galleryItemBackground 的属性，而该属性即为系统内置的 Gallery 背景样式。第二个参数代表默认值，也就是说当第一个参数不合法时，会直接应用第二个参数的值*/

```java
        galleryItemBackground = attrs.getResourceId(
                R.styleable.GalleryEx_android_galleryItemBackground, 0);
        /*attrs 调用 recycle()方法是为了之后可以重复利用 attrs 来取得属性值*/
        attrs.recycle();
    }
    /*返回图片总数，所以返回 images.length*/
    @Override
    public int getCount() {
        return images.length;
    }
    /*在此 getItem()与 getItemId()不重要，所以直接返回 position 参数即可*/
    @Override
    public Object getItem(int position) {
        return position;
    }

    @Override
    public long getItemId(int position) {
        return position;
    }
    /*系统会按照 getCount()方法返回的总数来决定调用 getView()方法的次数。第一
    调用 getView()方法时，position 参数会传入 0，代表要呈现索引为 0 的图片，而
    imageView.setImageResource(images[0]);就代表将索引为 0 的图片显示在 ImageView
    组件上，最后将 imageView 返回*/
    @Override
    public View getView(int position, View convertView, ViewGroup parent) {

        ImageView imageView;
        /* convertView 是上次调用 getView()方法所取得的 View，保存下来的目的是为了
        重复利用，应先检查 convertView 是否为 null，如果是，产生一个新的 ImageView；
        如果不是，则沿用之前旧的 View 并转换成 ImageView */
        if (convertView == null) {
            imageView = new ImageView(context);
                /*ImageView 组件的宽与高都为 150*/
            imageView.setLayoutParams(new Gallery.LayoutParams(150, 150));
                /*调整 ImageView 组件的尺寸以符合目的位置的尺寸*/
            imageView.setScaleType(ImageView.ScaleType.FIT_XY);
                /*ImageView 组件应用 galleryItemBackground 所指定的背景样式*/
            imageView.setBackgroundResource(galleryItemBackground);
        } else {
            imageView = (ImageView)convertView;
```

```
                }
                imageView.setImageResource(images[position]);
                return imageView;
            }
    }

    @Override
    public void onCreate(Bundle savedInstanceState) {
        super.onCreate(savedInstanceState);
        setContentView(R.layout.main);
        findViews();
    }

    private void findViews() {
        imageView = (ImageView)findViewById(R.id.imageView);
        gallery = (Gallery) findViewById(R.id.gallery);
        /*调用 setAdapter()方法设置 Gallery 组件可供浏览的图片与对应的索引，而这些设置声
明在 ImageAdapter 类*/
        gallery.setAdapter(new ImageAdapter(this));
        /*调用 setOnItemClickListener()方法向 OnItemClickListener 注册，
OnItemClickListener 会监听用户是否单击 Gallery 组件的图片，如果是，会自动调用
onItemClick()方法并传递 position 参数。position 参数代表被单击图片的索引，利用
Toast 将 position 值呈现出来，并用以取得 images 数组内所存储的图片资源 ID*/
        gallery.setOnItemClickListener(new OnItemClickListener() {
            public void onItemClick(AdapterView<?> parent, View v, int position, long id) {
                Toast.makeText(GalleryExActivity.this, "" + position, Toast.LENGTH_SHORT).show();
                /*调用 setImageResource()方法并搭配图片资源 ID 即可显示该 ID 所代表的
            图片于 ImageView 组件上*/
                imageView.setImageResource(images[position]);
            }
        });
    }
}
```

5.4.2 GridView

GridView 组件属于 ViewGroup，是以 2D 方式显示其内容，开发者经常使用 GridView 来呈现缩图方便用户浏览，其程序编写方式很像前述的 Gallery 组件。

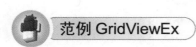
范例 GridViewEx

范例(如图 5-10 所示)说明：
单击 GridView 组件上的图片后会以 Toast 显示该图片的索引。

UI 高级设计 05

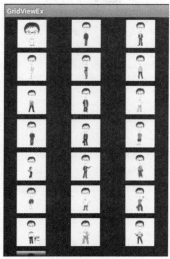

图 5-16

范例创建步骤如下：

STEP 1 准备图形文件并放在项目的 res/drawable 目录内。

STEP 2 在 res/layout/main.xml 文件内新建 GridView 组件：

```
<GridView xmlns:android="http://schemas.android.com/apk/res/android"
    android:id="@+id/gridView"
    android:layout_width="match_parent"
    android:layout_height="match_parent"
    android:columnWidth="90dp"
    android:numColumns="auto_fit"
    android:verticalSpacing="10dp"
    android:horizontalSpacing="10dp"
    android:gravity="center"
/>
<!--
android:columnWidth="90dp" 指定每一列的宽度为 90 像素
android:numColumns="auto_fit" 自动指定列数
android:verticalSpacing="10dp" 指定二行垂直间距为 10 像素
android:horizontalSpacing="10dp" 指定二列水平间距为 10 像素
android:gravity="center" 指定单元格对齐方式为居中对齐
-->
```

STEP 3 GridViewExActivity.java 文件内插入下列程序代码。

```
public class GridViewExActivity extends Activity {
    private GridView gridView;
    /*继承 BaseAdapter 类并改写 getCount()、getItem()、getItemId()、getView()方法,
```

141

系统会自动调用这些方法*/
```java
private class ImageAdapter extends BaseAdapter {
    private Context context;
    /*images 数组存储 res/drawable 所有图形文件的资源 ID*/
    private Integer[] images = {
        R.drawable.p01, R.drawable.p02, R.drawable.p03, R.drawable.p04,
        R.drawable.p05, R.drawable.p06, R.drawable.p07, R.drawable.p08,
        R.drawable.p09, R.drawable.p10, R.drawable.p11, R.drawable.p12,
        R.drawable.p13, R.drawable.p14, R.drawable.p15, R.drawable.p16,
        R.drawable.p17, R.drawable.p18, R.drawable.p19, R.drawable.p20,
        R.drawable.p21, R.drawable.p22
    };
    public ImageAdapter(Context context) {
        this.context = context;
    }
    /* getCount()方法返回图片总数，所以返回 images.length */
    @Override
    public int getCount() {
        return images.length;
    }
    /*在此 getItem()与 getItemId()不重要，所以直接返回 position 参数即可*/
    @Override
    public Object getItem(int position) {
        return position;
    }
    @Override
    public long getItemId(int position) {
        return position;
    }
    /*系统会按照 getCount()方法返回的总数来决定调用 getView()方法的次数。第一次调用
    getView()方法时，position 参数会传入 0，代表要呈现索引为 0 的图片，而
    imageView.setImageResource(images[0]); 就代表将索引为 0 的图片显示在
    ImageView 组件上，最后将 imageView 返回*/
    @Override
    public View getView(int position, View convertView, ViewGroup parent) {
        ImageView imageView;
        /* convertView 是上次调用 getView()方法所取得的 View，保存下来的目的是
        为了重复利用，应先检查 convertView 是否为 null，如果是，产生一个新的
        ImageView；如果不是，则沿用之前旧的 View 并转换成 ImageView */
        if (convertView == null) {
            imageView = new ImageView(context);
            /* ImageView 组件的宽与高都为 85*/
            imageView.setLayoutParams(new GridView.LayoutParams(85, 85));
```

```java
        /*调整 ImageView 组件的尺寸以符合目的位置的尺寸*/
        imageView.setScaleType(ImageView.ScaleType.FIT_XY);
    } else {
        imageView = (ImageView) convertView;
    }
    imageView.setImageResource(images[position]);
    return imageView;
    }
}

@Override
public void onCreate(Bundle savedInstanceState) {
    super.onCreate(savedInstanceState);
    setContentView(R.layout.main);
    findViews();
}

private void findViews() {
    gridView = (GridView)findViewById(R.id.gridView);
    /*调用 setAdapter()方法设置 GridView 组件可供浏览的图片与对应的索引,而这些设置声明在 ImageAdapter 类*/
    gridView.setAdapter(new ImageAdapter(this));
    /*调用 setOnItemClickListener()方法向 OnItemClickListener 注册,
    OnItemClickListener 会监听用户是否单击 GridView 组件的图片,如果是,会自动呼
    叫 onItemClick()方法并传递 position 参数。position 参数代表被单击图片的索引,利
    用 Toast 将 position 值呈现出来*/
    gridView.setOnItemClickListener(new OnItemClickListener() {
        @Override
        public void onItemClick(AdapterView<?> parent, View v, int position, long id) {
            Toast.makeText(GridViewExActivity.this, "" + position, Toast.LENGTH_SHORT).show();
        }
    });
    }
}
```

5.5 ListView

ListView 组件属于 ViewGroup,以列表方式呈现内容,如果内容过长,用户可以翻阅画面来浏览,此组件非常适合用来呈现大量数据。

5.5.1 ListActivity

如果只想要条列项目让用户单击，之后做对应事件处理，使用 ListActivity 组件最为恰当，程序编写也很简单，而 ListActivity 就是使用 ListView 来显示数据。

 范例 ListActivityEx

范例说明：
- 单击项目文字后会以 Toast 显示该文字，如图 5-17 所示。
- 直接输入英文字母可以将开头字母相同的项目筛选出来，如图 5-18 所示。原始数据不需要按照英文字母排列也可达到同样效果。

图 5-17

图 5-18

范例创建步骤如下：

STEP 1 创建 res/layout/listview_item.xml 作为 ListView 组件的 layout 文件，并在其内创建 TextView 组件，然后呈现在 ListView 组件上。

```
<!-- android:layout_width 设为 match_parent 而非 wrap_content 是因为设为 match_parent，只要单击该列范围内的区域即可；设为 wrap_content 则须单击字的区域方有反应。例如在上图 5-17 中，如果设为 wrap_content，想要单击 Activity 项目就必须单击 Activity 这个字方有反应 -->
<TextView xmlns:android="http://schemas.android.com/apk/res/android"
    android:layout_width="match_parent"
    android:layout_height="match_parent"
    android:padding="10dp"
    android:textSize="16sp" />
```

STEP 2 在 res/values/strings.xml 文件内添加 string-array，存储着各选项文字。

```
<string-array name="glossary">
```

```xml
            <item>Activity</item>
            <item>ADB</item>
            <item>Application</item>
            <item>Broadcast Receiver</item>
            <item>Canvas</item>
            <item>Content Provider</item>
            <item>Dalvik</item>
            <item>DDMS</item>
            <item>Drawable</item>
            <item>Intent</item>
            <item>Layout Resource</item>
            <item>Manifest File</item>
            <item>Nine-patch image</item>
            <item>OpenGL ES</item>
            <item>Resources</item>
            <item>Service</item>
            <item>SurfaceView</item>
            <item>Theme</item>
            <item>View</item>
            <item>Viewgroup</item>
            <item>Widget</item>
    </string-array>
```

STEP 3 在 MyListActivity.java 文件内添加下列程序代码。

```java
public class MyListActivity extends ListActivity {

    @Override
    public void onCreate(Bundle savedInstanceState) {
        super.onCreate(savedInstanceState);
         //可以取得 strings.xml 文件内名为 glossary 的字符串数组
        String[] glossary = getResources().getStringArray(R.array.glossary);

         /*调用 ArrayAdapter 构造函数以创建选项的内容与样式，glossary 所代表的选项会应用
R.layout.listview_item 样式*/
        setListAdapter(new ArrayAdapter<String>(this, R.layout.listview_item, glossary));
        ListView listView = getListView();

         /*输入英文字母可以将开头字母相同的项目筛选出来*/
        listView.setTextFilterEnabled(true);

         /* listView 调用 setOnItemClickListener()方法向 OnItemClickListener 注册，
OnItemClickListener 会监听用户是否单击 ListView 组件上的项目，如果是，会自
动调用 onItemClick()方法并传递 view 参数。view 参数代表被单击的 View 组件，在此
为 TextView 组件，调用 getText()方法可以取得文字，最后以 Toast 呈现*/
```

```
        listView.setOnItemClickListener(new OnItemClickListener() {
            @Override
            public void onItemClick(AdapterView<?> parent, View view,
                    int position, long id) {
                Toast.makeText(getApplicationContext(), ((TextView) view).getText(),
                        Toast.LENGTH_SHORT).show();
            }
        });
    }
}
```

5.5.2 ListView

ListView 组件与 ImageView、TextView 相同，都是 widget，也都可以使用 layout 文件来配置页面。如果 ListView 上要显示多个组件(例如 ImageView、TextView)，可以自定义类并利用其属性分别引用到对应的组件，下面范例将有详细说明。

 范例 ListViewEx

范例(如图 5-19 所示)说明：
- ListView 的每一行都有 ImageView 与 TextView 组件分别呈现图片和文字。
- 单击项目后会以 Toast 显示该文字。

图 5-19

范例创建步骤如下：

STEP 1 在 res/layout/main.xml 文件内新建 ListView 组件，之后会放在 ListView 组件上以显示文字。

```xml
<ListView
    android:id="@+id/listView"
    android:layout_width="match_parent"
    android:layout_height="wrap_content" />
```

STEP 2 创建 res/layout/listview_item.xml 作为 ListView 组件的 layout 文件，并在其内创建 ImageView 与 TextView 组件，然后呈现在 ListView 组件上。

```xml
<ImageView android:id="@+id/imageView"
    android:layout_width="48dp"
    android:layout_height="48dp"
    android:padding="6dp" />

<TextView android:id="@+id/textView"
    android:layout_gravity="center_vertical"
    android:layout_width="wrap_content"
    android:layout_height="wrap_content"
    android:padding="6dp" />
```

STEP 3 在 MyListView.java 文件内添加下列程序代码。

```java
public class MyListView extends Activity {
    private ListView listView;
    /*继承 BaseAdapter 类并改写 getCount()、getItem()、getItemId()、getView()方法，系统会自动调
    用这些方法*/
    private class ImageTextAdapter extends BaseAdapter {
        private LayoutInflater layoutInflater;
        /* images 数组存储 res/drawable 所有图形文件的资源 ID */
        private Integer[] images = {
            R.drawable.p01, R.drawable.p02, R.drawable.p03, R.drawable.p04,
            R.drawable.p05, R.drawable.p06, R.drawable.p07, R.drawable.p08,
            R.drawable.p09, R.drawable.p10, R.drawable.p11, R.drawable.p12
        };
        /* ViewHolder 类的 imageView 与 textView 属性会在之后分别存储 ListView 的子组件 */
        private class ViewHolder {
            ImageView imageView;
            TextView textView;
        }

        public ImageTextAdapter(Context context) {
            /*调用 getSystemService()方法取得 LayoutInflater 对象，可以通过该对象
            取得指定 layout XML 文件内容后初始化成 View 对象*/
            layoutInflater =
                (LayoutInflater)context.getSystemService(LAYOUT_INFLATER_SERVICE);
        }
```

```java
/* getCount()方法返回图片总数，所以返回 images.length */
@Override
public int getCount() {
    return images.length;
}
/*在此 getItem()与 getItemId()不重要，所以直接返回 position 参数即可*/
@Override
public Object getItem(int position) {
    return position;
}

@Override
public long getItemId(int position) {
    return position;
}

/*系统会按照 getCount()方法返回的总数来决定调用 getView()方法的次数*/
@Override
public View getView(int position, View convertView, ViewGroup parent) {
    ViewHolder viewHolder;

    /* convertView 是上次调用 getView()所取得的 View，当 convertView 不为
   null，就不需要再次调用 inflate()方法重新产生它*/
    if (convertView == null) {
        convertView = layoutInflater.inflate(R.layout.listview_item, null);
        /*创建 viewHolder 对象并利用其属性分别存储 TextView 与 ImageView 对象的引用*/
        viewHolder = new ViewHolder();
        /*只要是 View 组件都可调用 findViewById()方法取得其子组件*/
        viewHolder.textView = (TextView)convertView.findViewById(R.id.textView);
        viewHolder.imageView = (ImageView)convertView.findViewById(R.id.imageView);
        /*利用 setTag()方法将 convertView 与 viewHolder 创建 View 组件之间的层次，
       换句话说，convertView 为父组件，viewHolder 属性所引用的对象为子组件*/
        convertView.setTag(viewHolder);
    } else {
        viewHolder = (ViewHolder) convertView.getTag();
    }
    viewHolder.imageView.setImageResource(images[position]);
    viewHolder.textView.setText("image " + (position + 1));
    return convertView;
    }
}

@Override
public void onCreate(Bundle savedInstanceState) {
    super.onCreate(savedInstanceState);
    setContentView(R.layout.main);
```

```
            findViews();
        }
        private void findViews() {
            listView = (ListView)findViewById(R.id.listView);
            /*调用 setAdapter()方法设置 ListView 组件可供浏览的图片与文字,而这些设置声明在
ImageTextAdapter 类*/
            listView.setAdapter(new ImageTextAdapter(this));
            /*调用 setOnItemClickListener()方法向 OnItemClickListener 注册,
OnItemClickListener 会监听用户是否单击 ListView 组件上的项目,如果是,会自动
调用 onItemClick()方法并传递 view 参数。view 参数代表被单击的 View 组件,在此为
LinearLayout 组件*/
            listView.setOnItemClickListener(new OnItemClickListener() {
                @Override
                public void onItemClick(AdapterView<?> parent, View view,
                    int position, long id) {
                    /*取得 LinearLayout 上索引为 1 的子组件,该组件为 TextView*/
                    TextView textView = (TextView) ((LinearLayout)view).getChildAt(1);
                    /*调用 getApplicationContext()方法可以取得 Context 对象*/
                    Toast.makeText(getApplicationContext(), textView.getText(),
                        Toast.LENGTH_SHORT).show();
                }
            });
        }
    }
```

5.6 自定义 View 组件与 2D 绘图

Android 提供 2D 绘图功能,开发者所需包为 android.graphics。自定义 View 组件与 2D 绘图说明如下。

(1) 继承 View 类并改写 onDraw():想绘图必须有一个可显示的组件可以提供绘制,要取得该组件最简单的方式就是自定义类(例如 GeometricView 类)去继承 View 类,并且改写 onDraw(),将想要绘制的图形置入 onDraw()方法内,如下列程序代码。

```
public class GeometricView extends View {
    public GeometricView(Context context){
        super(context);
        //初始化组件
    }
    @Override
    protected void onDraw(Canvas canvas) {
```

```
        //绘制图形的程序代码
    }
}
```

利用构造函数产生该类的对象时，系统便会自动调用改写好的 onDraw() 方法并完成图形绘制；而产生的对象便属于 View 组件的一种。

(2) 通过 layout 文件作页面配置：如果想要像其他 UI 组件一样通过 layout 文件来作页面配置，还必须创建 GeometricView(Context, AttributeSet)构造函数并将 AttributeSet(该组件的 XML 属性)参数传递给父类对应的构造函数，如下列程序代码。

```
public class GeometricView extends View {
    public GeometricView(Context context){
        super(context);
        //初始化组件
    }

    public GeometricView(Context context, AttributeSet attrs){
        super(context, attrs); //attrs 就是 layout 文件内为该组件设置的属性
        //初始化组件
    }
    @Override
    protected void onDraw(Canvas canvas) {
        //绘制图形的程序代码
    }
}
```

以 layout 文件创建此组件会自动调用 GeometricView(Context, AttributeSet)构造函数，而 layout 文件的设置方式如下：

```
<org.draw2D.GeometricView
    android:id="@+id/geomView"
    android:layout_width="match_parent"
    android:layout_height="wrap_content" />
```

(3) 调用 View.invalidate() 重绘组件：如果自定义组件产生后想要重新绘制该组件上面的图形，可以调用 View.invalidate()，系统会先删掉原来的图形然后自动调用 onDraw()，以重新绘制此 View 的内容。

范例(如图 5-20 所示)说明：
按下"按我就变色"按钮会让下面的几何图形随机变色。

UI 高级设计   05

图 5-20

Draw2D/src/org/draw2D/GeometricView.java

```java
9.      public class GeometricView extends View {
10.         private ShapeDrawable[] shapes;
11.         public GeometricView(Context context) {
12.             super(context);
13.             makeShapes();
14.         }
15.
16.         public GeometricView(Context context, AttributeSet attrs){
17.             super(context, attrs);
18.             makeShapes();
19.         }
20.         public void makeShapes(){
21.             shapes = new ShapeDrawable[4];
22.             shapes[0] = new ShapeDrawable(new OvalShape());
23.             shapes[0].setBounds(10, 10, 100, 150);
24.             shapes[1] = new ShapeDrawable(new OvalShape());
25.             shapes[1].setBounds(120, 10, 260, 100);
26.             shapes[2] = new ShapeDrawable(new RectShape());
27.             shapes[2].setBounds(10, 170, 100, 310);
28.             shapes[3] = new ShapeDrawable(new RectShape());
29.             shapes[3].setBounds(120, 170, 260, 260);
30.         }
31.
32.         @Override
33.         protected void onDraw(Canvas canvas) {
34.             for (ShapeDrawable shape : shapes) {
35.                 int r = (int) (256 * Math.random());
36.                 int g = (int) (256 * Math.random());
```

151

```
37.            int b = (int) (256 * Math.random());
38.            shape.getPaint().setARGB(255, r, g, b);
39.            shape.draw(canvas);
40.        }
41.    }
42. }
```

9 行：要自行产生 UI 组件最简单的方式就是先继承 View 类。

10 行：如果需要绘制个性化的几何图形，可以利用 ShapeDrawable 类。

11 行：此构造函数方便直接在 Activity 使用程序代码创建 GeometricView 组件。

16 行：通过 layout 文件创建 GeometricView 组件会调用此构造函数。

20 行：调用 makeShapes()以绘制 4 个几何图形。

22 行：OvalShape 对象代表椭圆形。

23 行：调用 setBounds()设置该图形的四周边界。

26 行：RectShape 对象代表矩形。

33 行：改写 onDraw()以绘制自定义图形。系统会传入组件上的 Canvas(画布)对象供绘图之用。

35~37 行：随机出现红、绿、蓝颜色的值。

38 行：调用 ShapeDrawable 的 getPaint()会取得 Paint 对象，再调用 setARGB()可以设置颜色的 alpha 值与红、绿、蓝 3 原色。

39 行：调用 ShapeDrawable 的 draw()会将图形绘制在画布上。

Draw2D/src/org/draw2D/Draw2D.java

```
8.    public class Draw2D extends Activity {
9.        private Button btnSubmit;
10.       private GeometricView geomView;
11.       @Override
12.       public void onCreate(Bundle savedInstanceState) {
13.           super.onCreate(savedInstanceState);
14.           setContentView(R.layout.main);
15.           findViews();
16.
17.           //GeometricView geomView = new GeometricView(this);
18.           //setContentView(geomView);
19.       }
20.       public void findViews(){
21.           btnSubmit = (Button)findViewById(R.id.btnSubmit);
22.           btnSubmit.setOnClickListener(new OnClickListener() {
23.               @Override
24.               public void onClick(View v) {
25.                   geomView.invalidate();
26.               }
```

27.	});	
28.	geomView = (GeometricView)**findViewById**(R.id.geomView);	
29.	}	
30.	}	

17~18 行：通过此两行程序代码可以直接在 Activity 创建 GeometricView 组件。

25 行：调用 invalidate()，系统会先删掉原来在 GeometricView 组件上的画布，然后自动调用 onDraw() 并传送新的画布以便重新绘制。

28 行：此范例通过 layout 文件来创建 GeometricView 组件，所以可以利用 findViewById() 来取得对应对象。

5.7 补间动画

补间动画(tweening，就是 in between 的意思)是指填补两个图形之间的变化，让第一个图形逐渐改变成第二个图形。Android 提供位移、缩放、旋转、透明化等补间动画的功能，而这些功能可以应用在大部分的 UI 组件上，让开发者可以很简单地在 UI 组件中加上动画，使得 UI 画面更加丰富活泼，并提高用户与操作画面的交互性。建议在 XML 文件内设置补间动画的各种效果，而不要直接以程序代码编写；这点其实就与之前所述将页面设置放在 layout 文件内的观念相同，因为使用 XML 文件来设置，可以增加可读性与重复利用性。一般而言，会将设置补间动画的 XML 文件放在 Android 项目的 res/anim 目录内。设置补间动画常用到的 XML 属性大都定义在 Animation 类内，说明如表 5-5 所示。

表 5-5
Animation 类的 XML 属性

属 性 名 称	说 明	属 性 值
android:duration	动画播放的时间，单位为毫秒。对应的方法为 setDuration(long)	整数，不可为负值
android:interpolator[①]	指定动画的运行效果。设置位移补间动画后，还需要指定整个位移过程的效果是 accelerate(加速)还是 decelerate(减速)。对应的方法为 setInterpolator(Interpolator)	linear_interpolator(线性) ● accelerate_interpolator(加速) ● decelerate_interpolator(减速) ● anticipate_interpolator(先退后进) ● overshoot_interpolator(冲过头) ● bounce_interpolator(反弹) ● cycle_interpolator(以曲线方式加快重复次数)[②] 默认为 linear_interpolator

① 所有的动画特效参见 R.anim 类的参数。
② 参见 CycleInterpolator 类说明。

(续表)

Animation 类的 XML 属性

属性名称	说　　明	属 性 值
android:repeatCount	动画重复播放次数。对应的方法为 setRepeatCount(int)	整数。-1 代表无限重复播放 默认为 0(不重复播放)
android:repeatMode	动画重复播放模式。对应的方法为 setRepeatMode(int)	● restart：重新播放 ● reverse：反向播放 默认为 restart
android:startOffset	设置主动画开始后多久才运行此动画，单位为毫秒。要播放多个动画时，可以使用这个属性来指定各个动画播放的相对时间。对应的方法为 setStartOffset(long)	整数，不可为负值

除了前述 Animation 类所定义的 XML 属性外，在设置补间动画时还需要了解其他相关 XML 属性，常用的属性分类说明如表 5-6 所示。

表 5-6

通用 XML 属性

属 性 名 称	说　　明	属 性 值
android:pivotX / android:pivotY	组件的哪个位置会发生补间动画，指定该位置的 X 轴/ Y 轴坐标①	浮点数 ● 50%代表在该组件的中间位置 ● 50 代表在父组件的 50%的位置，也就是父组件的中间位置
android:fromXDelta / android:fromYDelta	位移开始时组件的 X 轴/ Y 轴坐标	
android:toXDelta / android:toYDelta	位移结束时组件的 X 轴/ Y 轴坐标	

与缩放有关的 XML 属性

属 性 名 称	说　　明	属 性 值
android:fromXScale / android:fromYScale	缩放开始时组件的水平/ 垂直尺寸	浮点数 ● 1.0 代表维持原来尺寸 ● <1.0 代表缩小 ● >1.0 代表放大
android:toXScale / android:toYScale	缩放结束时组件的水平/ 垂直尺寸	

与旋转有关的 XML 属性

属 性 名 称	说　　明	属 性 值
android:fromDegrees	旋转开始时组件的角度	浮点数 ● <0.0 代表逆时针 ● >0.0 代表顺时针
android:toDegrees	旋转结束时组件的角度	

---

① 原点(0,0)在左上角，向右则 X 轴的值增加；向下则 Y 轴的值增加。

(续表)

与透明化有关的 XML 属性

属性名称	说　明	属性值
android:fromAlpha	动画开始时组件的透明度	介于 0.0~1.0 之间的浮点数 ● 0.0 代表完全透明 ● 1.0 代表完全不透明
android:toAlpha	动画结束时组件的透明度	

Android 提供许多与补间动画设置有关的 XML 标签以方便开发者快速实现补间动画的功能，每个标签都会对应到 Android.view.animation 包内的类，说明如下[①]。

- <translate>：提供位移补间动画的功能，对应的类为 TranslateAnimation。
- <scale>：提供缩放补间动画的功能，对应的类为 ScaleAnimation。
- <rotate>：提供旋转补间动画的功能，对应的类为 RotateAnimation。
- <alpha>：提供透明化补间动画的功能，对应的类为 AlphaAnimation。
- <set>：使用此标签可以将数个补间动画设置成同一群组一同播放，而且可以共享相同的设置，对应的类为 AnimationSet。

使用补间动画所需使用到的方法说明如表 5-7 所示。

表 5-7

AnimationUtils 类
public static Animation loadAnimation (Context context, int id)
加载资源目录内的动画配置文件
● context：加载资源文件需给予指定的 Context 对象，一般为 Activity 对象 ● id：动画配置文件对应的资源 ID
public void startAnimation(Animation animation)
开始运行指定的动画
animation：要播放的 Animation 对象

范例 TweenAnimEx

范例(如图 5-21 所示)说明：

- 按下"振动"按钮，EditText 会左右快速摇晃。
- 勾选中间部分的 Spinner，可以选择"一二三四"文字要播放的特效。
- 单击下面部分的 Spinner，可以选择"百战不殆"文字要播放的特效。

[①] TranslateAnimation、ScaleAnimation、RotateAnimation、AlphaAnimation、AnimationSet 等类都是 Animation 的子类。

图 5-21

TweenAnimEx/res/layout/main.xml

```
56.    <ViewFlipper android:id="@+id/fpText"
57.        android:layout_width="match_parent"
58.        android:layout_height="wrap_content"
59.        android:flipInterval="1500"
60.        android:layout_marginTop="10dp"
61.        android:layout_marginBottom="10dp" >
62.        <TextView
63.            android:layout_width="match_parent"
64.            android:layout_height="wrap_content"
65.            android:gravity="center_horizontal"
66.            android:textSize="20sp"
67.            android:text="@string/anim_text1"/>
68.        <TextView
69.            android:layout_width="match_parent"
70.            android:layout_height="wrap_content"
71.            android:gravity="center_horizontal"
72.            android:textSize="20sp"
73.            android:text="@string/anim_text2"/>
74.        <TextView
75.            android:layout_width="match_parent"
76.            android:layout_height="wrap_content"
77.            android:gravity="center_horizontal"
78.            android:textSize="20sp"
79.            android:text="@string/anim_text3"/>
80.    </ViewFlipper>
```

56 行：ViewFlipper 是一个简易的动画组件，可以添加多个 View 组件进来。62~79 行共添加 3 个 TextView 组件，不过 1 次仅能显示 1 个。ViewFlipper 可以设置每个子组件显示出来的间隔时间。

59 行： 设置每隔 1500 毫秒显示 1 个子组件。

TweenAnimEx/src/org/tweenAnimEx/TweenAnimEx.java

```
36.     private void findViews_anim01() {
37.         etUserName = (EditText)findViewById(R.id.etUserName);
38.         btnSubmit = (Button)findViewById(R.id.btnSubmit);
39.         btnSubmit.setOnClickListener(new OnClickListener() {
40.             @Override
41.             public void onClick(View v) {
42.                 Animation anim = AnimationUtils.loadAnimation(
43.                             TweenAnimEx.this, R.anim.anim_edittext);
44.                 etUserName.startAnimation(anim);
45.             }
46.         });
47.     }
```

42~43 行：载入 res/anim/anim_edittext.xml 动画配置文件。

44 行：将 etUserName(EditText 组件)应用并播放 anim 所代表的动画效果。

```
49.     private void findViews_anim02() {
50.         tvAnim = (TextView)findViewById(R.id.tvAnim);
51.         spInter = (Spinner)findViewById(R.id.spInter);
52.         spInter.setOnItemSelectedListener(new OnItemSelectedListener() {
53.             @Override
54.             public void onItemSelected(AdapterView parent,
55.                     View view, int pos, long id) {
56.                 View parentView = (View)tvAnim.getParent();
57.                 TranslateAnimation anim = new TranslateAnimation(
58.                     0.0f,
59.                     parentView.getWidth() - parentView.getPaddingLeft() -
60.                     parentView.getPaddingRight() - tvAnim.getWidth(),
61.                     0.0f, 0.0f);
62.                 anim.setDuration(1000);
63.                 anim.setStartOffset(300);
64.                 anim.setRepeatMode(Animation.RESTART);
65.                 anim.setRepeatCount(Animation.INFINITE);
66.
67.                 int inter_id = android.R.anim.accelerate_interpolator;
68.                 switch (pos) {
69.                     case 0:
70.                         inter_id = android.R.anim.accelerate_interpolator;
71.                         break;
72.                     case 1:
73.                         inter_id = android.R.anim.decelerate_interpolator;
74.                         break;
75.                     case 2:
```

```
76.                         inter_id = android.R.anim.accelerate_decelerate_interpolator;
77.                         break;
78.                     case 3:
79.                         inter_id = android.R.anim.anticipate_interpolator;
80.                         break;
81.                     case 4:
82.                         inter_id = android.R.anim.overshoot_interpolator;
83.                         break;
84.                     case 5:
85.                         inter_id = android.R.anim.anticipate_overshoot_interpolator;
86.                         break;
87.                     case 6:
88.                         inter_id = android.R.anim.bounce_interpolator;
89.                         break;
90.                 }
91.                 anim.setInterpolator(AnimationUtils.loadInterpolator(
92.                     TweenAnimEx.this, inter_id));
93.                 tvAnim.startAnimation(anim);
94.             }
95.
96.             @Override
97.             public void onNothingSelected(AdapterView parent) {}
98.         });
99.     }
```

57 行：TranslateAnimation (float fromXDelta, float toXDelta, float fromYDelta, float toYDelta) 属于位移补间动画，构造函数的 4 个参数分别代表：fromXDelta–位移开始时组件的 X 轴坐标；toXDelta–位移结束时组件的 X 轴坐标；fromYDelta–位移开始时组件的 Y 轴坐标；toYDelta–位移结束时组件的 Y 轴坐标。

58~61 行：只有第 2 个参数值不是 0.0 代表组件作水平移动，而 parentView.getWidth()-parentView.getPaddingLeft()-parentView.getPaddingRight()-tvAnim.getWidth()是指"父组件的宽度减父组件左边填充宽度减父组件右边填充宽度减动画组件宽度"，剩下的宽度即为位移的宽度；这代表组件位移时，从 X 轴 0.0 的位置向右移至不会超出父组件的极限，因为不希望动画组件向右位移时被父组件遮蔽。

62 行：动画持续播放 1000 毫秒。

63 行：主动画开始 300 毫秒后才运行此动画。

64 行：设置重复播放(Animation.RESTART)动画。

65 行：设置无限(Animation.INFINITE)重复播放动画。

68~90 行：根据 Spinner 选择结果决定要采用何种内置的 interpolator 特效。android.开头代表系统内置。

```
101.     private void findViews_anim03() {
102.         fpText = (ViewFlipper)findViewById(R.id.fpText);
103.         fpText.startFlipping();
```

```
104.        spAnim = (Spinner)findViewById(R.id.spAnim);
105.        spAnim.setOnItemSelectedListener(new OnItemSelectedListener() {
106.            @Override
107.            public void onItemSelected(AdapterView parent,
108.                    View view, int pos, long id) {
109.
110.                int anim_in = R.anim.translate_up_in;
111.                int anim_out = R.anim.translate_up_out;
112.                switch (pos) {
113.                    case 0:
114.                        anim_in = R.anim.translate_up_in;
115.                        anim_out = R.anim.translate_up_out;
116.                        break;
117.                    case 1:
118.                        anim_in = R.anim.translate_left_in;
119.                        anim_out = R.anim.translate_left_out;
120.                        break;
121.                    case 2:
122.                        anim_in = android.R.anim.fade_in;
123.                        anim_out = android.R.anim.fade_out;
124.                        break;
125.                    case 3:
126.                        anim_in = R.anim.abstract_in;
127.                        anim_out = R.anim.abstract_out;
128.                        break;
129.                }
130.                fpText.setInAnimation(AnimationUtils.loadAnimation(
131.                        TweenAnimEx.this, anim_in));
132.                fpText.setOutAnimation(AnimationUtils.loadAnimation(
133.                        TweenAnimEx.this, anim_out));
134.            }
135.
136.            @Override
137.            public void onNothingSelected(AdapterView parent) {}
138.        });
139.    }
```

112~129 行：根据 Spinner 选择结果决定要采用何种自定义特效配置文件(放在该项目的 res/anim 目录内)。

5.8 Drawable 动画

Drawable 动画(Drawable animation)就是利用 ImageView 组件加载 res/drawable 资源目录内

的图形文件后，按照一定顺序与时间播放这一连串的图片，就像一般传统动画一样。可以使用 XML 文件设置 Drawable 动画播放的图片顺序与时间，另外还需要搭配 AnimationDrawable 类才能播放动画。

 范例 DrawableAnimationEx

范例(如图 5-22 所示)说明：
触摸画面，ImageView 组件会播放动画，再触摸则会停止播放。

图 5-22

范例创建步骤如下：

 在 res/layout/main.xml 文件内新建 ImageView 组件。

```xml
<ImageView
    android:id="@+id/imageView"
    android:layout_width="wrap_content"
    android:layout_height="wrap_content"
    android:layout_alignParentTop="true"
    android:layout_centerHorizontal="true"
    android:layout_marginTop="110dp" />
```

 创建 image_set.xml 文件，放在 res/drawable 目录内。

```xml
<animation-list xmlns:android="http://schemas.android.com/apk/res/android"
    android:oneshot="false">
    <!--每个 item 标签代表一张图片，duration 设置该图片播放时间，单位毫秒-->
    <item android:drawable="@drawable/android01" android:duration="200" />
    <item android:drawable="@drawable/android02" android:duration="200" />
    <item android:drawable="@drawable/android03" android:duration="200" />
</animation-list>
```

UI 高级设计

在 DrawableAnimationExActivity.java 文件内添加下列程序代码。

```java
public class DrawableAnimationExActivity extends Activity {
    AnimationDrawable animDrawable;

    @Override
    public void onCreate(Bundle savedInstanceState) {
        super.onCreate(savedInstanceState);
        setContentView(R.layout.main);
        /*取得 ImageView 组件当作播放 Drawable 动画的容器*/
        ImageView imageView = (ImageView) findViewById(R.id.imageView);
        /*将 ImageView 组件的背景设置指向 image_set.xml(Drawable 动画配置文件)*/
        imageView.setBackgroundResource(R.drawable.image_set);
        /*ImageView 对象调用 getBackground()方法会返回 AnimationDrawable 对象*/
        animDrawable = (AnimationDrawable) imageView.getBackground();
    }

    public boolean onTouchEvent(MotionEvent event) {
        if (event.getAction() == MotionEvent.ACTION_DOWN) {
            /*调用 isRunning()方法会检查动画是否在播放,如果没播就开始播放;如果已经播放,则停止播放*/
            if (!animDrawable.isRunning()) {
                /*开始播放动画*/
                animDrawable.start();
            } else {
                /*停止播放动画*/
                animDrawable.stop();
            }
            return true;
        }
        return super.onTouchEvent(event);
    }
}
```

第 6 章
Activity生命周期与平板电脑设计概念

本章学习目标：

- Activity 生命周期
- Activity 间传递数据
- 通知信息
- Broadcast
- Service 生命周期
- 平板电脑 UI 设计概念

6.1 Activity 生命周期

Activity 的主要功能就是创建一个可以让用户与移动设备交互的画面，而前两章所提及的 UI 组件就是放在 Activity 所呈现的画面上。从 Activity 画面的出现到用户操作再到整个 Activity 结束，会历经 7 个阶段，这 7 个阶段就是 Activity 的 7 个生命周期。如果开发者希望每个阶段都能按照他的意思运行对应的程序，就必须将想要运行的程序安排在对应的阶段。Android 为了方便开发者能够轻易指定每个阶段要运行什么程序，而创建了 Activity 类，并在其内定义了与生命周期有关的 7 种方法，如表 6-1 所示。

表 6-1

方 法	说 明	可否移除程序	下一个阶段
onCreate()	当 Activity 第一次被创建时，会调用此方法。通常会将下列程序置入： 创建 UI 画面：例如，调用 setContentView() 以载入 layout 文件内容 初始化 UI 组件：例如，调用 findViewById() 以取得对应的 UI 组件	否	onStart()
onStart()	当 Activity 画面准备要呈现时会调用此方法	否	onResume()
onResume()	当 Activity 将要与用户交互之前会调用此方法	否	onPause()
onPause()	当前的 Activity 画面要被其他 Activity 画面所替换，也就是其他 Activity 将要与用户交互时会调用此方法。因为当前的 Activity 即将进入暂停状态，所以应该在此阶段释放与此 Activity 有关的资源(例如停止 GPS 运行)，以免耗费 CPU、内存或电力等资源	是	• 当 Activity 画面被替换，会调用 onStop() • 如果之后又恢复此 Activity 画面，会调用 onResume()
onStop()	当 Activity 画面被替换会调用此方法	是	• 如果 Activity 要结束，会调用 onDestroy() • 如果恢复此 Activity 到可以与用户交互的状态，会调用 onRestart()
onRestart()	当 Activity 从 onStop() 状态要恢复到 onStart() 状态时会调用此方法	否	onStart()
onDestroy()	Activity 要准备结束之前会调用此方法	是	已经是最后阶段，所以没有下一个阶段

上述每一个方法都代表 Activity 生命周期的一个阶段。开发者可以自行创建一个类继承 Activity 类，并按照开发上的需要而改写对应的方法。只要 Activity 执行到指定阶段，就会自动

Activity 生命周期与平板电脑设计概念

调用被改写的对应方法以达成目的。

现在的智能手机,大多具有多任务(Multi-Task)功能,可以使用手机听音乐的同时也运行其他多个应用程序。这种多任务功能虽然方便,但也有缺点,那就是每多运行一个应用程序,就必须多消耗系统的内存。内存是有限的,所以当同时运行的程序越多,系统整体运行就会越慢,甚至不稳定。为了兼顾应用程序的正常运行与内存的有效利用,Android 有自己的一套程序管理模式,其中最重要的部分就是前述 Activity 生命周期的管理。按照前面表 6-1 可以绘制出图 6-1[①]Activity 生命周期的示意图。

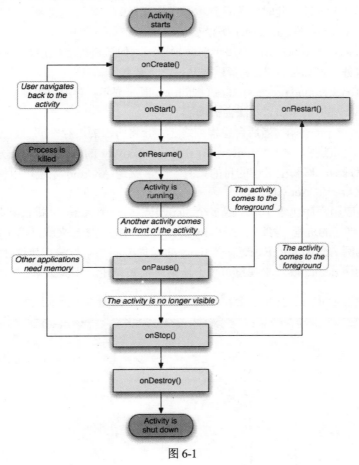

图 6-1

其实 Activity 的 7 种方法就决定了一个 Activity 的完整生命周期,但是 Activity 不一定会执行所有方法,而有些方法不一定只执行一次[②]。一个 Activity 会历经哪些方法,大多是由 Android 系统按照用户操作与系统资源使用情况加以控制。将 Activity 历经过的方法结合起来,其实代表的就是一个完整的程序流程[③],也可以称作进程(process)。但如前述,进程是由 Android 系统掌控,所以只要当内存不足,Android 系统可能会随时终止(kill)某一个 Activity 进程。当然

---

① 请参见 http://developer.android.com/training/basics/activity-lifecycle/starting.html。
② 这就像人类的生命周期中有生、老、病、死,而"病"这个阶段可能重复多次。
③ Activity 属于 MVC 架构中的 Controller,负责与 UI 交互,并处理与流程控制有关的程序。

Android 系统不会胡乱地终止 Activity 进程，而会按照进程的重要程度(importance hierarchy)来决定欲终止的进程；而最不重要的进程，最容易被终止，由重要至不重要的顺序排列如下：

(1) 前台进程(foreground process)：用户正在使用的进程，而且该进程的画面正显示在屏幕上，这种进程几乎不会被终止，除非现存内存空间少得可怜，而且没有其他不够重要的进程可以终止，前台进程才可能被终止。Activity 的 onCreate()、onStart()、onResume()方法被调用时，该 Activity 就会进入到前台进程。

(2) 可视进程(visible process)：虽然不是前台进程，但用户仍可看到该进程所呈现的画面，例如按电源键会打开键盘锁，此时仍可看到主画面的背景图，解开键盘锁后仍会回到该进程。进入到这个进程，Activity 的 onPause() 方法会被调用。

(3) 服务进程(service process)：不属于前两项进程，但该进程启动后会保持执行状态，以持续对用户提供服务，例如播放 MP3 音乐或从网络下载数据。

(4) 后台进程(background process)：此进程的画面既未显示，对用户也没有直接影响，进入到这个进程，Activity 的 onStop() 方法会被调用。

(5) 空进程(empty process)：当后台进程被终止，会将所占的内存空间释放出来，该进程就会由后台进程进入到空进程，但 Activity 仍存在(只要 Activity 的 onDestroy() 方法没被调用，Activity 就不会被移除)。移到空进程的目的只有一个，就是 Android 系统可以快速回复到前台进程，而不需要重新产生 Activity。

一个 Android 应用程序包含了一个或多个 Activity，每一个 Activity 提供某些功能，而且每一个 Activity 都有自己的进程，将每个 Activity 历经过的状态连接起来就是一个完整的 Android 应用程序流程。范例 ActivityLifeEx 改写 Activity 生命周期的 7 个方法，建议在 Eclipse 开发环境下设置断点与利用 debug 模式来观察每个方法何时会被调用，步骤如下：

STEP 1 双击欲设置断点的行号，会出现蓝色的断点，如图 6-2 所示。

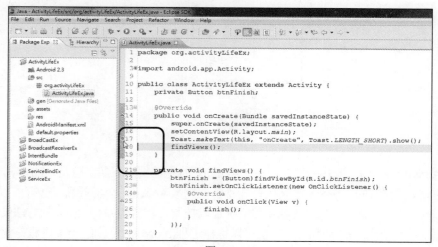

图 6-2

STEP 2 对着项目 ActivityLifeEx 右击选择 Debug As│Android Application 即可启动 debug 模式，如图 6-3 所示。

Activity 生命周期与平板电脑设计概念 06

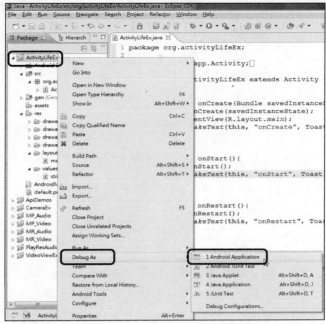

图 6-3

STEP 3 启动 debug 模式后会停在当初设置的断点，如图 6-4 所示。

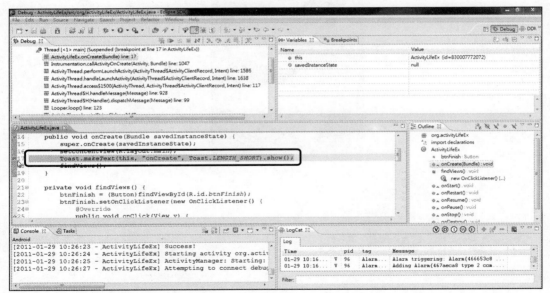

图 6-4

STEP 4 如果希望也能在实机上启动 debug 模式，必须设置 manifest 文件，在 <application> 标签内加上 android:debuggable="true"，如图 6-5 所示。

167

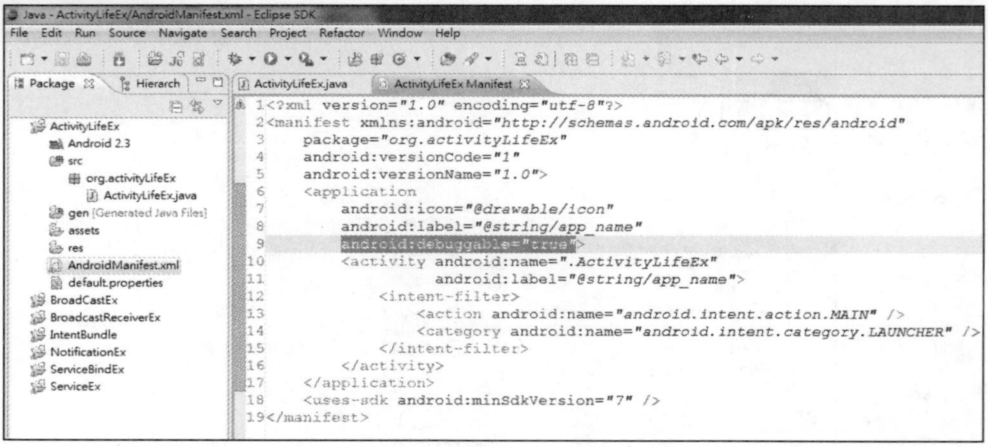

图 6-5

STEP 5 启动 debug 模式后，无论模拟器或实机都会弹出 Waiting For Debugger 的对话窗口，如图 6-6 所示，请勿按下 Force Close 按钮，否则会终止 debug 模式。过一秒钟后该对话窗口将会自动关闭。

图 6-6

6.2 Activity 间传递数据

一个 Android 应用程序可能包含多个 Activity，要从一个 Activity 切换到另一个 Activity，必须通过 Intent[①]，因为 Intent 存储切换时所需要的重要信息，图 6-7 为 Activity01 通过 Intent

---

① 想要启动一个 Activity、发送 Broadcast 给欲接收的 BroadcastReceiver、或打开 Service 都需要 Intent 对象，这是因为 Intent 对象存储发送端的重要信息，接收端则需要根据这些重要信息来进行接下来的处理。可参见 API 文件中 Intent 类的说明。

切换到 Activity02 的示意图。

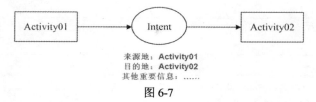

图 6-7

最简单的切换方式说明如下：

(1) 先自定义一个 Intent 对象。

　　Intent intent = new Intent();

(2) 调用 setClass() 并设置起始 Activity 与目的 Activity 类。

　　intent.setClass(Activity01.this, Activity02.class);

(3) 调用 startActivity() 并将 Intent 对象置入将要前往的目的 Activity。

　　startActivity(intent);

(4) 别忘了在 Manifest 文件内加上 Activity 设置(有多少个 Activity 就必须设置多少组<activity>标签)并且将其中一个 Activity 设为首页(<action android:name="android.intent.action.MAIN" />，可参见 IntentBundle 范例的 manifest 文件)。

如果想要在两个 Activity 切换时附带额外数据，可以将该项数据存储在 Bundle 上。Bundle 依附在 Intent 上，是一个专门用来存储添加数据的对象。Bundle 非常类似 Map，也是使用 key-value pairs 方式来存储数据，所以使用起来非常方便，图 6-8 为 Bundle 示意图。

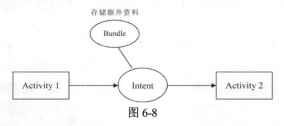

图 6-8

Activity 间传递数据所需使用到的方法说明如表 6-2 所示。

表 6-2

Intent 类
public Intent setClass(Context packageContext, Class<?> cls) 指定一个欲执行的类。会返回相同的 Intent 对象，以方便调用下一个方法(chaining multiple calls) ● packageContext：实现此类的对象，在此指 Activity 对象 ● cls：目的类名称
public Intent putExtras(Bundle extras) 新建 Bundle 到 Intent 内 　extras：欲新建的额外数据

(续表)

Intent 类
public Bundle getExtras()
取得 Bundle
Bundle 类
public void putString(String key, String value)
新建一个字符串及对应的键值到 Bundle 内
● key：数据对应的键值
● value：数据的值
public String getString (String key)
根据键值取得对应数据的值
key：数据对应的键值
Activity 类
public void startActivity (Intent intent)
打开一个指定的 Activity
intent：要打开指定 Activity 所需要使用的 Intent
public Intent getIntent ()
取得 Intent

 范例 IntentBundle

范例说明：

- 在第 1 页(也就是第 1 个 Activity)，如图 6-9 所示，输入完成绩后按下"确定"按钮会将成绩信息带到下一页(也就是第 2 个 Activity)，并将总分、平均分等计算结果呈现在第 2 页，如图 6-10 所示。
- 按下第 2 页的"返回"按钮会回到第 1 页。
- 如果未输入成绩即按下"确定"按钮，会出现成绩输入错误信息。

图 6-9

图 6-10

Activity 生命周期与平板电脑设计概念

IntentBundle/src/org/intentBundle/Score.java

```
22.     private void findViews(){
23.         etChinese = (EditText)findViewById(R.id.etChinese);
24.         etEnglish = (EditText)findViewById(R.id.etEnglish);
25.         etMath = (EditText)findViewById(R.id.etMath);
26.         btnSubmit = (Button)findViewById(R.id.btnSubmit);
27.         btnSubmit.setOnClickListener(new OnClickListener() {
28.             @Override
29.             public void onClick(View v) {
30.                 Intent intent = new Intent();
31.                 intent.setClass(Score.this, Result.class);
32.                 Bundle bundle = new Bundle();
33.                 bundle.putString("chinese", etChinese.getText().toString());
34.                 bundle.putString("english", etEnglish.getText().toString());
35.                 bundle.putString("math", etMath.getText().toString());
36.                 intent.putExtras(bundle);
37.                 startActivity(intent);
38.             }
39.         });
40.     }
```

31 行：调用 setClass()并设置起始 Activity 为 Score 对象，目的 Activity 为 Result 类。

33~35 行：通过 putString()按照顺序将用户输入的中文、英文、数学成绩存入 Bundle 内。

36 行：调用 putExtras()将 Bundle 对象存入 Intent 内。

37 行：根据 Intent 打开指定的 Activity，此范例会打开 Result 页面。

IntentBundle/src/org/intentBundle/Result.java

```
39.     private void showResults(){
40.         NumberFormat nf = NumberFormat.getInstance();
41.         nf.setMaximumFractionDigits(2);
42.         Bundle bundle = this.getIntent().getExtras();
43.         double chinese = Double.parseDouble(
44.                         bundle.getString("chinese"));
45.         double english = Double.parseDouble(
46.                         bundle.getString("english"));
47.         double math = Double.parseDouble(
48.                         bundle.getString("math"));
49.         double sum = chinese + english + math;
50.         double average = sum/3;
51.         tvSum.setText(getString(R.string.totalScores) + nf.format(sum));
52.         tvAverage.setText(getString(R.string.average) + nf.format(average));
53.         if(average >= 85)
54.             tvResult.setText(getString(R.string.top));
55.         else if(average >= 75)
56.             tvResult.setText(getString(R.string.high));
```

```
57.        else
58.            tvResult.setText(getString(R.string.general));
59.    }
```

41 行：将数字格式化成最多只有两位小数。

42 行：调用 getIntent()取得 Intent 对象，再调用 getExtras()取得 Bundle 对象。

43~49 行：调用 getString()并根据键值取得 Bundle 内对应的成绩后转换成 double 类型，之后再算出总分与平均分。

53~58 行：根据平均分的高低显示可以取得多少奖学金。

6.3 通知信息

应用程序通常会利用通知信息来告知用户重要信息或警示信息。为了不干扰用户目前的操作画面，通知信息利用状态栏来呈现简易的图标与消息正文，待用户向下拖曳其信息后会更进一步呈现详细内容，单击该内容可以打开其他 Activity，功能非常丰富。如果想要将通知信息显示在状态栏(status bar)上，步骤如下。

STEP 1 取得 NotificationManager 对象：必须取得 NotificationManager 对象才能在状态栏上发送通知信息。调用 Activity 的 getSystemService()并指定 NOTIFICATION_SERVICE 即可取得 NotificationManager 对象。

```
ntfMgr = (NotificationManager) getSystemService(NOTIFICATION_SERVICE);
```

STEP 2 创建 Notification 对象：调用 Notification.Builder 构造函数来创建 Notification.Builder 对象，再调用 getNotification()即可创建 Notification 对象。不过 API Level 16 将 getNotification()列为 deprecated(已经弃用)，并建议改为调用 build()方法。

```
Notification notification =
    new Notification.Builder(context).getNotification();
```

STEP 3 设置通知信息标题、图标与内容：一开始通知信息会呈现在状态栏上(如图 6-11 所示)，向下拖曳后会呈现详细信息在信息面板上(如图 6-12 所示)。设置这些信息内容与图标的方式如下。

```
Notification notification = new Notification.Builder(this)
// 通知信息在状态栏的文字
.setTicker(getString(R.string.tickerText))
// 通知信息在信息面板的标题
.setContentTitle(getString(R.string.contentTitle))
// 通知信息在信息面板的内容文字
.setContentText(getString(R.string.contentText))
// 通知信息的图标
```

```
.setSmallIcon(android.R.drawable.ic_menu_info_details)
.getNotification();
```

STEP 4 进行发送操作：最后调用 NotificationManager.notify()发送信息。

```
// id 代表通知信息的 ID
ntfMgr.notify(id, notification);
```

STEP 5 删除通知信息：如果想删除指定的通知信息，可以调用 NotificationManager.cancel() 进行删除操作，cancelAll()则会删除此应用程序之前发出的所有通知信息。

```
ntfMgr.cancel(id);
```

如果希望用户单击信息面板后会自动删除状态栏上的通知信息，可以调用 setAutoCancel(true)。

STEP 6 单击信息面板后欲直接回到原有的 Activity，而不打开新的，可在 manifest 文件中，将该 Activity 加上 android:launchMode="singleTop"属性设置。

```
<activity android:launchMode="singleTop" > ... </activity>
```

Notification.Builder 功能在 API Level 11 才开始支持，所以 minSdkVersion 必须设置在 11 以上。

```
<uses-sdk android:minSdkVersion="11" />
```

 范例 NotificationEx

范例说明：
- 按下"出现信息"按钮会将通知信息显示在状态栏上，如图 6-11 所示。
- 向下拖曳状态栏会显示信息面板，指定的通知信息会呈现在该面板上，如图 6-12 所示。
- 按下"删除信息"按钮会将通知信息删除。

图 6-11

图 6-12

范例创建步骤如下：

在 NotificationEx.java 文件内添加下列程序代码。

```java
public class NotificationEx extends Activity {
    // 自定义常数，代表通知信息的 ID
    private final static int NOTIFICATION_ID = 0;
    private Button btnNtf, btnCancel;
    private NotificationManager ntfMgr;

    @Override
    public void onCreate(Bundle savedInstanceState) {
        super.onCreate(savedInstanceState);
        setContentView(R.layout.main);
        findViews();
        // 调用 getSystemService()并指定 NOTIFICATION_SERVICE
        // 可取得 NotificationManager 对象，为了之后发送通知信息
        ntfMgr = (NotificationManager) getSystemService(NOTIFICATION_SERVICE);
    }

    private void findViews() {
        btnNtf = (Button)findViewById(R.id.btnNtf);
        btnNtf.setOnClickListener(new OnClickListener() {
            @Override
            public void onClick(View v) {
                showNotification();
            }
        });
        btnCancel = (Button)findViewById(R.id.btnCancel);
        btnCancel.setOnClickListener(new OnClickListener() {
            @Override
            public void onClick(View v) {
                // 删除指定 ID 的通知信息
                ntfMgr.cancel(NOTIFICATION_ID);

                // 调用 cancelAll()会删除此应用程序之前发出的所有通知信息,
                // 但不会删除其他应用程序发出的通知信息
                // ntfMgr.cancelAll();
            }
        });
    }

    private void showNotification() {
        // 在 manifest 文件，将此 Activity 加上 android:launchMode="singleTop"属性，
        // 单击信息面板后会直接回到目前 Activity，而不会打开新的
        Intent intent = new Intent(this, NotificationEx.class);
```

// 调用 getActivity()取得打开通知信息所需使用的 PendingIntent
// PendingIntent 代表不会立即打开 Intent 所指定的画面，
// 而是等待之后符合一定条件后才打开。
PendingIntent pendingIntent = PendingIntent.getActivity(
    this, 0, intent, 0);
// 利用 Notification.Builder 创建 Notification 对象
Notification notification = new Notification.Builder(this)
    // 通知信息在状态栏的文字
    .setTicker(getString(R.string.tickerText))
    // 通知信息在信息面板的标题
    .setContentTitle(getString(R.string.contentTitle))
    // 通知信息在信息面板的内容文字
    .setContentText(getString(R.string.contentText))
    // 通知信息的图标
    .setSmallIcon(android.R.drawable.ic_menu_info_details)
    // 单击信息面板后会自动删除状态栏上的通知信息
    .setAutoCancel(true)
    // 等待用户向下拨动状态栏后单击信息面板上的信息才会打开指定 Activity 的画面
    .setContentIntent(pendingIntent)
    // API Level 16 开始支持 build()，并建议不要使用 getNotification()
    .getNotification();
// 调用 notify()发出通知信息
ntfMgr.notify(NOTIFICATION_ID, notification);
    }
}
```

## 6.4 Broadcast

### 6.4.1 单纯接收 Broadcast

Android 系统有许多事件是以 Broadcast(广播)方式传递，例如移动设备收到短信，即是以 Broadcast 方式传递。可以在 manifest 文件设置欲拦截的 Broadcast，并作对应处理，步骤如下：

  自定义类继承 BroadcastReceiver，并改写 onReceive()方法。

```
public class MyReceiver extends BroadcastReceiver {
    @Override
    public void onReceive(Context context, Intent intent){
        //改写内容
    }
}
```
```

STEP 2 可以在 manifest 文件设置欲拦截带有何种 action 信息(会被存储在 Intent 对象内)的 Broadcast，并注册指定的 BroadcastReceiver。

```
<receiver android:name="MyReceiver">
  <intent-filter>
    <action android:name="欲拦截的 action" />
  </intent-filter>
</receiver>
```

STEP 3 当系统发出的 Broadcast 被拦截时，会自动调用被改写的 onReceive()[①]以作对应处理。

接收 Broadcast 所需使用到的方法说明如表 6-3 所示。

表 6-3

BroadcastReceiver 类
public abstract void onReceive (Context context, Intent intent)
当 BroadcastReceiver 接收到指定的 Broadcast 时会自动调用此方法
● context：Context 对象，通常为现行 Activity
● intent：接收到的 Broadcast 所对应的 Intent 对象

范例 BroadcastReceiverEx

范例说明：
- 一开始显示正在等待短信的信息，如图 6-13 所示。
- 如果收到短信就会以 Toast 消息框显示"接收到短信"，如图 6-14 所示。

图 6-13

图 6-14

---

① 此类会因为特定事件而被系统自动调用，而且需要被改写的方法即被称为 callback method。

BroadcastReceiverEx/AndroidManifest.xml

```
2.    <manifest xmlns:android="http://schemas.android.com/apk/res/android"
3.        package="org.broadcastReceiverEx"
4.        android:versionCode="1"
5.        android:versionName="1.0">
6.     <uses-sdk android:minSdkVersion="8" />
7.     <uses-permission android:name="android.permission.RECEIVE_SMS" />
8.     <application android:icon="@drawable/icon" android:label="@string/app_name">
9.         <activity android:name=".SMSEx" android:label="@string/app_name">
10.            <intent-filter>
11.                <action android:name="android.intent.action.MAIN" />
12.                <category android:name="android.intent.category.LAUNCHER" />
13.            </intent-filter>
14.        </activity>
15.        <receiver android:name="SMSReceiver">
16.            <intent-filter>
17.                <action android:name="android.provider.Telephony.SMS_RECEIVED" />
18.            </intent-filter>
19.        </receiver>
20.    </application>
21. </manifest>
```

7 行：因为拦截短信可以拦截到短信内容，会触及到用户隐私，所以须加上<uses-permission android:name="android.permission.RECEIVE_SMS" />，让用户决定是否同意安装此应用程序。

15 行：注册指定的 BroadcastReceiver–SMSReceiver。

16~18 行：设置要拦截短信。

<uses-permission>[①]：系统会要求如果应用程序要使用某些对用户有较大影响的功能时(例如：通信功能–影响通信费用、个人信息–影响个人隐私、定位功能–高耗电)必须在 manifest 文件内使用 <uses-permission> 标签以取得用户同意。所谓取得用户同意是指用户在安装时会显示该应用程序使用到的上述特定功能，如图 6-15 所示，让用户决定是否要安装。如果应用程序使用到上述功能却没有设置<uses-permission>，执行该功能时会产生 Exception 而导致无法执行。

---

① 所有的 permission 参见 Manifest.permission 类的常量。

图 6-15

BroadcastReceiverEx/src/org/broadcastReceiverEx/SMSReceiver.java

```
8.     public class SMSReceiver extends BroadcastReceiver {
9.         @Override
10.        public void onReceive(Context context, Intent intent){
11.            Toast.makeText(context,
12.                R.string.receivedSMS, Toast.LENGTH_LONG).show();
13.        }
14.    }
```

8~14 行：此项目的 manifest 文件设置拦截短信的 Broadcast 并注册 SMSReceiver，而 SMSReceiver 继承 BroadcastReceiver 并改写 onReceive()；所以当移动设备接收到短信时，系统会自动调用 10 行的 onReceive()，以 Toast 消息框显示"接收到短信"。

6.4.2 自行发送与接收 Broadcast

前面主要说明拦截系统的 Broadcast，其实开发者也可以自行发出 Broadcast，然后再拦截该 Broadcast。另外除了在 manifest 文件设置欲拦截的 Broadcast 与注册 BroadcastReceiver，其实也可以直接在程序代码内完成这些操作，说明如下：

(1) 自定义类继承 BroadcastReceiver，并改写 onReceive()。

(2) 调用 Context 的 registerReceiver()直接注册 BroadcastReceiver 而不在 manifest 文件内注册。

```
registerReceiver(receiver, filter);
```

(3) 调用 Context 的 sendBroadcast()，自行发出 Broadcast。

```
sendBroadcast (intent);
```

(4) 系统自动调用已注册 BroadcastReceiver 的 onReceive()。

(5) 调用 Context 的 unregisterReceiver() 解除 BroadcastReceiver 的注册①。

> unregisterReceiver(receiver);

自行发送与接收 Broadcast 所需使用到的方法说明如表 6-4 所示。

表 6-4

Context 类
public abstract Intent registerReceiver (BroadcastReceiver receiver, IntentFilter filter)
注册指定的 BroadcastReceiver 与欲拦截的 Broadcast ● receiver：BroadcastReceiver 会处理对应发出的 Broadcast ● filter：设置欲拦截的 Broadcast
public abstract void sendBroadcast (Intent intent)
发出带有 Intent 对象的 Broadcast ● intent：Broadcast 所带有的 Intent 对象
public abstract void unregisterReceiver (BroadcastReceiver receiver)
解除对指定 BroadcastReceiver 的注册 ● receiver：欲解除的 BroadcastReceiver。此 receiver 所有之前已经注册的 filter 都会被移除

 范例 BroadCastEx

范例说明：

- 按下"发送 Broadcast"按钮会发出 Broadcast，对应的 receiver 会接收到并以 Toast 消息框显示"接收到 Broadcast"，如图 6-16 所示。
- 按下"删除receiver"按钮，会以 Toast 消息框显示"删除receiver"。再按下"发送Broadcast"按钮则不再显示"接收到 Broadcast"。如果再按下"删除 receiver"按钮，则会以 Toast 消息框显示"receiver 已被删除"，如图 6-17 所示。

图 6-16

图 6-17

---

① 如果已经解除 BroadcastReceiver 的注册，之后再调用 unregisterReceiver() 会产生 IllegalArgumentException 异常。

BroadCastEx/src/org/broadCastEx/BroadCastEx.java

```java
14.    public class BroadCastEx extends Activity {
15.        private static final String BROADCAST_ACTION = "org.broadCastEx.BroadCastEx";
16.        private Button btnSend, btnCancel;
17.        private MyReceiver receiver;
18.
19.        @Override
20.        public void onCreate(Bundle savedInstanceState) {
21.            super.onCreate(savedInstanceState);
22.            setContentView(R.layout.main);
23.            findViews();
24.            registerMyReceiver();
25.        }
26.
27.        private class MyReceiver extends BroadcastReceiver{
28.            @Override
29.            public void onReceive(Context context, Intent intent) {
30.                Toast.makeText(BroadCastEx.this, R.string.brocastReceived,
31.                        Toast.LENGTH_SHORT).show();
32.            }
33.        }
34.
35.        private void registerMyReceiver() {
36.            Toast.makeText(BroadCastEx.this, R.string.brocastRegister,
37.                    Toast.LENGTH_SHORT).show();
38.            IntentFilter filter = new IntentFilter(BROADCAST_ACTION);
39.            receiver = new MyReceiver();
40.            registerReceiver(receiver, filter);
41.        }
42.
43.        private void findViews() {
44.            btnSend = (Button) findViewById(R.id.btnSend);
45.            btnSend.setOnClickListener(new OnClickListener() {
46.                @Override
47.                public void onClick(View v) {
48.                    Toast.makeText(BroadCastEx.this, R.string.brocastSend,
49.                            Toast.LENGTH_SHORT).show();
50.                    sendBroadcast(new Intent(BROADCAST_ACTION));
51.                }
52.            });
```

```
53.        btnCancel = (Button) findViewById(R.id.btnCancel);
54.        btnCancel.setOnClickListener(new OnClickListener() {
55.            @Override
56.            public void onClick(View v) {
57.                try{
58.                    unregisterReceiver(receiver);
59.                    Toast.makeText(BroadCastEx.this, R.string.brocastCancel,
60.                            Toast.LENGTH_SHORT).show();
61.                }catch(IllegalArgumentException e){
62.                    Toast.makeText(BroadCastEx.this, R.string.brocastCancelled,
63.                            Toast.LENGTH_SHORT).show();
64.                }
65.            }
66.        });
67.    }
68.
69.    public void onDestroy(){
70.        super.onDestroy();
71.        try{
72.            unregisterReceiver(receiver);
73.        }catch(IllegalArgumentException e){
74.            Toast.makeText(BroadCastEx.this, R.string.brocastCancelled,
75.                    Toast.LENGTH_SHORT).show();
76.        }
77.    }
78. }
```

15 行：自定义 BROADCAST_ACTION 字符串常数。

27~33 行：MyReceiver 继承 BroadcastReceiver 并改写 onReceive()。拦截到对应的 Broadcast 时，系统会自动调用 onReceive()。

38 行：创建 IntentFilter 对象，并指定拦截带有 BROADCAST_ACTION 信息的 Broadcast。

40 行：注册指定的 BroadcastReceiver–MyReceiver 与欲拦截的 Broadcast。

50 行：发出带有 BROADCAST_ACTION 信息的 Broadcast。

58 行：解除对 MyReceiver 的注册。如果之后再发出对应的 Broadcast，便无法接收。

61 行：解除 receiver 后若再调用 unregisterReceiver()，对相同的 receiver 进行删除的操作便会产生 IllegalArgumentException 异常。

6.5 Service 生命周期

Service(服务)虽然与 Activity 都属于 Context，但最大的不同点就是 Activity 有 UI 画面供用户操作，但 Service 却没有 UI 画面，所以用户无法直接与 Service 交互，而只能通过 Activity 的 UI 间接与 Service 交互(例如打开或关闭 Service)。如果想要在后台持续运行程序(例如播放音乐、扫描病毒、下载文件等)，Service 是最好的选择。Service 可以说是 Android 应用程序背后的无名英雄。

想在应用程序使用 Service 功能，就如同使用 Activity 一样，必须在项目的 manifest 文件内设置，使用的标签为<service>。

```
<service android:enabled="true" android:name=".TimerService" />
```

- android:enabled：可否启动此 Service，也就是系统可否产生 Service 对象实体。
- android:name：Service 类名称。

可通过下列两种方式打开 Service：

(1) 直接调用 Context 的 startService() 打开指定的 Service。
(2) 调用 Context 的 bindService() 连接 Service，若 Service 尚未打开，会自动打开。

与 Service 有关的类及其常用方法如表 6-5 所示。

表 6-5

Service 类
public void onCreate()
第一次启动 Service 时会调用此方法
public int onStartCommand(Intent intent, int flags, int startId)
每次只要调用 startService() 打开 Service 就会自动调用此方法；反过来说，若不调用 startService()就不会调用此方法。例如，调用 bindService() 打开 Service，系统就不会自动调用 onStartCommand()   • intent：调用 startService(Intent) 时所传递过来的 Intent 对象   • flags：额外添加的信息，可能为 0、START_FLAG_REDELIVERY 或 START_FLAG_RETRY   • startId：给予打开 Service 操作一个标识符，并将此标识符传递过来，方便之后通过 stopSelfResult(int)[①]来停止对应的 Service
public abstract IBinder onBind (Intent intent)
调用 Context 的 bindService() 会打开 Service，并调用此方法。会返回 IBinder 对象以方便其他 client 与 Service 交互，所以 IBinder 是与 Service 沟通的桥梁。例如，Activity 要与 Service 沟通就需要通过 IBinder，而该 Activity 就被称为 Service 的 client   intent：调用 bindService() 时所传递过来的 Intent 对象，为了要连接 Service

---

① 本书不会探讨 stopSelfResult(),若读者有兴趣可以参见 http://developer.android.com/reference/android/app/Service.html#stopSelfResult(int)说明。

(续表)

Service 类
public boolean onUnbind(Intent intent)
调用 Context 的 unbindService(),而且没有其他 client 连接 Service 时会调用此方法,以解除与 Service 的连接
intent:调用 bindService() 时所传递过来的 Intent 对象,为了要连接 Service
public void onDestroy()
当 Service 准备结束时会调用此方法
Context 类
public ComponentName startService (Intent intent)
要求打开指定的 Service
intent:在 Intent 对象内指定欲打开的 Service
public boolean stopService (Intent intent)
要求停止指定的 Service
intent:在 Intent 对象内指定欲停止的 Service
public abstract boolean bindService (Intent intent, ServiceConnection conn, int flags)
连接指定的 Service,若该 Service 不存在,则自动创建一个
• intent:在 Intent 对象内指定欲连接的 Service
• conn:当 service 启动或结束时,会利用此对象来接收信息
• flags:连接 Service 时所指定的选项
public void unbindService (ServiceConnection conn)
解除与指定 Service 之间的连接
conn:之前调用 bindService()时所提供的 ServiceConnection 对象,通过该对象可以解除与对应 Service 之间的连接
ServiceConnection 界面
public abstract void onServiceConnected (ComponentName name, IBinder service)
成功创建与 Service 间的连接时会调用此方法
• name:Service 的组件名称
• service:与 Service 沟通的桥梁
public abstract void onServiceDisconnected (ComponentName name)
失去与 Service 间的连接时会调用此方法。不过这方法通常不会被调用,因为失去连接通常代表 Service 进程已经结束,而执行此方法的进程属于同一进程,既然已经结束,所以就无法调用此方法
name:Service 的组件名称

6.5.1 调用 startService()打开 Service

从调用 Context 的 startService()打开指定的 Service 到该 Service 结束会历经如图 6-18 所示的过程:

(1) 直接调用 Context 的 startService()以显式打开 Service。
(2) 系统打开 Service 后会自动调用该 Service 的 onCreate()。
(3) 接下来调用该 Service 的 onStartCommand()[①]。

---

① 在 Android 2.0 版之前会调用 onStart(),但是在 Android 2.0 以后改成调用 onStartCommand()而不再调用 onStart()。

(4) 如果调用 Context 的 stopService()，会终止该 Service；此时系统会调用该 Service 的 onDestroy()方法并关闭 Service。

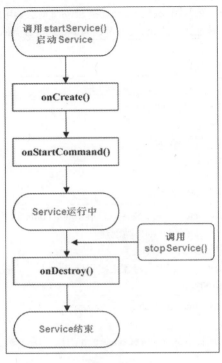

图 6-18

建议在范例 ServiceEx 对应的方法添加断点后使用 debug 模式来观察 Service 生命周期。

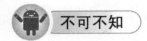

无论调用多少次 startService() 来打开 Service，只要调用一次 stopService() 即可停止该 Service。

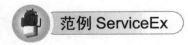

范例说明：
- 输入 3 秒后按下"确定"按钮，如图 6-19 所示。会打开 Service，并会在状态栏显示图标与"3 秒后播放音乐"的信息，显示"停止"按钮，并将"确定"按钮隐藏，如图 6-20 所示。
- 按下"停止"按钮会将 Service 停止并将状态栏上的图标与信息移除。

Activity 生命周期与平板电脑设计概念

图 6-19

图 6-20

范例创建步骤如下：

STEP 1 在 TimerSet.java 文件内添加下列程序代码。

```java
public class TimerSet extends Activity {
    private EditText etTimer;
    private Button bt_Submit, bt_Stop;
    private TimerReceiver receiver;
    private int sec;
    // isActive 是一个控制变量，代表 Service 是否启动
    private boolean isActive = false;

    @Override
    public void onCreate(Bundle savedInstanceState) {
        super.onCreate(savedInstanceState);
        setContentView(R.layout.main);
        findViews();
        resetLayout(isActive);
        registerTimerReceiver();
    }

    private void findViews() {
        etTimer = (EditText) findViewById(R.id.etTimer);
        bt_Submit = (Button) findViewById(R.id.bt_Submit);
        // 按下"确定"按钮会获取输入的秒数、打开 Service 并将秒数传递过去
        bt_Submit.setOnClickListener(new OnClickListener() {
            @Override
            public void onClick(View v) {
                // 获取用户输入的秒数并转换成整数
```

```java
            sec = Integer.parseInt(etTimer.getText().toString());
            // 创建 Intent 对象,并设置欲打开 TimerService
            Intent intent = new Intent(TimerSet.this, TimerService.class);
            Bundle bundle = new Bundle();
            // 将用户输入的秒数存储到 Bundle 对象
            bundle.putInt("sec", sec);
            intent.putExtras(bundle);
            // 调用此方法以启动指定的 Service
            startService(intent);

            isActive = true;
            resetLayout(isActive);
        }
    });

    bt_Stop = (Button) findViewById(R.id.bt_Stop);
    // 按下 "停止" 按钮会将 Service 停止
    bt_Stop.setOnClickListener(new OnClickListener() {
        @Override
        public void onClick(View v) {
            Intent intent = new Intent(TimerSet.this, TimerService.class);
            // 停止指定的 Service
            stopService(intent);

            isActive = false;
            resetLayout(isActive);
        }
    });
}

// 检查 Service 是否启动,如果已启动,显示 "停止" 按钮,让用户可以停止 Service;
// 否则显示 "确定" 按钮,让用户可以启动 Service
private void resetLayout(boolean isActive) {
    if (isActive) {
        bt_Submit.setVisibility(View.INVISIBLE);
        bt_Stop.setVisibility(View.VISIBLE);
    } else {
        bt_Submit.setVisibility(View.VISIBLE);
        bt_Stop.setVisibility(View.INVISIBLE);
    }
}

private void registerTimerReceiver() {
    // 设置只拦截会发送指定字符串的 Broadcast
    IntentFilter filter = new IntentFilter(TimerService.TIMER_ACTION);
```

Activity 生命周期与平板电脑设计概念

```java
        receiver = new TimerReceiver();
        // 注册 BroadcastReceiver，当欲拦截的 Broadcast 发送过来时，
        // 会调用对应的 onReceive()
        registerReceiver(receiver, filter);
    }

    private class TimerReceiver extends BroadcastReceiver {
        @Override
        public void onReceive(Context context, Intent intent) {
            // 创建 MediaPlayer 对象并调用 start()，
            // 播放在项目"res/raw"目录内的音乐文件
            MediaPlayer mp = MediaPlayer.create(TimerSet.this, R.raw.ring);
            mp.start();

            Toast.makeText(TimerSet.this, R.string.msg_playMusic,
                Toast.LENGTH_SHORT).show();
        }
    }

    public void onDestroy() {
        super.onDestroy();
        // 解除 BroadcastReceiver 的注册
        unregisterReceiver(receiver);
    }
}
        // 调用 notify()发出通知信息
        ntfMgr.notify(NOTIFICATION_ID, notification);
    }
}
```

STEP 2 在 TimerService.java 文件内添加下列程序代码。

```java
public class TimerService extends Service {
    private final static int NOTIFICATION_ID = 0;
    public final static String TIMER_ACTION = "org.serviceEx.TimerService";
    private Timer timer;
    private int sec;
    private NotificationManager ntfMgr;

    @Override
    // 第一次启动 Service 时会调用 onCreate()
    public void onCreate() {
        super.onCreate();
    }
```

```java
@Override
// 以 startService()方式启动会调用 onStartCommand()
public int onStartCommand(Intent intent, int flags, int startId) {
    super.onStartCommand(intent, flags, startId);
    // 取得之前用户输入的秒数
    Bundle bundle = intent.getExtras();
    sec = bundle.getInt("sec");

    // 创建 Timer,并将用户输入的秒数设置成要延迟启动的时间;
    // 时间一到便会执行 TimerTask 的 run()内容— 传送指定的 Broadcast
    TimerTask task = new TimerTask() {
        @Override
        public void run() {
            sendBroadcast(new Intent(TIMER_ACTION));
        }
    };
    timer = new Timer();
    timer.schedule(task, sec * 1000);

    // 创建 NotificationManager 并调用 showNotification()发送通知信息
    ntfMgr = (NotificationManager) getSystemService(NOTIFICATION_SERVICE);
    showNotification();

    // 返回 START_STICKY 可以保证再次创建新的 Service 时仍会调用 onStartCommand()
    return START_STICKY;
}

// 详细说明可参见 NotificationEx 范例
private void showNotification() {
    Intent intent = new Intent(this, TimerSet.class);
    PendingIntent pendingIntent = PendingIntent.getActivity(this, 0,
        intent, 0);
    Notification notification = new Notification.Builder(this)
        .setTicker(sec + getString(R.string.msg_playAfterSec))
        .setContentTitle(getString(R.string.msg_preparePlay))
        .setContentText(getString(R.string.msg_speakerOn))
        .setSmallIcon(android.R.drawable.ic_menu_info_details)
        .setAutoCancel(true).setContentIntent(pendingIntent)
        .getNotification();
    ntfMgr.notify(NOTIFICATION_ID, notification);
}

@Override
// Service 准备结束时会调用此方法
public void onDestroy() {
```

```
    super.onDestroy();
    // 停止 Timer 与其指定的操作进程
    timer.cancel();

    // 取消之前在状态栏上显示的信息
    ntfMgr.cancelAll();

    Toast.makeText(this, R.string.msg_serviceOver, Toast.LENGTH_SHORT)
        .show();
}
@Override
// onBind()将于 ServiceBindEx 范例说明
public IBinder onBind(Intent intent) {
    return null;
}
}
```

6.5.2 调用 bindService()连接 Service

从调用 Context 的 bindService()连接 Service 开始到该 Service 结束会历经如图 6-21 所示的过程。

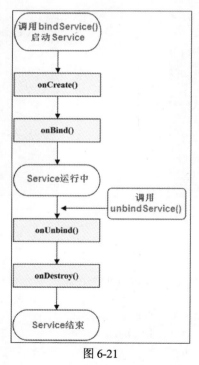

图 6-21

(1) 调用 Context 的 bindService() 会连接 Service，若 Service 尚未打开，会自动打开。
(2) 打开 Service 后会调用 Service 的 onCreate()，但是不会调用 onStartCommand()。

(3) 调用 onBind() 并返回 IBinder 对象。

(4) 调用 ServiceConnection 的 onServiceConnected()，并将步骤 3 返回的 IBinder 对象当作参数传递给 onServiceConnected()，通过 IBinder 可以取得与 Service 的联系，例如直接取得 Service 对象。

(5) 若调用 Context 的 unbindService()，系统会调用该 Service 的 onUnbind() 以解除与 Service 的连接，随后会自动调用 Service 的 onDestroy() 关闭 Service。

建议在范例 ServiceBindEx 对应的方法添加断点后使用 debug 模式来观察 Service 生命周期。

 范例 ServiceBindEx

范例说明：

- 按下"连接 Service"按钮，会连接 Service 并显示"Service 已经完成连接"的信息，而且会显示"播放音乐"、"停止播放"按钮，如图 6-22 所示。
- 按下"结束 Service 连接"按钮，会结束 Service 连接并显示"Service 已经结束连接"的信息，而且会隐藏"播放音乐"、"停止播放"按钮，如图 6-23 所示。

图 6-22

图 6-23

范例创建步骤如下：

STEP 1 在 ServiceSet.java 文件内添加下列程序代码。

```java
public class ServiceSet extends Activity {
    private Button bt_connect, bt_disconnect;
    private Button bt_play, bt_stop;
    private TextView tv_msg;
    private boolean isBound;
    private MusicService musicService;
```

```java
@Override
public void onCreate(Bundle savedInstanceState) {
    super.onCreate(savedInstanceState);
    setContentView(R.layout.main);
    findViews();
}

private void findViews() {
    tv_msg = (TextView) findViewById(R.id.tv_msg);
    bt_connect = (Button) findViewById(R.id.bt_connect);
    // 按下"连接 Service"按钮,会调用 doBindService()以连接 Service
    bt_connect.setOnClickListener(new OnClickListener() {
        @Override
        public void onClick(View v) {
            doBindService();
        }
    });

    bt_disconnect = (Button) findViewById(R.id.bt_disconnect);
    // 按下"结束 Service 连接"按钮,会调用 doUnbindService()以结束 Service 连接
    bt_disconnect.setOnClickListener(new OnClickListener() {
        @Override
        public void onClick(View v) {
            doUnbindService();
        }
    });

    bt_play = (Button) findViewById(R.id.bt_play);
    bt_play.setOnClickListener(new OnClickListener() {
        @Override
        public void onClick(View v) {
            String msg = musicService.play();
            tv_msg.setText(msg);
        }
    });

    bt_stop = (Button) findViewById(R.id.bt_stop);
    bt_stop.setOnClickListener(new OnClickListener() {
        @Override
        public void onClick(View v) {
            String msg = musicService.stop();
            tv_msg.setText(msg);
        }
    });
```

```java
        // 一开始隐藏"播放音乐"、"停止播放"按钮
        bt_play.setVisibility(View.INVISIBLE);
        bt_stop.setVisibility(View.INVISIBLE);
    }

    void doBindService() {
        if (!isBound) {
            Intent intent = new Intent(ServiceSet.this, MusicService.class);
            // 连接 intent 所指定的 Service
            // serviceCon 是实现 ServiceConnection 接口的对象
            // Context.BIND_AUTO_CREATE 代表只要连接到 Service，就会自动创建该 Servive
            // isBound 代表是否与 Service 连接，一旦连接就设置为 true
            bindService(intent, serviceCon, Context.BIND_AUTO_CREATE);
            isBound = true;
        }
    }

    void doUnbindService() {
        // 先检查是否与 Service 连接，如果是，则调用 unbindService()，
        // 解除与该 Service 间的连接，并将 isBound 设为 false
        if (isBound) {
            unbindService(serviceCon);
            isBound = false;
            bt_play.setVisibility(View.INVISIBLE);
            bt_stop.setVisibility(View.INVISIBLE);
            tv_msg.setText(R.string.msg_serviceDisconnected);
        }
    }

    private ServiceConnection serviceCon = new ServiceConnection() {
        @Override
        // 成功创建与 Service 间的连接时会调用此方法并传入 IBinder 对象
        public void onServiceConnected(ComponentName className, IBinder service) {
            musicService = ((MusicService.ServiceBinder) service).getService();
            tv_msg.setText(R.string.msg_serviceConnected);
            // 显示"播放音乐"、"停止播放"按钮
            bt_play.setVisibility(View.VISIBLE);
            bt_stop.setVisibility(View.VISIBLE);
        }

        @Override
        // 失去与 Service 间的连接时会调用此方法，但并未移除 ServicConnection，
        // 当 Service 再次运行时，onServiceConnected()会再次被调用
        public void onServiceDisconnected(ComponentName className) {
            musicService = null;
```

```
            tv_msg.setText(R.string.msg_serviceLostConnection);
        }
    };

    @Override
    public void onDestroy() {
        super.onDestroy();
        doUnbindService();
    }
}
```

STEP 2　在 MusicService.java 文件内添加下列程序代码。

```
public class MusicService extends Service {
    private final IBinder binder = new ServiceBinder();

    // 假设 MusicService 提供音乐播放服务
    public String play() {
        String msg = getString(R.string.msg_musicPlay);
        return msg;
    }

    public String stop() {
        String msg = getString(R.string.msg_musicStop);
        return msg;
    }

    @Override
    public void onCreate() {
        super.onCreate();
    }

    @Override
    // 为了让连接 Service 的 client 可以取得 IBinder 对象
    public IBinder onBind(Intent intent) {
        Toast.makeText(this, "onBind", Toast.LENGTH_SHORT).show();
        return binder;
    }

    public class ServiceBinder extends Binder {
        // 调用 getService()可以取得 Service，
        // 这样一来 client 就可以与 Service 交互
        MusicService getService() {
            return MusicService.this;
        }
    }

    @Override
```

```
// 没有任何 client 连接 Service 时会调用此方法
public boolean onUnbind(Intent intent) {
    Toast.makeText(this, "onUnbind", Toast.LENGTH_SHORT).show();
    return true;
}

@Override
public void onDestroy() {
    super.onDestroy();
}
}
```

不可不知

如果要在 Android 设备上强制关闭 Service 可以通过下列步骤(以 Android 4.0 模拟器为例)：主画面按 Menu 按键，选择"管理应用程序">"运行中"，会显示运行中的 Service 列表，如图 6-24 所示。

点击欲停止 Service 所属的程序(例如 ServiceEx)后，会显示如图 6-25 对应的 Service(例如 TimerService)；按下"停止"按钮即可停止该 Service。

图 6-24

图 6-25

6.6 平板电脑 UI 设计概念

单单就程序设计而言，平板电脑(tablet 或 pad)与智能手机程序设计概念可以说几乎相同，最大的差别点在于 UI 设计，这是因为平板电脑与手机尺寸不同。目前市场上平板电脑的尺寸至少在 7 英寸以上，而且主流尺寸为 10 英寸；而手机尺寸目前都不超过 5 英寸，这意味着平板电脑屏幕尺寸比手机屏幕尺寸要大许多，所以 UI 设计也必然有所不同。如果将适用于手机的应用程序直接安装在平板电脑上，基本上可以运行，只是 UI 画面可能会有点"丑"，如图 6-26 所示。

图 6-26

从 Android 3.0(API level 11)开始，Android 系统正式进入平板电脑时代，而且引进了新的 UI 组件——Fragment。该组件提供类似窗格(pane)的功能，让开发者可以将平板电脑这种较大的画面拆分成两块以上；不仅可以各自显示自己的画面，也可以实现交互效果。不过在当时 Android 手机与平板电脑使用不同的 API，所以手机无法使用 Fragment 组件。但是从 Android 4.0(API level 14)开始，不论平板电脑与手机，都是使用相同的 API；换句话说，Android 手机也可以使用 Fragment 组件。除了 Fragment 组件外，如果想要让 UI 画面呈现更具弹性，Android 官方网站也建议善用 ActionBar 组件[①]，以下将会详细说明这两个重要组件的应用。

6.6.1 Fragment 生命周期

创建 Fragment 非常类似于创建 Activity，只要创建 Fragment 类的子类，并改写关于 Fragment 生命周期的方法即可让系统在特定时候调用对应的生命周期方法。另外要注意的是，Fragment 组件必须依附在 Activity 上，而且 Fragment 生命周期直接受到 Activity 生命周期的影响。所以当一个 Activity 进入到 onPause()状态，所有依附的 Fragment 都会进入到 onPause()状态；当该 Activity 进入到 onDestroy()状态，所有依附的 Fragment 也都会进入到 onDestroy()状态。当 Activity 正在运行时，开发者可以添加 Fragment，也可以移除它。

关于 Fragment 生命周期的整体流程参见图 6-27，并同时参见方法说明。建议如前面图 6-2 一样，在 Eclipse 开发环境下设置断点与利用 debug 模式来观察每个生命周期的方法何时会被调用。

当 Fragment 将被添加在 Activity 上，一直到画面呈现出来让用户可以与其交互，会历经下

---

① 参见 http://developer.android.com/guide/practices/tablets-and-handsets.html。

列 6 个方法。

(1) onAttach()：Fragment 第一次添加在 Activity 时会调用此方法。

(2) onCreate()：在此初始化 Fragment。

(3) onCreateView()：可以在此初始化 Fragment 的 UI。

(4) onActivityCreated()：调用此方法代表该 Fragment 所属的 Activity 已经创建完毕(完成 Activity.onCreate())。

(5) onStart()：调用此方法代表 Fragment 画面准备要呈现。

(6) onResume()：调用此方法代表 Fragment 画面准备要与用户交互。

当 Fragment 画面准备要离开，直到离开所属 Activity 会历经下列 5 个方法。

(1) onPause()：当 Activity 进入暂停状态(Activity.onPause())或是 Fragment 准备要离开 Activity 会调用此方法。

(2) onStop()：当 Activity 进入停止状态(Activity.onStop()) 或是 Fragment 因离开 Activity 而停止时会调用此方法。

(3) onDestroyView()：Fragment 的画面(View)确定离开 Activity 时会调用此方法，此时可以清除与 Fragment 画面有关的资源。

(4) onDestroy()：调用此方法代表 Fragment 不再被使用。

(5) onDetach()：Fragment 确定完全离开 Activity 时会调用此方法。

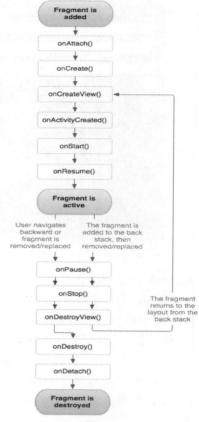

图 6-27

 范例 FragmentEx

范例说明：

- 按下 Add Fragment 按钮会创建 Fragment 并添加在画面下方(Fragment A)，如图 6-28 所示，之后用户无法按下 Add Fragment 按钮，而 Detach Fragment 按钮则呈现可使用状态。
- 按下 Detach Fragment 按钮会移除 Fragment 画面，而且 Detach Fragment 按钮会变成 Attach Fragment 按钮，如图 6-29 所示。按下 Attach Fragment 按钮会把 Fragment 画面添加回去。
- 按下 Finish Activity 按钮会结束 Activity。

Activity 生命周期与平板电脑设计概念

图 6-28

图 6-29

范例创建步骤如下：

STEP 1 在 res/layout/myactivity.xml 文件内新建 FrameLayout，之后会当作 Fragment 画面的父组件。

```xml
<FrameLayout
    android:id="@+id/frameLayout"
    android:layout_width="match_parent"
    android:layout_height="wrap_content" />
```

STEP 2 创建 res/layout/myfragment.xml 作为 Fragment 的 layout 文件。

```xml
<LinearLayout xmlns:android="http://schemas.android.com/apk/res/android"
    android:layout_width="match_parent"
    android:layout_height="match_parent"
    android:orientation="vertical" >

    <TextView
        android:id="@+id/textView"
        android:layout_width="wrap_content"
        android:layout_height="wrap_content"
        android:layout_gravity="center"
        android:layout_margin="50dp"
        android:text="My Fragment"
        android:textColor="#FFFF00"
        android:textSize="20sp" />

</LinearLayout>
```

197

在 MyActivity.java 文件内添加下列程序代码。FragmentManager 专门用来管理 Fragment。FragmentTransaction 则是用来添加、移除、替换 Fragment，最后要调用 FragmentTransaction 的 commit()以确定执行。

```java
public class MyActivity extends Activity {
    private Button bt_detach;
    private Button bt_finish;
    private Button bt_add;

    @Override
    public void onCreate(Bundle savedInstanceState) {
        super.onCreate(savedInstanceState);
        setContentView(R.layout.myactivity);
        findViews();
    }

    private void findViews() {
        bt_add = (Button) findViewById(R.id.bt_add);
        bt_add.setOnClickListener(new OnClickListener() {

            @Override
            public void onClick(View v) {
                /* 调用 getFragmentManager()取得 FragmentManager 实体引用 */
                FragmentManager fm = getFragmentManager();

                /* 调用 beginTransaction()取得 FragmentTransaction 实体引用 */
                FragmentTransaction ft = fm.beginTransaction();

                /* 调用 findFragmentById()寻找 R.id.frameLayout 指定的 FrameLayout 上
                   有没有添加 Fragment */
                Fragment fragment = fm.findFragmentById(R.id.frameLayout);

                /* 如果 Fragment 不存在，就产生新的 Fragment 并显示"Fragment A"，
                   而且添加在指定的 container 上 */ ①
                if (fragment == null) {
                    String title = "Fragment A";
                    MyFragment myFragment = new MyFragment(title);

                    /* 将 Fragment 添加在 R.id.frameLayout 指定的 FrameLayout 上 */
                    ft.add(R.id.frameLayout, myFragment);

                    /* 将 FragmentTransaction 所做的操作以 Stack(堆栈)方式记录，
                       方便用户按下移动设备返回键时可以返回上个操作 */
                    ft.addToBackStack(null);
```

---

① 这里所谓的 container(容器)属于 ViewGroup 类型，也就是我们所熟悉的 Layout，例如：FrameLayout、LinearLayout、RelativeLayout 等。

```java
            /* 确定执行 FragmentTransaction 指定的操作 */
            ft.commit();

            /* 既然已经新建且添加了 Fragment，
               就可以将 Add Fragment 按钮设为 disabled 状态 */
            bt_add.setEnabled(false);

            /* 将 Detach Fragment 按钮设为 enabled 状态 */
            bt_detach.setEnabled(true);
        }
    }
});

bt_detach = (Button) findViewById(R.id.bt_detach);

/* 一开始还没有任何 Fragment，所以将 Detach Fragment 按钮设为 disabled 状态 */
bt_detach.setEnabled(false);
bt_detach.setOnClickListener(new OnClickListener() {

    @Override
    public void onClick(View v) {
        FragmentManager fm = getFragmentManager();
        FragmentTransaction ft = fm.beginTransaction();
        Fragment fragment = fm.findFragmentById(R.id.frameLayout);

        /* 按下 Detach Fragment 按钮：
           1. 如果已经有 Fragment，而且也已经添加上来，就将其移除，并记录此过程，
              按钮上的显示变为 Attach Fragment。
           2. 如果已经有 Fragment，但没有添加上来，就将其添加，并记录此过程，
              按钮上的显示变为 Detach Fragment。         */
        if (fragment != null && !fragment.isDetached()) {
            ft.detach(fragment);
            ft.addToBackStack(null);
            ft.commit();
            bt_detach.setText("Attach Fragment");
        } else if (fragment != null && fragment.isDetached()) {
            ft.attach(fragment);
            ft.addToBackStack(null);
            ft.commit();
            bt_detach.setText("Detach Fragment");
        }
    }
});

bt_finish = (Button) findViewById(R.id.bt_finish);
bt_finish.setOnClickListener(new OnClickListener() {
```

```java
        @Override
        /* 按下 Finish Activity 按钮会结束 Activity */
        public void onClick(View v) {
            MyActivity.this.finish();
        }
    });
}

/* 以下为 Activity 生命周期的各个方法,已在前面有过说明,所以省略 */

@Override
public void onStart() {
    super.onStart();
}
// ......
}
```

> **STEP 4** 在 MyFragment.java 文件内添加下列程序代码。通过改写 onCreateView()来显示 Fragment 的画面。

```java
public class MyFragment extends Fragment {
    String title;
    TextView textView;

    public MyFragment() {
    }

    public MyFragment(String title) {
        this.title = title;
    }

    @Override
    /* Fragment 的画面是 Activity 画面的一部分,必须通过改写 onCreateView()来显示
Fragment 的画面 */
    public View onCreateView(LayoutInflater inflater, ViewGroup container,
            Bundle savedInstanceState) {
        super.onCreateView(inflater, container, savedInstanceState);

        /* 为了产生 Fragment 的 View(也就是画面),调用 inflate()取得指定 layout 文件内容,有下列 3 个参数:
           1. R.layout.myfragment 所代表的 layout 文件将成为 Fragment 的 View
           2. container 是 Activity 所设置的 ViewGroup,
              MyActivity.java 文件 ft.add(R.id.frameLayout, myFragment),代表
           container 为 FrameLayout
           3. false 代表不要将产生的 View 添加在 container 上,这是因为
```

```
            ft.add(R.id.frameLayout, myFragment)已经将 Fragment 的 View 添加在
            container(在此为 FrameLayout)上，所以不需要多此一举  */
            View view = inflater.inflate(R.layout.myfragment, container, false);

            /*  找到 View 内的 TextView 子组件  */
            textView = (TextView) view.findViewById(R.id.textView);
            textView.setText(title);
            return view;
        }
    /*  以下为 Fragment 生命周期的各个方法，已在前面有说明，所以省略  */
        @Override
        public void onAttach(Activity activity) {
            super.onAttach(activity);
        }
        // ......
    }
```

6.6.2　Activity 画面拆分

　　用户操作平板电脑时，一般都是横向(landscape 模式)，所以画面较宽，如果整个 Activity 只显示一个画面会十分单调且不美观，所以最好将 Activity 拆分成两个以上画面(multi-pane)，每一个画面都由一个 Fragment 负责，如图 6-30 的左图部分[①]；当单击左半部选项，会在右半部呈现对应内容，而且这两个 Fragment 都归属于同一个 Activity 控制。

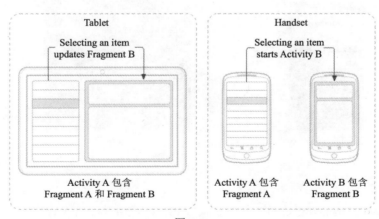

图 6-30

　　智能手机大多是垂直(portrait 模式)使用，手机屏幕已比平板电脑小，又是垂直模式时，能显示的画面更窄，如果将画面拆分成多块，可以显示文字、图片的空间会被压缩。手机等小屏幕设计的准则就是画面简单不复杂，才能方便用户触控操作，所以一个 Activity 显示一个画面

---

①　参见 http://developer.android.com/guide/topics/fundamentals/fragments.html#Design。

即可，此时有两种设计方式：

(1) 直接使用 Activity 来控制画面。

(2) 使用 Fragment 来控制画面，然后再将该 Fragment 添加在 Activity 上，如图 6-30 的右图部分，Activity A 上添加 Fragment A，单击 Fragment A 上的选项后会打开新的 Activity B 并添加 Fragment B，在 Fragment B 上显示对应内容。

方式 2 的设计模式更具弹性，因为 Fragment 可以添加在任何指定的 Activity 上，更容易达到重复利用与模块化，也是 Android 官方网站所建议的设计模式。

 范例 Fragment_dual

范例说明：

- 移动设备横向时(模拟器按 Ctrl+F11 可以改变横向/垂直状态)，导航栏在左侧，单击导航栏项目会将对应 detail 内容显示在右侧，也就是 dual-pane(双窗格)模式，如图 6-31 所示。
- 移动设备垂直时，整个画面只显示导航栏，单击导航栏项目会将对应内容显示在下一页，如图 6-32 所示。

图 6-31

图 6-32

范例创建步骤如下：

STEP 1 res/layout/main.xml 文件用在非横向模式，换句话说，当移动设备垂直时，Activity 会自动应用此 layout 文件。

```
<FrameLayout xmlns:android="http://schemas.android.com/apk/res/android"
    android:layout_width="match_parent"
    android:layout_height="match_parent" >

    <!-- Activity 的 layout 文件可以使用<fragment>标签与 class 属性来添加指定的 Fragment -->
```

```
        <fragment
            android:id="@+id/master"
            android:layout_width="match_parent"
            android:layout_height="match_parent"
            class="org.fragment.MasterFragment" />

</FrameLayout>
```

STEP 2 创建 res/layout/detail_fragment.xml 作为 DetailFragment 的 layout 文件。ImageView 显示公园图标，TextView 显示 detail 内容。

```
<ScrollView xmlns:android="http://schemas.android.com/apk/res/android"
    android:layout_width="wrap_content"
    android:layout_height="wrap_content" >

    <LinearLayout
        android:layout_width="match_parent"
        android:layout_height="wrap_content"
        android:orientation="vertical" >

        <ImageView
            android:id="@+id/imageView"
            android:layout_width="wrap_content"
            android:layout_height="wrap_content"
            android:layout_gravity="center"
            android:layout_margin="8dp" />

        <TextView
            android:id="@+id/textView"
            android:layout_width="wrap_content"
            android:layout_height="wrap_content"
            android:layout_gravity="center"
            android:padding="8dp" />

    </LinearLayout>

</ScrollView>
```

STEP 3 res/**layout-land**/main.xml 文件用在横向模式。

```
<LinearLayout xmlns:android="http://schemas.android.com/apk/res/android"
    android:layout_width="match_parent"
    android:layout_height="match_parent"
    android:orientation="horizontal" >

    <fragment
        android:id="@+id/master"
        android:layout_width="0px"
```

```xml
        android:layout_height="match_parent"
        android:layout_weight="1"
        class="org.fragment.MasterFragment" />

    <!-- Fragment 的父组件，android:background 属性设置背景样式 -->
    <FrameLayout
        android:id="@+id/detail"
        android:layout_width="0px"
        android:layout_height="match_parent"
        android:layout_weight="2"
        android:background="?android:attr/detailsElementBackground" />

</LinearLayout>
```

STEP 4 在 MasterActivity.java 文件内添加下列程序代码。MasterActivity 载入 main.xml 所代表的 layout 文件。

```java
public class MasterActivity extends Activity {

    @Override
    protected void onCreate(Bundle savedInstanceState) {
        super.onCreate(savedInstanceState);
        setContentView(R.layout.main);
    }
}
```

STEP 5 在 MasterFragment.java 文件内添加下列程序代码。MasterFragment 是 ListFragment 的子类，在此用来当作导航栏。

```java
public class MasterFragment extends ListFragment {
    boolean isDualPane;
    int position;

    @Override
    public void onActivityCreated(Bundle savedInstanceState) {
        super.onActivityCreated(savedInstanceState);
        /* Resort.PARKS 存储多个 Park(公园)对象，
           用循环取得各公园名称后存入 List 当作之后导航栏的项目文字 */
        ArrayList<String> parkNames = new ArrayList<String>();
        for (Park park : Resort.PARKS) {
            parkNames.add(park.getName());
        }

        /* simple_list_item_activated_1 内置样式可以更改被选择选项的背景色 */
        setListAdapter(new ArrayAdapter<String>(getActivity(),
```

```java
                        android.R.layout.simple_list_item_activated_1, parkNames));

        /* R.id.detail 是 res/layout-land/main.xml 的组件，
           如果是横向模式，将有两个画面(dual-pane) */
        View detailFrame = getActivity().findViewById(R.id.detail);
        isDualPane = detailFrame != null && detailFrame.getVisibility() == View.VISIBLE;

        if (savedInstanceState != null) {
            /* 将存储的选项位置取出 */
            position = savedInstanceState.getInt("position", 0);
        }

        if (isDualPane) {
            /* 设置成单选模式 */
            getListView().setChoiceMode(ListView.CHOICE_MODE_SINGLE);

            /* 根据导航栏被选择的项目位置，在另一个 Fragment 呈现对应 detail 内容 */
            showDetail(position);
        }
    }

    @Override
    /* 画面被切换就会调用此方法，需要存储选项位置方便之后取出 */
    public void onSaveInstanceState(Bundle outState) {
        super.onSaveInstanceState(outState);
        outState.putInt("position", position);
    }

    @Override
    public void onListItemClick(ListView l, View v, int position, long id) {
        showDetail(position);
    }

    /* 根据导航栏被选择的项目位置，在另一个 Fragment 呈现对应 detail 内容 */
    void showDetail(int position) {
        this.position = position;

        /* 如果是 dual-pane 模式，就直接在另一个 Fragment 呈现对应 detail 内容 */
        if (isDualPane) {
            /* 根据被选择的项目位置，将该选项设置为选择状态 */
            getListView().setItemChecked(position, true);

            DetailFragment detailFragment = (DetailFragment)
                    getFragmentManager().findFragmentById(R.id.detail);

            /* detail 的 Fragment 不存在或即使存在但 position 与被选择选项的位置不符合，
               就创建新的 DetailFragment 来呈现 detail 内容 */
```

```
            if (detailFragment == null || detailFragment.getIndex() != position) {
                detailFragment = new DetailFragment(position);
                    FragmentTransaction ft = getFragmentManager().beginTransaction();

                    /* 将新的 DetailFragment 替换旧的，添加在 R.id.detail 位置上 */
                    ft.replace(R.id.detail, detailFragment);

                    /* 设置换页的动画模式 */
                    ft.setTransition(FragmentTransaction.TRANSIT_FRAGMENT_FADE);
                    ft.commit();
            }
        }

        /* 如果不是 dual-pane 模式，打开新的 Activity 并在之后添加 Fragment 后呈现对应内容 */
        else {
            Intent intent = new Intent();
            intent.setClass(getActivity(), DetailActivity.class);
            intent.putExtra("position", position);
            startActivity(intent);
        }
    }
}
```

STEP 6 在 DetailActivity.java 文件内添加下列程序代码。如果不是横向模式，就创建 DetailActivity，并在其上添加 DetailFragment 来呈现 detail 内容。

```
public class DetailActivity extends Activity {

    @Override
    protected void onCreate(Bundle savedInstanceState) {
        super.onCreate(savedInstanceState);
        Configuration configuration = getResources().getConfiguration();

        /* 如果是横向模式，就结束此 Activity */
        if (configuration.orientation == Configuration.ORIENTATION_LANDSCAPE) {
            finish();
            return;
        }

        /* 将 MasterFragment 传来的 position 转给新创建的 DetailFragment，
            并将 DetailFragment 加在此 Activity 上 */
        Bundle bundle = getIntent().getExtras();
        int position = bundle.getInt("position");
        DetailFragment detailFragment = new DetailFragment(position);
        FragmentTransaction ft = getFragmentManager().beginTransaction();
```

```
            ft.add(android.R.id.content, detailFragment).commit();
    }
}
```

STEP 7 在 DetailFragment.java 文件内添加下列程序代码。DetailFragment 在此专门用来呈现公园的 detail 内容。

```java
public class DetailFragment extends Fragment {
    private int position;

    public DetailFragment() {
    }

    /* 创建 DetailFragment 实体，position 代表导航栏被选择选项的位置 */
    public DetailFragment(int position) {
        this.position = position;
    }

    public int getIndex() {
        return position;
    }

    @Override
    /* 返回的 View 会成为 DetailFragment 的画面，container 是 DetailFragment 的父组件 */
    public View onCreateView(LayoutInflater inflater, ViewGroup container,
            Bundle savedInstanceState) {
        super.onCreateView(inflater, container, savedInstanceState);

        /* 如果没有 container，代表不是 dual-pane，也就不需要 DetailFragment，因此返回 null */
        if (container == null) {
            return null;
        }

        /* 调用 inflate()取得指定 layout 文件内容，有下列 3 个参数：
        1. R.layout.detail_fragment 所代表的 layout 文件将成为 Fragment 的 View
        2. container 是 DetailFragment 的父组件
        3. 因为已经将 Fragment 的 View 添加在 container 上，所以设为 false，不需要再添加 */
        View view = inflater.inflate(R.layout.detail_fragment, container, false);

        /* 取得 ImageView，并按照选项 position 显示对应的图标 */
        ImageView imageView = (ImageView) view.findViewById(R.id.imageView);
        imageView.setImageResource(Resort.PARKS[position].getImageId());

        /* 取得 TextView，并按照选项 position 显示对应的内容 */
        TextView textView = (TextView) view.findViewById(R.id.textView);
        textView.setText(Resort.PARKS[position].getDescription());
        return view;
    }
}
```

6.6.3　ActionBar

Android 3.0(API level 11)开始提供 ActionBar 组件，它是极具弹性且容易使用的组件。

(1) 类似功能列表：提供 Button、Spinner、Tab 等功能让用户可以简易地操控应用程序。

(2) 适用于各种宽窄画面：ActionBar 会根据画面的宽窄自动调整显示方式，如果希望应用程序能够同时在平板电脑与手机上运行，一定要善用此组件。

(3) 只要会使用 Options Menu 就一定会使用 ActionBar，因为不需要更改 Options Menu 的程序代码，只要将 manifest 文件的 minSdkVersion 或 targetSdkVersion 设置成 API level 11 以后版本而且 theme 设为 Theme.Holo(默认)或其子系，系统即可自动将 Options Menu 转为 ActionBar；Menu Item 转为 Action Button。

ActionBar 外观大致如图 6-33 所示，并说明如下[①]。

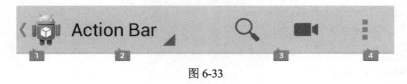

图 6-33

(4) 应用程序图标(App icon)：这个位置默认摆放应用程序的识别图标，也可以按照需求改变成其他图标。

(5) View 控制器(View controller)：这个部分可以让开发者放入不同的 View，例如下拉菜单或 Tab 控制器。

(6) Action 按钮(Action buttons)：让用户方便操作的按钮。Options Menu 的 Menu Item 会自动转换成 Action 按钮。

(7) 延伸栏(Action overflow)：如果移动设备没有 Menu 键(平板电脑一般没有此键)而且按钮数量过多无法完全显示在 ActionBar 上，就会显示延伸栏，单击后会显示剩余的按钮。如果移动设备有 Menu 键，就不会显示延伸栏，因为可以按下 Menu 键来显示剩余的按钮。

1. 下拉列表功能

也可以在 ActionBar 上放置下拉菜单，如图 6-34 所示。

图 6-34

---

① 参见 http://developer.android.com/design/patterns/actionbar.html。

Activity 生命周期与平板电脑设计概念

创建步骤如下:

STEP 1 调用 setNavigationMode(ActionBar.NAVIGATION_MODE_LIST)设置浏览模式为菜单模式。

STEP 2 创建 SpinnerAdapter 以提供下拉菜单的选项。

STEP 3 实现 ActionBar.OnNavigationListener 的 onNavigationItemSelected()方法,完成用户单击下拉菜单选项时的事件处理。

STEP 4 调用 setListNavigationCallbacks(spinnerAdapter, onNavigationListener)方法设置 SpinnerAdapter 与 OnNavigationListener 对象,让下拉菜单运作顺利。

2. 分割 ActionBar

如果应用程序在 Android 4.0 或更高版本上运行,可以启动自动分割机制(Split action bar),只要在 manifest 文件的<activity>标签(只针对单一 Activity),或<application>标签(针对一个应用程序的所有 Activity)加上 android:uiOptions="splitActionBarWhenNarrow"设置即可。当屏幕不够宽时(例如手机的 Portrait 模式)会将摆放不下的项目放到屏幕底部,如图 6-35 所示。例如,ActionBar 同时有下拉菜单与众多按钮,当屏幕不够宽时会将下拉菜单保留在上部,但按钮放在底部。

图 6-35

 范例 ActionBarEx

范例说明:
- 移动设备垂直时,ActionBar 的下拉菜单在上而按钮栏在下,如图 6-36 所示。移动设备横向时,下拉菜单与按钮栏都在上,而且因为空间足够,所以按钮的文字部分也会显示,如图 6-37 所示。

- 按下 ADD 按钮会增加一个 EditText 组件，按下 DELETE 按钮会删除第一个 EditText 组件。
- 无论单击下拉菜单选项或单击按钮，都会将对应文字以 Toast 方式显示。

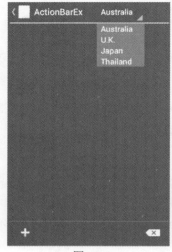

图 6-36　　　　　　　　　　　　　　　图 6-37

范例创建步骤如下：

STEP 1　在 res/AndroidManifest.xml 文件内新建 android:uiOptions="splitActionBarWhenNarrow"，当屏幕不够宽时，会将摆放不下的项目放到屏幕底部。

```
<application
    android:icon="@drawable/ic_launcher"
    android:label="@string/app_name"
    android:uiOptions="splitActionBarWhenNarrow" >
    <activity
        android:name=".ActionBarExActivity"
        android:label="@string/app_name" >
        <intent-filter>
            <action android:name="android.intent.action.MAIN" />
            <category android:name="android.intent.category.LAUNCHER" />
        </intent-filter>
    </activity>
</application>
```

STEP 2　在 res/menu/mymenu.xml 文件内新建 MenuItem，就会转换成 Action 按钮。

```
<menu xmlns:android="http://schemas.android.com/apk/res/android" >
    <item
```

```xml
            android:id="@+id/add"
            android:icon="@android:drawable/ic_input_add"
            android:title="@string/add"
            android:showAsAction="ifRoom|withText"/>
        <!-- ifRoom 代表只有空间足够才会显示此项目，如不足够，会将项目放在延伸栏。
                withText 代表显示文字(但只有在 Landscape 模式) -->

    <item
            android:id="@+id/delete"
            android:icon="@android:drawable/ic_input_delete"
            android:title="@string/delete"
            android:showAsAction="ifRoom|withText"/>
</menu>
```

STEP 3 在 ActionBarExActivity.java 文件内添加下列程序代码。

```java
public class ActionBarExActivity extends Activity {
    LinearLayout linearLayout;
    ActionBar actionBar;
    @Override
    public void onCreate(Bundle savedInstanceState) {
        super.onCreate(savedInstanceState);
        setContentView(R.layout.main);
        linearLayout = (LinearLayout)findViewById(R.id.linearLayout);

        /* 取得 ActionBar 实体引用 */
        actionBar = getActionBar();

        /* 设置为 true，才能单击 ActionBar 上的 App 图标 */
        actionBar.setDisplayHomeAsUpEnabled(true);

        setListNavigation();
    }

    /* 让 ActionBar 有下拉菜单功能 */
    private void setListNavigation() {
        /* 设置浏览方式为 list 方式 */
        actionBar.setNavigationMode(ActionBar.NAVIGATION_MODE_LIST);

        final String[] places = {"Australia", "U.K.", "Japan", "Thailand"};
        ArrayAdapter<String> adapterPlace =
            new ArrayAdapter<String>(this,
                    android.R.layout.simple_spinner_item, places);

        /* 利用匿名内部类实现 OnNavigationListener 的 onNavigationItemSelected()
```

```java
            方法，当 Spinner 选项改变时会调用该方法并取得选项位置 */
            OnNavigationListener navigationListener = new OnNavigationListener() {
                @Override
                public boolean onNavigationItemSelected(int itemPosition, long itemId) {
                    Toast.makeText(getApplicationContext(), places[itemPosition],
                Toast.LENGTH_SHORT).show();
                    return true;
                }
            };

            /* 设置 ActionBar 的浏览功能：adapterPlace 内容将成为列表选项，
               navigationListener 负责监听是否改变选项 */
            actionBar.setListNavigationCallbacks(adapterPlace, navigationListener);
    }

    @Override
    public boolean onCreateOptionsMenu(Menu menu) {
        MenuInflater inflater = getMenuInflater();
        inflater.inflate(R.menu.mymenu, menu);
        return true;
    }

    @Override
    public boolean onOptionsItemSelected(MenuItem item) {
        switch (item.getItemId()) {
        /* 单击 ActionBar 上的 App 图标就回到 ActionBarExActivity 的页面 */
        case android.R.id.home:
            Intent intent = new Intent(this, ActionBarExActivity.class);
            /* 设置为 FLAG_ACTIVITY_CLEAR_TOP，
               代表欲打开的 Activity 如果存在就直接打开已存在的，
               并将其他 Activity 都关闭 */
            intent.addFlags(Intent.FLAG_ACTIVITY_CLEAR_TOP);
            startActivity(intent);

        /* 按下 ADD 按钮会增加一个 EditText 组件 */
        case R.id.add:
            EditText editText = new EditText(this);
            editText.setHint("please input");
            editText.requestFocus();
            LinearLayout.LayoutParams layoutParams =
                new LinearLayout.LayoutParams(
                        LinearLayout.LayoutParams.MATCH_PARENT,
                        LinearLayout.LayoutParams.WRAP_CONTENT);
            linearLayout.addView(editText, layoutParams);
```

```
                Toast.makeText(this, item.getTitle(), Toast.LENGTH_SHORT).show();
                break;
            /* 按下 DELETE 按钮会删除第一个 EditText 组件 */
            case R.id.delete:
                if(linearLayout.getChildCount() > 0) {
                    linearLayout.removeViewAt(0);
                }
                Toast.makeText(this, item.getTitle(), Toast.LENGTH_SHORT).show();
                break;
            default:
                return super.onOptionsItemSelected(item);
        }
        return true;
    }
}
```

6.6.4 Tabs

如果开发者想要加上 Tab 导航方式，可以使用 ActionBar 的 Tab 功能，而避免使用 TabWidget[①](而且 TabActivity 已列为已弃用，因为 ActionBar 可以适用于各种屏幕尺寸。创建 Tab 的步骤如下：

STEP 1 实现 ActionBar.TabListener 界面以处理用户单击 Tab 的事件。

STEP 2 通过 newTab()初始化 Tab 后调用 setText()/setIcon()设置 Tab 文字与图标，并调用 setTabListener()设置已实现好的 TabListener。

STEP 3 ActionBar 调用 addTab()新建各个 Tab。

范例 TabsEx

范例(如图 6-38 所示)说明：
- 单击 Tab 会显示对应内容。
- Tab 栏超过页面就可以使用 scroll 方式翻阅。
- 如果 Tab 已经被选择又再次被单击，会以 Toast 显示 Reselected!。

---

① 参见 http://developer.android.com/guide/topics/ui/actionbar.html#Tabs。

图 6-38

范例创建步骤如下：

STEP 1 在 TabsExActivity.java 文件内添加下列程序代码。

```
public class TabsExActivity extends Activity {
    @Override
    protected void onCreate(Bundle savedInstanceState) {
        super.onCreate(savedInstanceState);
        ActionBar actionBar = getActionBar();

        /* 不显示 ActionBar 的图标与标题 */
        actionBar.setDisplayShowHomeEnabled(false);
        actionBar.setDisplayShowTitleEnabled(false);

        /* 设置浏览模式为 Tab 模式 */
        actionBar.setNavigationMode(ActionBar.NAVIGATION_MODE_TABS);
        /* 调用 newTab()取得 ActionBar.Tab 实体引用，
           setText()设置 Tab 文字，
           setIcon()设置 Tab 图标，
           setTabListener()设置 Tab 是否被单击的监听器 */
        actionBar.addTab(actionBar.newTab()
                .setText("Tab 1")
                .setIcon(R.drawable.ic_action_star)
                .setTabListener(new MyTabListener<Tab1>(this, "tab1", Tab1.class)));

        actionBar.addTab(actionBar.newTab()
                .setText("Tab 2")
                .setIcon(R.drawable.ic_action_photo)
                .setTabListener(new MyTabListener<Tab2>(this, "tab2", Tab2.class)));
```

```java
            actionBar.addTab(actionBar.newTab()
                    .setText("Tab 3")
                    .setIcon(R.drawable.ic_action_video)
                    .setTabListener(new MyTabListener<Tab3>(this, "tab3", Tab3.class)));

            actionBar.addTab(actionBar.newTab()
                    .setText("Tab 4")
                    .setIcon(R.drawable.ic_action_mail)
                    .setTabListener(new MyTabListener<Tab4>(this, "tab4", Tab4.class)));
    }

    public class MyTabListener<T extends Fragment> implements ActionBar.TabListener {
        private final Activity activity;
        private final String tag;
        private final Class<T> clz;
        private final Bundle bundle;
        private Fragment fragment;

        public MyTabListener(Activity activity, String tag, Class<T> clz) {
            this(activity, tag, clz, null);
        }

        public MyTabListener(Activity activity, String tag, Class<T> clz, Bundle bundle) {
            this.activity = activity;
            this.tag = tag;
            this.clz = clz;
            this.bundle = bundle;
        }

        /* 当 Tab 选择时调用 */
        public void onTabSelected(Tab tab, FragmentTransaction ft) {
            /* 如果 Fragment 不存在就创建，如果存在就直接添加以呈现给用户 */
            if (fragment == null) {
                fragment = Fragment.instantiate(activity, clz.getName(), bundle);
                ft.add(android.R.id.content, fragment, tag);
            } else {
                ft.attach(fragment);
            }
        }

        /* 当 Tab 结束选择时调用 */
        public void onTabUnselected(Tab tab, FragmentTransaction ft) {
            /* 如果 Fragment 已经存在，就直接移除 */
            if (fragment != null) {
                ft.detach(fragment);
            }
```

```
        }
        /* 当 Tab 已经选择又再次单击时调用 */
        public void onTabReselected(Tab tab, FragmentTransaction ft) {
            Toast.makeText(activity, "Reselected!", Toast.LENGTH_SHORT).show();
        }
    }
}
```

STEP 2 在 Tab1.java 文件内添加下列程序代码。Tab1 类继承 Fragment，用来呈现单击 Tab 后的内容。Tab2、Tab3、Tab4 等类的功能亦相同，不再赘述。

```
public class Tab1 extends Fragment {
    @Override
    public View onCreateView(LayoutInflater inflater, ViewGroup container,
            Bundle savedInstanceState) {

        /* Fragment 应用 R.layout.tab1 的 layout 文件 */
        View view = inflater.inflate(R.layout.tab1, container, false);
        return view;
    }
}
```

第 7 章

数据访问

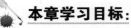

本章学习目标：

- Android 数据访问概论
- Assets
- Resources
- Shared Preferences
- Internal Storage
- External Storage

7.1 Android 数据访问概论

应用程序在运行过程中往往需要存储数据以方便之后取得，Android 提供许多数据访问的机制方便开发者按照应用程序的需求将数据存储至特定地方后再通过特定方式取得该数据内容。有些数据访问机制可以让应用程序将数据存储后仅提供给自己访问；也有机制是可以将应用程序存储的数据开放给其他应用程序来访问。Android 提供的数据访问机制有下列几种。

- assets：访问应用程序本身 assets 目录内的资源。
- resources：访问应用程序本身 res 目录内的资源。
- shared preferences：访问首选文件的内容。
- internal storage：访问移动设备的内存(memory)内容。
- external storage：访问移动设备外部存储媒体(例如 SD 卡)的内容。
- SQLite 数据库：访问 Android 内置数据库的内容。

除了 SQLite 数据库的访问较一般数据访问复杂许多，将在 SQLite 一章说明外，其余数据访问方式将于本章详细说明。

7.2 Assets

虽然一般比较少见，但有时候会需要直接使用应用程序本身的源文件(raw data，例如 txt 文件①)；如果有这种需求，不妨将源文件放在 assets 目录内。在 assets 目录内的文件不需要使用 resource ID 来访问，换句话说，不必通过 R.java 所定义的 resource ID 来取得该源文件内容。这是因为取得 assets 目录内的文件内容是通过较低级(lower-level)的 IO(Input/Output，输入与输出)方式，几乎与 Java IO 取得文件内容的观念相同。不过要取得 assets 目录内的文件内容，必须通过 AssetManager 类的功能，而且不能修改文件内容。

 范例 AssetEx

范例(如图 7-1 所示)说明：

取得 assets 目录内 txt 文件后，将其内容呈现在画面上。

---

① 如果要取得的源文件属于多媒体文件(例如 video 或 audio 文件)，则可将该文件存储在 res/raw 目录内，然后按照 7.3 节介绍的方式来取得该文件。

218

图 7-1

AssetEx/src/org/assetEx/AssetEx.java

```
10.     public class AssetEx extends Activity {
11.         private TextView tvAsset;
12.
13.         @Override
14.         public void onCreate(Bundle savedInstanceState) {
15.             super.onCreate(savedInstanceState);
16.             setContentView(R.layout.main);
17.             findViews();
18.         }
19.
20.         private void findViews() {
21.             tvAsset = (TextView)findViewById(R.id.tvAsset);
22.             try {
23.                 InputStream is = getAssets().open("dearJohn.txt");
24.                 int size = is.available();
25.                 byte[] buffer = new byte[size];
26.                 is.read(buffer);
27.                 is.close();
28.                 String text = new String(buffer);
29.                 tvAsset.setText(text);
30.             } catch (IOException e) {
31.                 Log.w("AssetEx", e);
32.             }
33.         }
34.     }
```

23 行：通过调用 Context 的 getAssets() 可以取得 AssetManager 对象；再通过 AssetManager 对象调用 open()并指定 dearJohn.txt 文件名，可以取得对应的 InputStream 对象，接下来就属于 Java IO 的观念了。

24 行：调用 InputStream.available() 可以取得该数据流内的数据大小，单位是 byte，在此是指 dearJohn.txt 文本文件的大小。

25 行：创建一个与文本文件大小相同的 byte 数组当作 buffer(暂存区)，用以暂存数据内容。

26 行：将读进来的文件内容暂存在 buffer 空间内。

28 行：将暂存空间内的数据转换成 String 类型。

29 行：最后将数据内容显示在 TextView 组件上。

7.3 Resources

所谓取得 resources 就是取得 res 目录内的资源，若要取得 res 目录内的资源，基本上都必须通过该资源的 resource ID(R.java 文件内会自动产生对应的标识符)来取得，举例如下：

- 取得文本文件内的文本：例如通过 R.string.styled_text 取得 strings.xml 文件内 styled_text 所代表的文本。

    String string = getString(***R.string.styled_text***);

- 取得 res/drawable 目录内的图形文件：例如通过 R.drawable.photo 取得 res/drawable/photo.jpg 图形文件。

    Drawable drawable = getResources().getDrawable(***R.drawable.photo***);

- 取得 res/raw 目录内的影音文件：例如通过 R.raw.ring 取得 res/raw/ring.mp3 音乐文件。

    MediaPlayer.create(context, ***R.raw.ring***);

- 取得 res/layout 目录内的 layout 文件或 UI 组件：例如通过 R.layout.main 取得 res/layout/main.xml 文件；R.id.tvGetText 取得 layout 文件内 tvGetText 对应的 TextView 组件。

    setContentView(***R.layout.main***);
    TextView tvGetText = (TextView) findViewById(***R.id.tvGetText***);

 范例 ResourceEx

范例(如图 7-2 所示)说明：

(1) 通过 R.string.styled_text 取得 strings.xml 文件内 styled_text 所代表的文本并呈现在 TextView 组件上。

- 调用 getText() 会连同该文本的格式信息也取得，所以可以呈现加入下划线、粗体、斜体字等文本格式。

- 调用 getString() 则会略过格式信息，而单纯只取得文本内容，所以呈现的文本即无粗体、斜体等文本格式。

(2) 通过 R.drawable.photo 取得 res/drawable/photo.jpg 图形文件。

(3) 通过 R.raw.ring 取得 res/raw/ring.mp3 音乐文件。

图 7-2

ResourceEx/res/values/strings.xml

```
1.   <?xml version="1.0" encoding="utf-8"?>
2.   <resources>
3.       <string name="app_name">ResourceEx</string>
4.       <string name="styled_text">
5.           <u>三国演义</u>，作者 <b>罗贯中</b>，号<i>湖海散人</i>
6.       </string>
7.   </resources>
```

4 行：styled_text 是第 5 行文本的标识符。

5 行：<u>、<b>、<i>分别代表加入下划线、粗体、斜体等文本格式。

ResourceEx/src/org/resourceEx/ResourceEx.java

```
10.  public class ResourceEx extends Activity {
11.      private TextView tvGetText;
12.      private TextView tvGetString;
13.      private TextView tvRes;
14.      private ImageView ivPhoto;
15.
16.      @Override
```

```
17.     protected void onCreate(Bundle savedInstanceState) {
18.         super.onCreate(savedInstanceState);
19.         setContentView(R.layout.main);
20.         findViews();
21.     }
22.
23.     private void findViews() {
24.         tvGetText = (TextView) findViewById(R.id.tvGetText);
25.         tvGetString = (TextView) findViewById(R.id.tvGetString);
26.         tvRes = (TextView) findViewById(R.id.tvRes);
27.         ivPhoto = (ImageView)findViewById(R.id.ivPhoto);
28.         CharSequence cs;
29.         String str;
30.         cs = getText(R.string.styled_text);
31.         tvGetText.setText(cs);
32.
33.         str = getString(R.string.styled_text);
34.         tvGetString.setText(str);
35.
36.         Resources res = this.getResources();
37.         cs = res.getText(R.string.styled_text);
38.         tvRes.setText(cs);
39.         ivPhoto.setImageDrawable(res.getDrawable(R.drawable.photo));
40.         MediaPlayer mp = MediaPlayer.create(ResourceEx.this, R.raw.ring);
41.         mp.start();
42.     }
43. }
```

30 行：调用 Context 的 getText()并搭配参数 R.string.styled_text 可以取得 res/values/strings.xml 文件内 styled_text 所代表的文本，而且不会失去文本格式。取得的文本，其数据类型为 CharSequence。

31 行：将取得的文本内容呈现在 tvGetText 所代表的 TextView 组件上。

33 行：调用 Context 的 getString()并搭配参数 R.string.styled_text 可以取得 res/values/strings.xml 文件内 styled_text 所代表的文本，但是会失去文本格式。取得的文本，其数据类型为 String。

36 行：调用 Context 的 getResources()可以取得 Resources 对象。

37 行：通过 Resources 对象调用 getText()、getString()，效果与 30、33 行相同，不再赘述。

39 行：调用 Resources 的 getDrawable()并搭配参数 R.drawable.photo 可以取得 Drawable 对象，该对象代表的即是 res/drawable/photo.jpg 图形文件信息。执行 ivPhoto.setImageDrawable() 并将前述返回的 Drawable 对象当作参数传递，即可将 photo.jpg 图形文件内容呈现在 ivPhoto 所代表的 ImageView 组件上。

40~41 行：调用 MediaPlayer 的 create()并搭配参数 R.raw.ring 可以取得 MediaPlayer 对象，再调用 start()代表播放 res/raw/ring.mp3 音乐文件。

7.4 Shared Preferences

SharedPreferences 类提供开发者一个可以通过 key-value pairs(键值组，类似 java.util.Map 访问数据的方式)而达到快速访问数据的功能，其数据会被存储在一个文件内，一般称为首选配置文件[①]。想要将数据存储在首选配置文件内，可按照下列步骤。

STEP 1 取得 SharedPreferences 对象：SharedPreferences 对象代表的就是首选配置文件，要取得该对象必须调用 Context 的 getSharedPreferences()。

SharedPreferences sharedPreferences = *getSharedPreferences*(name, mode);

STEP 2 取得 SharedPreferences.Editor 对象：要编辑首选配置文件内的数据，必须调用 SharedPreferences 的 edit()以取得 SharedPreferences.Editor 对象，通过该对象可以调用 putBoolean()、putFloat()、putInt()、putLong()、putString()以存储布尔值、整数、浮点数以及文本等类型的数据，最后必须调用 SharedPreferences.Editor 的 commit()将修改结果存回 SharedPreferences 对象。

sharedPreferences.*edit*()
.*putString*(key1, stringValue)
.*putBoolean*(key2, booleanValue)
.*putInt*(key3, intValue)
.commit();

STEP 3 存储后若想将数据从首选配置文件内取出，可以调用 SharedPreferences 对应的 getter 方法，说明如下：

String string = sharedPreferences.*getString*(key, defValue);

STEP 4 已经存储的数据也可以调用 SharedPreferences.Editor 的 remove() 将之卸载。

sharedPreferences.edit()
.*remove*(key)
.commit();

访问首选配置文件所需使用到的相关方法，说明如表 7-1 所示。

---

① 即使应用程序被关闭，再启动时，仍可访问存储在首选配置文件内的数据；除非将数据删除，或将该应用程序卸载。

表 7-1

Context 类
public abstract SharedPreferences getSharedPreferences (String name, int mode)
取得首选配置文件所代表的对象,数据类型为 SharedPreferences。 (1) name：首选设置文件名 (2) mode：共有下列几种 ● MODE_PRIVATE(只有应用程序本身可以访问) ● MODE_WORLD_READABLE(其他应用程序有读取权限) ● MODE_WRITEABLE(其他应用程序有写入权限)
SharedPreferences 界面
public abstract SharedPreferences.Editor edit()
取得 SharedPreferences.Editor 对象以编辑首选配置文件内的数据
public abstract String getString (String key, String defValue)[①]
取得文本类型的数据 ● key：数据的键值 ● defValue：默认值。当 key 所代表的数据不存在时，会返回默认值
public abstract boolean contains (String key)
检查首选配置文件内是否有 key 所代表的数据存在 key：首选配置文件内的键值
public abstract Map<String, ?> getAll ()
取得首选配置文件内的所有数据并存入 Map 后返回
SharedPreferences.Editor 界面
public abstract SharedPreferences.Editor putString (String key, String value)[②]
将文本类型的数据放入 SharedPreferences.Editor 对象内 ● key：数据的键值 ● value：数据值
public abstract SharedPreferences.Editor remove (String key)
卸载 key 所代表的一组数据 key：数据的键值
public abstract SharedPreferences.Editor clear ()
清空首选配置文件内的所有数据
public abstract boolean commit()
对 SharedPreferences.Editor 对象作任何修改后，都必须调用此方法将修改结果存回 SharedPreferences 对象

---

① 其余 getter 方法的功能与 getString()大致相同，仅差在数据类型，所以不再赘述。
② 其余 putBoolean()、putFloat、putInt()、putLong()等方法的功能与 putString()大致相同，仅差在放入数据的数据类型，所以不再赘述。

范例 SharedPrefEx

范例(如图 7-3 所示)说明：
- 输入完对应值后按下"存储首选设置"按钮，即可将输入的值存储在首选配置文件内。
- 按下"加载首选设置"按钮，即可将原来存储的首选设置值还原。
- 按下"恢复默认值"按钮，即可恢复成原始默认值。

图 7-3

SharedPrefEx/src/org/sharedPrefEx/SharedPrefEx.java

```
44.         //按下"存储首选设置"按钮
45.         btnSave.setOnClickListener(new OnClickListener() {
46.             @Override
47.             public void onClick(View v) {
48.                 SharedPreferences settings =
49.                     getSharedPreferences(prefName, Context.MODE_PRIVATE);
50.
51.                 String fileName = etFileName.getText().toString();
52.
53.                 boolean isAutoFocus = true;
54.                 if (rbYes.isChecked())
55.                     isAutoFocus = true;
56.                 else
57.                     isAutoFocus = false;
58.
59.                 int secAutoShoot = 0;
60.                 try{
61.                     secAutoShoot = Integer.parseInt(
62.                         etSecAutoShoot.getText().toString());
```

```
63.            }catch(NumberFormatException e){
64.                Toast.makeText(SharedPrefEx.this, getText(R.string.askForNumber),
65.                    Toast.LENGTH_SHORT).show();
66.            }
67.
68.            settings.edit()
69.                .putString("fileName", fileName)
70.                .putBoolean("isAutoFocus", isAutoFocus)
71.                .putInt("secAutoShoot", secAutoShoot)
72.                .commit();
73.
74.            Toast.makeText(SharedPrefEx.this, getText(R.string.prefSaved),
75.                Toast.LENGTH_SHORT).show();
76.        }
77.    });
```

49 行：调用 getSharedPreferences()并指定首选配置文件的名称以取得对应的 SharedPreferences 对象。MODE_PRIVATE 代表只有应用程序本身可以访问。

54~57 行：rbYes 代表是否自动对焦的"是"按钮，如果选择该按钮，就将 isAutoFocus 值设为 true，否则就设为 false。

61~62 行：获取用户输入的秒数后转换成整数值存储在 secAutoShoot 变量内。

63~66 行：如果用户输入的不是数字，会产生 NumberFormatException 异常，就以 Toast 消息框呈现错误信息。

68~72 行：调用 edit()会创建 SharedPreferences.Editor 对象，以方便修改 SharedPreferences 所存储的数据，执行 72 行的 commit() 会将修改结果存回 SharedPreferences 对象。按照数据类型的不同，可以分别调用 putString()、putBoolean()、putInt() 以存储字符串、布尔以及整数类型的数据。

```
79.    //按下"加载首选设置"按钮
80.    btnLoad.setOnClickListener(new OnClickListener() {
81.        @Override
82.        public void onClick(View v) {
83.            loadPref();
84.            Toast.makeText(SharedPrefEx.this, getText(R.string.prefLoaded),
85.                Toast.LENGTH_SHORT).show();
86.        }
87.    });
88.
89.    //按下"恢复默认值"按钮
90.    btnDefault.setOnClickListener(new OnClickListener() {
91.        @Override
92.        public void onClick(View v) {
93.            loadDefault();
```

```
94.                    Toast.makeText(SharedPrefEx.this, , getText(R.string.beenDefaultValue),
95.                        Toast.LENGTH_SHORT).show();
96.            }
97.        });
98.    }
99.
100.    //加载首选设置
101.    protected void loadPref() {
102.        SharedPreferences settings =
103.            getSharedPreferences(prefName, Context.MODE_PRIVATE);
104.
105.        String fileName = settings.getString("fileName", default_fileName);
106.        etFileName.setText(fileName);
107.
108.        boolean isAutoFocus = settings.getBoolean("isAutoFocus", default_isAutoFocus);
109.        if(isAutoFocus)
110.            rbYes.setChecked(true);
111.        else
112.            rbNo.setChecked(true);
113.
114.        int secAutoShoot = settings.getInt("secAutoShoot", default_secAutoShoot);
115.        etSecAutoShoot.setText(Integer.toString(secAutoShoot));
116.    }
117.
118.    //加载默认值
119.    protected void loadDefault() {
120.        etFileName.setText(default_fileName);
121.
122.        if(default_isAutoFocus)
123.            rbYes.setChecked(true);
124.        else
125.            rbNo.setChecked(true);
126.
127.        etSecAutoShoot.setText(Integer.toString(default_secAutoShoot));
128.    }
```

80~87 行：按下"加载首选设置"按钮会调用 101 行 loadPref()，并以 Toast 消息框呈现"已加载首选设置"信息。

90~97 行：按下"恢复默认值"按钮会调用 119 行 loadDefault()。

101~116 行：loadPref() 会将首选设置内容取出，并将结果呈现在各个对应的窗口组件上。

- 105~106 行：取得"照片默认存档名称"首选设置的值并呈现在 etFileName 所代表的 EditText 组件上。
- 108~112 行：取得"是否自动对焦"首选设置的值并决定选择哪一个 RadioButton。

- 114~115 行：取得"几秒后自动拍摄"首选设置的值并呈现在 etSecAutoShoot 所代表的 EditText 组件上。

119~128 行：loadDefault()会在画面上显示各个首选设置的默认值。

7.5 Internal Storage

应用程序可以指定将文件直接存储在移动设备的内存内，这种方式称为内部存储(internal storage)。这些存储在内存内的文件默认只有该应用程序可以访问，其他的应用程序一概不能访问，这类文件称为私有文件(private file)。当用户卸载(uninstall)应用程序后，存储在内存内对应的私有文件会一并被卸载。创建私有文件并存储至内存内的步骤如下：

STEP 1 调用 Context 的 openFileOutput() 可以打开指定文件，如果该文件不存在，就会自动创建对应文件并返回 FileOutputStream 对象供数据写入之用。

FileOutputStream fos = *openFileOutput*(fileName, mode);

STEP 2 调用 FileOutputStream 的 write()，将数据写入文件内。

fos.*write*(buffer);

STEP 3 调用 FileOutputStream 的 close()，关闭写入文件的数据串流。

fos.*close*();

如果要读取内部存储的文件，可按照下列步骤：

STEP 1 调用 Context 的 openFileInput() 可以打开指定文件，并返回 FileInputStream 对象供数据读取之用。如果该文件不存在，会产生 FileNotFoundException 异常。

FileInputStream fis = *openFileInput*(fileName);

STEP 2 调用 FileInputStream 的 read()，将数据从文件中读出。

fis.*read*();

STEP 3 调用 FileInputStream 的 close()，关闭读取文件的数据串流。

fis.*close*();

内部存储所需使用到的相关方法，说明如表 7-2 所示：

数据访问

表 7-2

Context 类

public abstract FileOutputStream openFileOutput (String name, int mode)

打开指定的私有文件并返回 FileOutputStream 对象，若该文件不存在，会自动创建一个

name：指定打开的文件名，不可以有路径分隔符("/")，也就是目录符号，否则会产生 IllegalArgumentException 异常

mode：共有下列 4 种

- MODE_PRIVATE(只有应用程序本身可以访问)
- MODE_APPEND(如果文件已经存在，将新写入的数据附加在后面)
- MODE_WORLD_READABLE(其他应用程序有读取权限)
- MODE_WRITEABLE(其他应用程序有写入权限)

public abstract FileInputStream openFileInput (String name)

打开指定的私有文件并返回 FileInputStream 对象，若该文件不存在，会产生 FileNotFoundException 异常

name：指定打开的文件名，不可以有路径分隔符("/")

 范例 InternalStorageEx

范例说明：

- 输入完文本后按下"存储文件"按钮，即可将输入的文本存储在程序指定的文件内，如图 7-4 所示。
- 按下"附加后存储"按钮，即可将新输入的文本附加在原来指定的文件内，如图 7-5 所示。
- 按下"打开文件"按钮，即可取得文件内容文本，如图 7-6 所示。

图 7-4

图 7-5

图 7-6

InternalStorageEx/src/org/internalStorageEx/InternalStorageEx.java

```
39.     //按下"存储文件"按钮
40.     btnSave.setOnClickListener(new OnClickListener() {
41.         @Override
42.         public void onClick(View v) {
43.             FileOutputStream fos = null;
44.             try {
45.                 fos = openFileOutput("note.txt", Context.MODE_PRIVATE);
46.                 fos.write(etNote.getText().toString().getBytes());
47.                 fos.close();
48.             } catch (IOException e) {
49.                 Log.e("InternalStorageEx", e.toString());
50.             }
51.             tvNote.setText(getText(R.string.fileSaved));
52.         }
53.     });
```

45 行：以输出方式打开 note.txt 文件，打开模式为仅有此应用程序可以访问。

46 行：调用 write() 将用户输入的文本写入 45 行指定的文件。

47 行：调用 close() 关闭与文件之间的数据串流。

51 行："已将输入存储"文本显示在指定的 TextView 上。

```
55.     //按下"附加后存储"按钮
56.     btnAppend.setOnClickListener(new OnClickListener() {
57.         @Override
58.         public void onClick(View v) {
59.             FileOutputStream fos = null;
60.             try {
61.                 fos = openFileOutput("note.txt", Context.MODE_APPEND);
62.                 fos.write(etNote.getText().toString().getBytes());
63.                 fos.close();
64.             } catch (IOException e) {
65.                 Log.e("InternalStorageEx", e.toString());
66.             }
67.             tvNote.setText(getText(R.string.fileAppended));
68.         }
69.     });
```

61 行：以输出方式打开 note.txt 文件，打开模式可以将新写入的数据附加在原文件内容的后面。

```
71.     //按下"打开文件"按钮
72.     btnOpen.setOnClickListener(new OnClickListener() {
73.         @Override
```

```
74.     public void onClick(View v) {
75.         FileInputStream fis = null;
76.         StringBuilder sb = new StringBuilder();
77.         try {
78.             fis = openFileInput("note.txt");
79.             InputStreamReader isr = new InputStreamReader(fis);
80.             BufferedReader br = new BufferedReader(isr);
81.             String str = "";
82.             while((str = br.readLine())!=null){
83.                 sb.append(str);
84.             }
85.             br.close();
86.             isr.close();
87.             fis.close();
88.         } catch (IOException e) {
89.             Log.e("InternalStorageEx", e.toString());
90.         }
91.         tvNote.setText(sb);
92.     }
93. });
```

78 行：以输入方式打开 note.txt 文件。

82~83 行：使用 BufferedReader 将文件内的文本一行行读入后附加在 StringBuilder 原来文本的后面，直到文件结尾才停止读取。

不可不知

内部私有文件会存储在：data/data/[项目包名称]/files 内，以 InternalStorageEx 范例运行所产生的 note.txt 文件为例，会存放在 data/data/org.internalStorageEx/files 目录内；如果在模拟器上运行，可以使用 DDMS 的 File Explorer 来查看该文件；如果在实机上运行，因为安全性考虑，所以没有浏览 data 目录内容的权限。

之前 SharedPrefEx 范例会产生 prefSet.xml 文件，存放在 data/data/org.sharedPrefEx/shared_prefs 目录内。

7.6 External Storage

应用程序也可以将文件存储在移动设备的可卸载存储媒体上(例如：SD 卡)，这种方式被称为外部存储(external storage)，这类文件被称作外部文件(external file)。外部文件可能会因为外部存储媒体被卸载而无法读取,也可能会被其他应用程序修改,属于安全性与私密性较低的文件。虽然外部文件可以被其他应用程序访问,但仍分成私有的外部文件与公开的外部文件。

(1) 私有的外部文件:会存储在/sdcard/Android/data/<package_name>/files/,其中<package_name>就是应用程序的包名称,让人很容易就能区别出该文件是专门提供给指定应用程序使用的。例如,一个应用程序的包名称为 org.externalStorageEx,将某一文件存储成私有的外部文件并指定目录类型为 DIRECTORY_MUSIC,该文件的路径应为/sdcard/Android/data/org.externalStorageEx/files/Music/music.mp3。从 API level 8 开始,私有的外部文件会随着应用程序被卸载而一同被卸载。

(2) 公开的外部文件:文件存储时可以指定欲存储之目录类型,而该目录会在存储媒体的根目录内。例如,欲将某一文件存储成公开的外部文件,并指定目录类型为 DIRECTORY_PICTURES,而存储媒体为 SD 卡,那么存储路径应为/sdcard/Pictures/photo.jpg。公开的外部文件不会随着应用程序被卸载而一同被卸载。

 不可不知

创建外部文件时可以指定欲存储之目录类型,定义在 Environment 类内,属于字符串常量,共有下列 9 种。

- DIRECTORY_ALARMS:用来存放闹钟铃声专用的音频文件(非一般音乐文件)。
- DIRECTORY_DCIM:用来存放移动设备拍摄下来的相片或视频文件。
- DIRECTORY_DOWNLOADS:用来存放用户下载的文件。
- DIRECTORY_MOVIES:用来存放电影文件。
- DIRECTORY_MUSIC:用来存放音乐文件。
- DIRECTORY_NOTIFICATIONS:用来存放通知铃声专用的音频文件(非一般音乐文件)。
- DIRECTORY_PICTURES:用来存放相片文件。
- DIRECTORY_PODCASTS:用来存放 podcasts[1]专用的音频文件(非一般音乐文件)。
- DIRECTORY_RINGTONES:用来存放来电铃声专用的音频文件(非一般音乐文件)。

创建外部文件的步骤如下。

STEP 1 检查可否访问外部存储媒体:如果文件存储在 SD 卡这种可卸载的外部存储媒体上,可能会因为被挂载在 PC 上或处于只读状态而无法顺利被 Android 应用程序访问。为了确定能够正常访问,第一步就是要调用 Environment 的 getExternalStorageState() 检查该存储媒体是否处于可被访问的状态[2]。

```
String state = Environment.getExternalStorageState();
if (state.equals(Environment.MEDIA_MOUNTED)) {
```

---

[1] podcasts 源自苹果电脑的 iPod 与 broadcast(广播)的混合词。中国香港也常直接称作 Podcasting,是指一种在因特网上发布文件并允许用户订阅 feed 以自动接收新文件的方法,或用此方法来制作的电台节目。参见维基百科 http://zh.wikipedia.org/zh-tw/%E6%92%AD%E5%AE%A2。

[2] 外部存储媒体有哪些状态,参见 API 文件关于 Environment 类的状态常量部分。

```
        //存储媒体处于可读写状态
    } else if (state.equals(Environment.MEDIA_MOUNTED_READ_ONLY)) {
        //存储媒体处于只读状态
    } else {
        //存储媒体处于无法读写状态
    }
```

STEP 2 创建私有的外部文件：调用 Context 的 getExternalFilesDir()[1]并指定欲存储的目录类型，会返回该目录的路径以方便创建私有的外部文件。

```
File path = getExternalFilesDir(Environment.DIRECTORY_PICTURES);
File file = new File(path, "photo.jpg");
```

STEP 3 创建公开的外部文件：调用 Environment 的 getExternalStoragePublicDirectory() 并指定欲存储的目录类型，会返回该目录的路径以方便创建公开的外部文件。

```
File path = Environment.getExternalStoragePublicDirectory(
            Environment.DIRECTORY_PICTURES);
File file = new File(path, "photo.jpg");
```

STEP 4 打开媒体扫描器：如果创建的外部文件属于多媒体文件，可以调用 MediaScannerConnection 的 scanFile() 启动媒体扫描器，扫描是否有新的文件。如果有，则将此新的文件添加到其他应用程序，如相片集(Gallery)、音乐(Music)的文件清单上[2]。

```
MediaScannerConnection.scanFile(context, paths, mimeTypes, onScanCompletedListener);
```

STEP 5 如果要将文件存放在 SD 卡上，必须允许应用程序将文件存放在外部存储媒体，所以必须在 manifest 文件作以下设置：

```
<uses-permission android:name="android.permission.WRITE_EXTERNAL_STORAGE" />
```

外部存储所需要使用到的相关方法，说明如表 7-3 所示。

表 7-3

Environment 类
public static String getExternalStorageState() 取得主要外部存储媒体(primary external storage)的当前状态
public static File getExternalStoragePublicDirectory(String type) 指定欲存储的目录类型，会返回该目录的路径以方便创建公开的外部文件 type：目录类型

---

[1] API level 7 或更早之前的版本则调用 getExternalStorageDirectory()。
[2] 如果不想让媒体扫描器扫描到新建的多媒体文件，可以在该文件名前加上点号.，例如.photo。

(续表)

Context 类
public abstract File getExternalFilesDir (String type) 指定欲存储的目录类型，会返回该目录的路径以方便创建私有的外部文件 type：目录类型

MediaScannerConnection 类
public static void scanFile(Context context, String[] paths, String[] mimeTypes, MediaScannerConnection.OnScanCompletedListener listener) 启动媒体扫描器，扫描是否有新的文件 ● context：Context 对象，通常为现行 Activity ● paths：欲扫描的路径 ● mimeTypes：欲扫描的 MIME type，如果设为 null，会从扩展名来推断 MIME type ● listener：实现 OnScanCompletedListener 的监听器

MediaScannerConnection.OnScanCompletedListener 界面
public abstract void onScanCompleted(String path, Uri uri) 媒体扫描器扫描完成时会调用此方法 ● path：被扫描到的文件之路径 ● uri：文件的 uri。如果扫描成功，文件会被添加到媒体数据库；uri 值可能为：content://media/external/images/media/1；如果扫描不到文件，则 uri 为 null

 范例 ExternalStorageEx

范例说明：
- 按下"存储至公开照片区"按钮，即可将 res 目录内的照片存储至指定的公开目录，如图 7-7 所示。
- 按下"存储至私有照片区"按钮，即可将 res 目录内的照片存储至指定的私有目录，如图 7-8 所示。
- 如果 SD 卡没有被挂载，或根本没有 SD 卡，会以 Toast 消息框显示"检测不到存储卡"。

图 7-7　　　　　　　　　　　图 7-8

数 据 访 问 07

ExternalStorageEx/src/org/externalStorageEx/ExternalStorageEx.java

```java
34.     private void findViews() {
35.         ivPhoto = (ImageView)findViewById(R.id.ivPhoto);
36.         ivPhoto.setImageDrawable(
37.                 getResources().getDrawable(R.drawable.photo));
38.         btnSavePublic = (Button)findViewById(R.id.btnSavePublic);
39.         btnSavePrivate = (Button)findViewById(R.id.btnSavePrivate);
40.         tvMsg = (TextView)findViewById(R.id.tvMsg);
41.
42.         //按下"存储至公开照片区"按钮
43.         btnSavePublic.setOnClickListener(new OnClickListener() {
44.             @Override
45.             public void onClick(View v) {
46.                 File path = Environment.getExternalStoragePublicDirectory(
47.                         Environment.DIRECTORY_PICTURES);
48.                 File file = new File(path, "photo.jpg");
49.                 createFile(file);
50.             }
51.         });
52.
53.         //按下"存储至私有照片区"按钮
54.         btnSavePrivate.setOnClickListener(new OnClickListener() {
55.             @Override
56.             public void onClick(View v) {
57.                 File path = getExternalFilesDir(Environment.DIRECTORY_PICTURES);
58.                 File file = new File(path, "photo.jpg");
59.                 createFile(file);
60.             }
61.         });
62.
63.     }
```

45~50 行：按下"存储至公开照片区"按钮后取得公开目录 Pictures 的路径，之后指定该路径与 photo.jpg 文件名来创建 file 对象，最后调用 65 行 createFile() 创建公开的外部文件。

56~60 行：按下"存储至私有照片区"按钮后取得私有目录 Pictures 的路径，之后指定该路径与 photo.jpg 文件名来创建 file 对象，最后调用 65 行 createFile()创建私有的外部文件。

```java
65.     protected void createFile(File file) {
66.         File parentPath = file.getParentFile();
67.         if (!isSDExist()){
68.             Toast.makeText(this,
69.                     R.string.SDCardNotFound, Toast.LENGTH_LONG).show();
70.             return;
```

```
71.        }
72.        try {
73.            if(!parentPath.exists())
74.                parentPath.mkdirs();
75.            if(file.exists())
76.                file.delete();
77.            InputStream is = getResources().openRawResource(R.drawable.photo);
78.            OutputStream os = new FileOutputStream(file);
79.            byte[] data = new byte[is.available()];
80.            is.read(data);
81.            os.write(data);
82.            tvMsg.setText(getString(R.string.saveFileTo) + file.toString());
83.            is.close();
84.            os.close();
85.        } catch (IOException e) {
86.            Log.e("ExternalStorageEx", e.toString());
87.        }
88.
89.        String[] paths = {file.toString()};
90.        callMediaScanner(paths);
91.    }
92.
93.    private boolean isSDExist() {
94.        String state = Environment.getExternalStorageState();
95.
96.        if (state.equals(Environment.MEDIA_MOUNTED))
97.            return true;
98.        else
99.            return false;
100.   }
101.
102.   private void callMediaScanner(String[] paths) {
103.       MediaScannerConnection.scanFile(this, paths, null,
104.           new MediaScannerConnection.OnScanCompletedListener() {
105.               public void onScanCompleted(String path, Uri uri) {
106.                   Log.i("ExternalStorageEx", "Scanned " + path + ":");
107.                   Log.i("ExternalStorageEx", "-> uri=" + uri);
108.               }
109.       });
110.   }
```

67~71 行：要创建外部文件前，先调用 93 行 isSDExist()检查可否访问 SD 卡，如果无法访问 SD 卡，则停止创建文件。

73~74 行：如果欲创建的目录不存在，先创建目录。

75~76 行：如果欲创建的文件已经存在，就将该文件删除，之后再重新创建新的文件。

77 行：打开 res/drawable/photo.jpg 图形文件。

78 行：创建输出连接到 file 所代表的文件。

79 行：根据图形文件的大小创建 data 暂存区。

80 行：将读入的 photo.jpg 图形文件暂存在 data 暂存区。

81 行：将 data 暂存区内的数据全部写入 file 所代表的文件。

89~90 行：调用 102 行 callMediaScanner()扫描 file 所代表的文件。

93~100 行：调用 isSDExist()会先检查 SD 卡的状态，如果可以访问 SD 卡，则返回 true；否则返回 false。

103 行：调用 scanFile()会扫描 paths 路径中的文件。

104~105 行：实现 OnScanCompletedListener 的 onScanCompleted()，当媒体扫描器扫描完所有指定的文件后会调用 onScanCompleted()。

106~107 行：执行 onScanCompleted() 内容的线程不是 Activity 的主要线程，所以无法在 onScanCompleted()内访问 UI 组件(例如：TextView 或 Toast)，而将路径与 URI 信息写入 log 文件。

第 8 章

移动数据库SQLite

本章学习目标：

- SQLite 数据库概论与数据类型
- 使用命令行创建数据库
- SQL 语法
- Android 应用程序访问 SQLite 数据库
- SQLite 新增功能
- SQLite 查询功能
- SQLite 修改与删除功能

8.1 SQLite 数据库概论与数据类型

8.1.1 SQLite 数据库概论

SQLite[①]是一个嵌入式的 SQL 数据库引擎(embedded SQL database engine)，而且属于关系数据库(relational database management system)。该数据库系统非常小(大约只需要 275K 大小空间)，属于轻量级的数据库系统，主要是以 C 语言写成，作者[②]不仅将其源代码公开，还捐出来成为公共财产；换句话说，无论个人或商业使用，都不需要支付任何费用，因此非常受到大众喜爱。SQLite 虽小，不仅支持 SQL 语法，而且关系数据库的功能齐全，所以非常受到小型、嵌入式等系统的青睐。SQLite 与其他数据库的最大差异点在于，SQLite 直接将数据库的数据存储在本机端，而非服务器端，而且将一个完整数据库所拥有的数据表(table)、index、trigger 以及 view 都存储成一个文件。Android 完全支持 SQLite 数据库系统，所以开发者可以将较大量、复杂的数据以有系统的方式直接存储至移动设备上，以方便应用程序访问与管理。

8.1.2 SQLite 数据类型[③]

大多数的 SQL 数据库引擎都使用静态数据类型，而这种方式会使得数据值的类型受限于它所存储的字段。SQLite 使用动态数据类型，数据值属于何种类型是根据数据值本身而非受限于所存储的字段，这样一来可以让 SQLite 兼容大多数使用了静态数据类型数据库的 SQL 语法。

1. 存储类型

每一个存储在 SQLite 数据库的值属于下列 5 种存储类型(storage class)的其中一种。
- NULL：代表空值。
- INTEGER：可存储有正负号的整数类型数据。
- REAL：可存储浮点数(也就是小数)类型的数据。
- TEXT：可存储文本类型的数据。
- BLOB：BLOB 全名为 Binary Large Object，可存储 binary 数据，例如图形文件。

SQLite 没有 Boolean 类型，而是将 Boolean 数据存储成整数 0(false)与 1(true)。SQLite 也没有 Date 或 Time 类型，当存储日期或时间数据时会自动调用对应的日期或时间函数[④]。

2. 近似类型

为了与其他数据库兼容，SQLite 支持近似类型(Type Affinity)，而且用于字段数据类型上，

---

① SQLite 数据库官方网站 http://www.sqlite.org/。关于 SQLite 数据库的限制说明参见 http://www.sqlite.org/limits.html。
② SQLite 是由 D. Richard Hipp 博士在 2000 年时设计出来，并于 2005 年赢得 Google O'Reilly Open Source Award。现在最新版为 SQLite version 3.7.10。
③ 参见 http://www.sqlite.org/datatype3.html。
④ 参见 http://www.sqlite.org/lang_datefunc.html。

这种近似类型只是建议该字段应该存储什么数据类型，但无强制性。下面说明 SQLite 的 5 种近似类型，以及其他数据库与 SQLite 数据类型间的转换。

- INTEGER：数据类型包含 INT 字符会转换成 INTEGER。
- REAL：数据类型包含 REAL、FLOA、DOUB 等字符会转换成 REAL。
- TEXT：数据类型包含 CHAR、CLOB、TEXT 等字符会转换成 TEXT。VARCHAR 类型因为包含 CHAR 这个字符，所以也会转换成 TEXT。
- NONE：数据类型包含 BLOB 或没有定义数据类型都会转换成 NONE。
- NUMERIC：不属于以上类型者会转换成 NUMERIC。

其他数据库类型转换成 SQLite 近似类型的对照如表 8-1 所示。

表 8-1

其他数据库数据类型名称	SQLite 近似类型
INT INTEGER TINYINT SMALLINT MEDIUMINT BIGINT UNSIGNED BIG INT INT2 INT8	INTEGER
CHARACTER(20) VARCHAR(255) VARYING CHARACTER(255) NCHAR(55) NATIVE CHARACTER(70) NVARCHAR(100) TEXT CLOB	TEXT
BLOB 无定义数据类型	NONE
REAL DOUBLE DOUBLE PRECISION FLOAT	REAL
NUMERIC DECIMAL(10,5) BOOLEAN DATE DATETIME	NUMERIC

3. 存储类型与近似类型

不要将"SQLite 存储类型"与"SQLite 近似类型"搞混,不过 INTEGER、REAL、TEXT 等近似类型大多存储成前述对应的存储类型。例如,INTEGER 近似类型大多存储成 INTEGER 类型。NONE 近似类型则完全根据数据值属于何种类型,就直接存储成该类型。NUMERIC 则会先试图存储成 INTEGER 类型,如果不行,则试图存储成 REAL 类型,如果还是不行,才会选择存储成 TEXT 类型。下面 SQL 语法将呈现数据存储在各种近似类型的字段时,会存储成何种类型。

```
-- 各字段的近似类型分别为 TEXT, NUMERIC, INTEGER, REAL, BLOB
CREATE TABLE t1(
    t  TEXT,
    nu NUMERIC,
    i  INTEGER,
    r  REAL,
    no BLOB
);

-- 插入的值会分别存储成 TEXT, INTEGER, INTEGER, REAL, TEXT 类型
INSERT INTO t1 VALUES('500.0', '500.0', '500.0', '500.0', '500.0');
SELECT typeof(t), typeof(nu), typeof(i), typeof(r), typeof(no) FROM t1;
text|integer|integer|real|text

-- 插入的值会分别存储成 TEXT, INTEGER, INTEGER, REAL, REAL 类型
DELETE FROM t1;
INSERT INTO t1 VALUES(500.0, 500.0, 500.0, 500.0, 500.0);
SELECT typeof(t), typeof(nu), typeof(i), typeof(r), typeof(no) FROM t1;
text|integer|integer|real|real

-- 插入的值会分别存储成 TEXT, INTEGER, INTEGER, REAL, INTEGER 类型
DELETE FROM t1;
INSERT INTO t1 VALUES(500, 500, 500, 500, 500);
SELECT typeof(t), typeof(nu), typeof(i), typeof(r), typeof(no) FROM t1;
text|integer|integer|real|integer

-- 空值不会受到近似类型影响
DELETE FROM t1;
INSERT INTO t1 VALUES(NULL,NULL,NULL,NULL,NULL);
SELECT typeof(t), typeof(nu), typeof(i), typeof(r), typeof(no) FROM t1;
null|null|null|null|null
```

8.2 使用命令行创建数据库

在说明如何编写 Android 应用程序访问 SQLite 数据库前,不妨先了解如何使用命令行来创

移动数据库 SQLite ⑧

建数据库，在 Android 系统下创建数据库的步骤如下：

STEP 1 先启动模拟器，然后在命令行先输入 adb shell 命令进入 Android 的 shell 环境。

>adb shell

如果同时打开多个模拟器，就需要指定模拟器序号(例如 5554)；如果不清楚打开了哪些模拟器，则可以使用 adb devices 命令来查询。

>adb devices
>adb -s emulator-5554 shell

STEP 2 进入 Android 的 shell 环境后，在 sdcard 目录下创建一个 databases 目录，以便之后存放测试用的数据库。databases 目录创建完毕后请进入该目录。

cd sdcard
mkdir databases
cd databases

STEP 3 输入 sqlite3 sites 命令会创建 sites 数据库[①]并进入到 SQLite 数据库命令行管理模式，如果想要了解 SQLite 命令，可输入.help 打开说明文件。

sqlite3 sites
sqlite> .help

8.3 SQL 语法

    SQL(Structured Query Language)最早是由 IBM 公司于 1970 年发展出来的一套专门用于数据库访问的语法。因为这套语法十分接近人类语言，所以易于了解与使用，因此之后各个数据库厂商，如 Oracle、Sybase，等也都纷纷推出可以执行 SQL 语法的关系数据库，使得 SQL 语法更被广泛地使用在关系数据库上。为了让 SQL 语法可以兼容于各个数据库系统，ISO(International Standards Organization)和 ANSI(American National Standards Institute)联合主导 SQL 语法标准化规范的制定。先后制定了 SQL-92[②](代表 1992 年制定)、SQL-1999、SQL-2003 的 SQL 标准语法。所以现在 SQL 就成为访问数据库的标准语法。

    SQL 语法不区分大小写，按照功能不同又可区分成下列 3 种语法。
- DDL(Data Definition Language，数据定义语法)——创建数据库、表的语法，其实就是在定义数据库、表。所以创建、修改数据库或表的语法就称作 DDL。

---

① 若该数据库文件已经存在，就会直接打开该数据库。
② SQLite 几乎完全支持 SQL-92 标准语法。

243

- DML(Data Manipulation Language,数据处理语法)——专门用来处理表内数据的语法。DML 主要有 4 大语法:INSERT、UPDATE、DELETE 和 SELECT。
- DCL(Data Control Language,数据控制语法)——设置数据库、表权限的语法。例如:GRANT(授权使用)、DENY(拒绝使用)、REVOKE(取消授权)。

因为表创建后就会不断地更改数据内容与查询数据,所以上述 3 种语法中又以 DML 语法最为重要。

下面说明如何使用 SQL 语法完成数据表的创建与新增、修改、删除、查询数据表内容。

8.3.1 创建数据表

假设要创建一个 sitesInfo 数据表要存储旅游地点的代号、地名、联系电话与地址等相关信息,可以执行下列 SQL 的 DDL 语法以创建对应数据表:

```
sqlite> CREATE TABLE sitesInfo (
    ...> id TEXT NOT NULL,
    ...> name TEXT NOT NULL,
    ...> phoneNo TEXT,
    ...> address TEXT,
    ...> PRIMARY KEY (id)
    ...> );
```

(1) TEXT 相当于 Java 的 String 类型,NOT NULL 代表不可为空值;换句话说就是一定要输入值。

(2) PRIMARY KEY (id) 代表将 id 字段设为 PK(Primary Key,主键),也就是这个字段内的值不会重复。

(3) 输入一行 SQL 语法后按 Enter 按钮会出现...>,这代表 SQL 语法尚未结束,若要结束整个语法,要加上";"。

(4) 执行完上述 SQL 语法后会创建 sitesInfo 数据表及对应字段,但各个字段内还没有填入对应的值,如表 8-2 所示。

表 8-2

sitesInfo			
id	name	phoneNo	address

创建完毕后,可以输入.tables(属于 SQLite 命令,而非 SQL 语法)以查询是否成功新增 sitesInfo 数据表,如果新增成功,就会显示该数据表名称。

```
sqlite> .tables
sitesInfo
```

如果想要知道当初创建 sitesInfo 数据表的语法,可以输入.schema sitesInfo 以查询创建数据表的 SQL 语法。

```
sqlite> .schema sitesInfo
CREATE TABLE sitesInfo (
id TEXT NOT NULL,
name TEXT NOT NULL,
phoneNo TEXT,
address TEXT,
PRIMARY KEY (id)
);
```

8.3.2 DML 语法

1. INSERT 语法

新增 1 个旅游地点数据至 sitesInfo 数据表内。新增的值若是字符串则需要加上单引号' '，不过 SQLite 也支持双引号" "。为了要让显示结果多样化，不妨先新增几个数据。

```
sqlite> INSERT INTO sitesInfo (id, name, phoneNo, address)
    ...> VALUES ('yangmingshan','阳明山公园','02-28613601','台北市北投区竹子湖路 1 之 20 号');
```

2. SELECT 语法

如果想查看刚刚新增的数据是否正确，可以使用 SELECT 语法，也称为查询语法。*代表所有字段的值都会列出。不妨使用.mode column SQLite 命令将呈现的方式改成字段模式，这样一来各个字段的数据都会对齐。

```
sqlite> SELECT * FROM sitesInfo;
yangmingshan|阳明山公园|02-28613601|台北市北投区竹子湖路 1 之 20 号
yushan|玉山公园管理处|049-2773121|南投县水里乡中山路一段 300 号
taroko|太鲁阁公园管理处|03-8621100|花莲县秀林乡 258 号
sqlite> .mode column
sqlite> SELECT * FROM sitesInfo;
yangmingshan    阳明山公园      02-28613601     台北市北投区竹子湖路 1 之 20 号
yushan          玉山公园管理处   049-2773121     南投县水里乡中山路一段 300 号
taroko          太鲁阁公园管理处 03-8621100      花莲县秀林乡 258 号
```

3. UPDATE 语法

如果想将"阳明山公园"改成"阳明山公园管理处"，可以使用 UPDATE 语法。下面 UPDATE 语法的意思为：将 id 为 yangmingshan 的数据行中 name 字段内的值改成"阳明山公园管理处"；其中 WHERE 与 Java 的 while 循环功能非常相似，会不断寻找符合条件的数据行直到所有数据行寻找完毕为止。如果不加上 WHERE 条件，当有多个数据时，所有数据行 name 字段的值都会被修改成"阳明山公园管理处"，所以要适时地加上 WHERE 条件。

修改数据后也可以使用 SELECT 语法检查是否正确修改。

```
sqlite> UPDATE sitesInfo SET name='阳明山公园管理处' WHERE id='yangmingshan';
sqlite> SELECT * FROM sitesInfo;
yangmingshan    阳明山公园管理处    02-28613601    台北市北投区竹子湖路 1 之 20 号
yushan          玉山公园管理处      049-2773121    南投县水里乡中山路一段 300 号
taroko          太鲁阁公园管理处    03-8621100     花莲县秀林乡 258 号
```

4. DELETE 语法

如果想将数据删除，可以使用 DELETE 语法。下面 DELETE 语法意思为：将 id 为 yangmingshan 的数据行从 sitesInfo 数据表中删除；如果不加上 WHERE 条件式，会将所有数据全部删除，请务必谨慎。

```
sqlite> DELETE FROM sitesInfo WHERE id='yangmingshan';
```

5. SELECT 语法再探讨

SELECT 语法可说是 SQL DML 语法中使用频率最高的，因为所有数据都已输入完毕后，就常常需要通过 SELECT 语法来查询所需的数据，接下来会介绍多种常用的 SELECT 语法。

将 sitesInfo 数据表内的数据列出，但只列出 name、phoneNo 等两个字段。

```
sqlite> SELECT name, phoneNo FROM sitesInfo;
阳明山公园管理处    02-28613601
玉山公园管理处      049-2773121
太鲁阁公园管理处    03-8621100
```

按照 id 字段作升序排序。ORDER BY 代表排序，默认是升序排序，加上 DESC 就变成降序排序。

```
sqlite> SELECT * FROM sitesInfo ORDER BY id;
taroko          太鲁阁公园管理处    03-8621100     花莲县秀林乡 258 号
yangmingsh      阳明山公园管理处    02-2861360     台北市北投区竹子湖路 1 之 20 号
yushan          玉山公园管理处      049-277312     南投县水里乡中山路一段 300 号

sqlite> SELECT * FROM sitesInfo ORDER BY id DESC;
yushan          玉山公园管理处      049-2773121    南投县水里乡中山路一段 300 号
yangmingsh      阳明山公园管理     02-28613601    台北市北投区竹子湖路 1 之 20 号
taroko          太鲁阁公园管理     03-8621100     花莲县秀林乡 258 号
```

将地名以"玉山"开头的数据列出，"%"代表要使用模糊查询，可以出现 0 个以上任意字符。

```
sqlite> SELECT * FROM sitesInfo WHERE name LIKE '玉山%';
yushan          玉山公园管理处      049-2773121    南投县水里乡中山路一段 300 号
```

8.4 Android 应用程序访问 SQLite 数据库

Android 应用程序自行创建的数据库可以被该应用程序内的任何类访问，但无法被其他应用程序所访问[①]。

如果想要在 SQLite 数据库系统内创建自己的数据库以及数据表，甚至对数据库作新增、修改、删除、查询的操作，参见下列范例与步骤。

```java
public class SitesDBHlp extends SQLiteOpenHelper {
    private static final String DATABASE_NAME = "sites";
    private static final int DATABASE_VERSION = 1;
    private static final String TABLE_NAME = "sitesInfo";
    private static final String TABLE_CREATE =
            "CREATE TABLE " + TABLE_NAME + " ( " +
            " id TEXT NOT NULL, " +
            " name TEXT NOT NULL, " +
            " phoneNo TEXT, " +
            " address TEXT, PRIMARY KEY (id)); ";
    public SitesDBHlp(Context context) {
        super(context, DATABASE_NAME, null, DATABASE_VERSION);
    }

    @Override
    public void onCreate(SQLiteDatabase db) {
        db.execSQL(TABLE_CREATE);
    }

    public void InsertDB() {
        SQLiteDatabase db = getWritableDatabase();
        //新增数据
    }

    public void UpdateDB() {
        //修改数据
    }

    public void DeleteDB() {
        //删除数据
    }
```

---

① 若要开放给其他应用程序访问，必须使用 Content Provider 的功能，可参见 http://developer.android.com/guide/topics/providers/content-providers.html。

```
public void QueryDB() {
    //查询数据
}
}
```

STEP 1 创建 SQLiteOpenHelper 的子类(例如：SitesDBHlp)。

STEP 2 定义该子类的构造函数：通过调用子类的构造函数(例如：SitesDBHlp(context))取得 SQLiteOpenHelper 对象，以方便创建、打开或管理数据库。此时数据库尚未真正创建，直到 getWritableDatabase() 或 getReadableDatabase() 被调用才会开始创建数据库。

STEP 3 改写 onCreate()：当数据库第一次被创建时(例如第一次调用 getWritableDatabase() 或 getReadableDatabase())，onCreate()会被调用，所以一般会在该方法内创建数据表。如果要创建数据表，会调用 execSQL(String sql)，而 sql 参数代表的就是想要创建数据表的 SQL 语法(参见 TABLE_CREATE 字符串内容)。

STEP 4 调用 getWritableDatabase()或 getReadableDatabase()：创建或打开数据库，并取得 SQLiteDatabase 对象，以便之后访问数据库内的数据。第一次调用 getWritableDatabase()或 getReadableDatabase()时，因为数据库尚未创建，所以会创建数据库并自动调用 onCreate()以创建相关数据表。数据库创建完毕后再调用 getWritableDatabase()或 getReadableDatabase()只会取得 SQLiteDatabase 对象，而不会再创建数据库，也不会再调用 onCreate()。

STEP 5 在 SQLiteOpenHelper 子类内增加其他方法：这些方法就是将来想要对数据库执行的操作(例如：新增、修改、删除、查询)，之后可以通过 SQLiteOpenHelper 子类对象来调用这些方法。

经过上面的说明可知欲访问 SQLite 数据库，必须使用 SQLiteOpenHelper 与 SQLiteDatabase 两个类所提供的功能，这两个类的说明如表 8-3 所示。

表 8-3

SQLiteOpenHelper 类
构 造 函 数
public SQLiteOpenHelper(Context context，String name, SQLiteDatabase.CursorFactory factory, int version) 创建 SQLiteOpenHelper 对象来打开与管理数据库，不过此时尚未创建数据库。只要没有调用 getWritableDatabase() 或 getReadableDatabase()，数据库就未被创建或打开 ● context：用来打开或创建数据库的 context 对象 ● name：数据库名称 ● factory：用来创建 cursor 对象，null 代表使用默认方式 ● version：数据库版本(起始值为 1)，如果数据库是较旧的版本，会调用 onUpgrade()以更新数据库

(续表)

方法
public void onCreate(SQLiteDatabase db) 第一次调用 getWritableDatabase() 或 getReadableDatabase()时，因为数据库尚未创建，所以会创建数据库并自动调用此方法。通常会在此方法内作初始化的操作，例如创建数据表 db：数据库
public abstract void onUpgrade(SQLiteDatabase db, int oldVersion, int newVersion) 当数据库需要被更新时(例如数据库版本有更新的情况)，此方法会被调用。一般会在此方法内重新创建数据表，以及完成更新数据表的操作 • db：目前使用中的数据库 • oldVersion：旧的数据库版本 • newVersion：新的数据库版本
public synchronized SQLiteDatabase getWritableDatabase() 创建、打开数据库后并返回数据库对象，以便之后访问数据库内的数据
public synchronized SQLiteDatabase getReadableDatabase() 创建、打开数据库后并返回数据库对象，在一般情况下，调用此方法与调用 getWritableDatabase() 功能相同，只有在少数情况下才会返回只读的数据库
public synchronized void close() 关闭任何已经打开数据库的 SQLiteOpenHelper 对象
SQLiteDatabase 类
public void execSQL(String sql) execSQL() 主要执行单一 SQL 语法(非查询语法)，例如 CREATE TABLE、DELETE、INSERT 等语法，但要注意，不支持以;(分号)作分隔的连续多段语法 sql：SQL 语法
public long insert(String table, String nullColumnHack, ContentValues values) 新增指定的 ContentValues 数据至数据表内。若该个数据新增成功，则返回对应的数据行 ID(row ID)；若新增失败，则返回-1 • table：数据表名称 • nullColumnHack：SQLite 不允许新增完全空白的数据行，如果未输入域值，则以 NULL 值填补 • values：欲新增的数据行
public int update(String table, ContentValues values, String whereClause, String[] whereArgs) 修改指定的 ContentValues 数据，并返回修改成功的个数 • table：数据表名称 • values：欲修改的数据行 • whereClause：是否要加上修改条件，与 SQL 语法 WHERE 条件式的功能相同(此参数内容不需再放入 WHERE 字符串) • whereArgs：可以在 SQL 语法的 WHERE 条件加上?并使用 whereArgs 内的参数值替换该?①
Public int delete (String table, String whereClause, String[] whereArgs) 删除指定数据，并返回删除成功的个数 • table：数据表名称 • whereClause：是否要加上删除条件，与 SQL 语法 WHERE 条件式的功能相同(此参数内容不需要再放入 WHERE 字符串) • whereArgs：可以在 SQL 语法的 WHERE 条件式加上?并使用 whereArgs 内的参数值替换该?

---

① 加上?的功能类似 java.sql.PreparedStatement 类所提供的功能(属于 JDBC 范围)。

(续表)

SQLiteDatabase 类
public Cursor rawQuery(String sql, String[] selectionArgs) 执行单纯的 SQL 查询语法，返回 Cursor 对象[①]，之后可通过 Cursor 对象取得结果数据 ● sql：SQL 的查询语法，但要注意结尾请勿加上 ";"。 ● selectionArgs：可以在 SQL 语法的 WHERE 条件加上?并使用 selectionArgs 内的参数值替换该?
public Cursor query(String table, String[] columns, String selection, String[] selectionArgs, String groupBy, String having, String orderBy) 将 SQL 查询语法拆解成多个部分，并分别以参数表示。执行完毕一样会返回 Cursor 对象 ● table：数据表 ● columns：以数组存储欲查询的域名。设为 null 会返回所有字段数据，但性能较差，所以不建议 ● selection：设置查询条件，效果与 SQL 语法 WHERE 条件的功能相同(此参数内容不需要再放入 WHERE 字符串)，如果值为 null，代表不设置条件，所以会返回所有数据行 ● selectionArgs：可以在参数 selection 内容加上?并使用 selectionArgs 内的值替换该? ● groupBy：效果与 SQL 语法 GROUP BY 的功能相同(此参数内容不需要再放入 GROUP BY 字符串) ● having：效果与 SQL 语法 HAVING 的功能相同(此参数内容不需要再放入 HAVING 字符串) ● orderBy：排序，效果与 SQL 语法 ORDER BY 的功能相同(此参数内容不需要再放入 ORDER BY 字符串)

8.5 SQLite 新增功能

欲新增一个数据至数据表内，最直接的方法就是调用 SQLiteDatabase 的 insert()，此方法已在前面表 8-3 中说明过，现在针对 values 这个参数(ContentValues 对象)加以深入说明。ContentValues 存储数据的方式非常类似 java.util.Map，也是通过调用 put() 并搭配 key-value pairs 来存储数据。因为这些数据要新增至数据表内，所以 key 必须为域名，而 value 则为该字段对应的值，新增数据的步骤如下：

STEP 1 创建、打开数据库并返回数据库对象。

```
SQLiteDatabase db = getWritableDatabase();
```

STEP 2 创建 ContentValues 对象存储欲新增的数据。

```
ContentValues values = new ContentValues();
```

STEP 3 新增数据到指定的字段内。

```
values.put("id", "yushan"); //id 字段欲新增"yushan"
values.put("name", "玉山公园管理处"); //name 字段欲新增"玉山公园管理处"
```

---

① Cursor 对象非常类似 java.sql.ResultSet 对象(属于 JDBC 范围)。

移动数据库 SQLite

values.*put*("phoneNo", "049-2773121"); //phoneNo 字段欲新增"049-2773121"
values.*put*("address", "中山路一段 300 号"); // address 字段欲新增"中山路一段 300 号"

STEP 4 执行新增命令后会返回执行结果，如果新增成功，则会返回该数据行的 ID；如果新增失败，则返回 -1。

long rowID = db.*insert*("sitesInfo", null, values);

 范例 InsertData

范例(如图 8-1 所示)说明：
- 用户输入完欲新增的各个字段数据后按下"新增"按钮。如果数据新增成功，就以 Toast 消息框呈现"数据新增成功"信息；如果数据新增失败，就以 Toast 消息框呈现"数据新增失败"信息。
- 按下"删除"按钮会将文字输入方块的内容删除。

图 8-1

其实每一个旅游地信息都可以使用一个对象来存储，本范例会创建一个 Site 类，每一个 Site 对象都会存储一个完整的旅游地点信息。

InsertData/src/org/insertData/SitesDBHlp.java

```
8.    public class SitesDBHlp extends SQLiteOpenHelper {
9.        private static final String DATABASE_NAME = "sites";
10.        private static final int DATABASE_VERSION = 1;
11.        private static final String TABLE_NAME = "sitesInfo";
12.        private static final String TABLE_CREATE =
```

```
13.                    "CREATE TABLE " + TABLE_NAME + " ( " +
14.                    " id TEXT NOT NULL, " +
15.                    " name TEXT NOT NULL, " +
16.                    " phoneNo TEXT, " +
17.                    " address TEXT, PRIMARY KEY (id)); ";
18.         private static final String COL_id = "id";
19.         private static final String COL_name = "name";
20.         private static final String COL_phoneNo = "phoneNo";
21.         private static final String COL_address = "address";
22.
23.         public SitesDBHlp(Context context) {
24.             super(context, DATABASE_NAME, null, DATABASE_VERSION);
25.         }
26.
27.         @Override
28.         public void onCreate(SQLiteDatabase db) {
29.             db.execSQL(TABLE_CREATE);
30.         }
31.
32.         @Override
33.         public void onUpgrade(SQLiteDatabase db, int oldVersion,
34.                 int newVersion) {
35.             db.execSQL("DROP TABLE IF EXISTS " + TABLE_NAME);
36.             onCreate(db);
37.         }
```

8 行：创建 SQLiteOpenHelper 的子类 SitesDBHlp。

24 行：调用父类 SQLiteOpenHelper 的构造函数以创建 helper 对象，之后将通过该对象进行数据库管理。

28~30 行：改写 onCreate()，并调用 execSQL()，搭配创建数据表的 SQL 语法以创建数据表。当数据库第一次被创建时，onCreate()会被调用。

33~37 行：改写 onUpgrade()。当数据库需要被更新时(例如数据库版本有更新的情况)，此方法会被调用。当此方法被调用时会删除已存在的数据表并调用 onCreate()重建该数据表。

```
39.         public long insertDB(Site site){
40.             SQLiteDatabase db = getWritableDatabase();
41.             ContentValues values = new ContentValues();
42.             values.put(COL_id, site.getId());
43.             values.put(COL_name, site.getName());
44.             values.put(COL_phoneNo, site.getPhoneNo());
45.             values.put(COL_address, site.getAddress());
46.             long rowId = db.insert(TABLE_NAME, null, values);
47.             db.close();
```

移动数据库 SQLite

```
48.            return rowId;
49.        }
```

39 行：调用 insertDB()会新增数据到指定的数据表内。

40 行：调用 getWritableDatabase()以打开并返回数据库对象，以便之后访问数据库内的数据。

41 行：创建 ContentValues 对象来存储数据。

42~45 行：调用 put()将欲新增的数据先存放在 ContentValues 对象内。

46 行：调用 insert()可以将一个 ContentValues 数据新增至指定数据表内，如果新增成功，则返回对应的数据行 ID，如果新增失败，则返回 -1。

InsertData/src/org/insertData/InsertData.java

```
27.     @Override
28.     public void onResume() {
29.         super.onResume();
30.         if(dbHlp == null)
31.             dbHlp = new SitesDBHlp(this);
32.     }
33.
34.     @Override
35.     public void onPause() {
36.         super.onPause();
37.         if(dbHlp != null){
38.             dbHlp.close();
39.             dbHlp = null;
40.         }
41.     }
```

28~32 行：调用 onResume()代表用户可以开始与程序画面交互，此时应该检查 dbHelp (SQLiteOpenHelper 对象)是否为 null，如果是，则创建新的对象实体，以方便之后连接 SQLite 数据库。

35~41 行：调用 onPause()代表本程序暂停，而改为运行其他程序；如果 dbHlp 不为 null，就调用 close()关闭数据库对象。将 dbHlp 对象设为 null 以方便之后检查该对象是否为 null(例如 30 行)，进而决定是否再创建新的 SQLiteOpenHelper 对象。

```
43.     private void findViews() {
44.         etId = (EditText)findViewById(R.id.etId);
45.         etName = (EditText)findViewById(R.id.etName);
46.         etPhoneNo = (EditText)findViewById(R.id.etPhoneNo);
47.         etAddress = (EditText)findViewById(R.id.etAddress);
48.         btnInsert = (Button)findViewById(R.id.btnInsert);
49.         btnClear = (Button)findViewById(R.id.btnClear);
50.
```

```
51.        //按下"新增"按钮
52.        btnInsert.setOnClickListener(new OnClickListener() {
53.            @Override
54.            public void onClick(View v) {
55.                String id = etId.getText().toString().trim();
56.                String name = etName.getText().toString().trim();
57.                String phoneNo = etPhoneNo.getText().toString().trim();
58.                String address = etAddress.getText().toString().trim();
59.                if(id.length() <= 0 || name.length() <= 0){
60.                    Toast.makeText(InsertData.this, getString(R.string.blank),
61.                            Toast.LENGTH_SHORT).show();
62.                    return;
63.                }
64.
65.                StringBuilder sb = new StringBuilder();
66.                Site site = new Site(id, name, phoneNo, address);
67.                long rowId = dbHlp.insertDB(site);
68.                if(rowId != -1){
69.                    sb.append(getString(R.string.insert_success));
70.                }else{
71.                    sb.append(getString(R.string.insert_fail));
72.                }
73.                Toast.makeText(InsertData.this, sb,
74.                        Toast.LENGTH_SHORT).show();
75.            }
76.        });
77.
78.        //按下"删除"按钮
79.        btnClear.setOnClickListener(new OnClickListener() {
80.            @Override
81.            public void onClick(View v) {
82.                etId.setText("");
83.                etName.setText("");
84.                etPhoneNo.setText("");
85.                etAddress.setText("");
86.            }
87.        });
88.    }
```

54~58 行：当按下"新增"按钮时，取得用户输入的数据。

59~63 行：检查用户是否输入旅游地代号(id)以及地名(name)，如果未输入，则要求重新输入。

67 行：新增一个旅游地点(site)的信息至数据表内，如果新增成功，则返回对应的数据行 ID，如果新增失败，则返回-1。

移动数据库 SQLite

68~74 行：如果返回不为 -1，则代表新增成功，以 Toast 显示"数据新增成功"；如果返回-1 代表新增失败，以 Toast 显示"数据新增失败"。

81~86 行：当"删除"按钮被按下时，将文字输入方块的内容删除。

8.6 SQLite 查询功能

SQLiteDatabase 类提供两种查询功能的方法，分别是 rawQuery() 与 query() 这两种方法(参数说明参见表 8-2)，差别在于：

- rawQuery()：接受的是单纯的 SQL 查询语法，所以直接将 SQL 查询语法传入即可。无论在任何平台上都可以使用标准的 SQL 语法，所以使用此方法的好处是只要为标准的 SQL 查询语法，大部分都可以直接移植过来使用，不需要进行太多修改。
- query()：必须先将 SQL 查询语法按照 query() 的参数来拆分。好处是更改查询语法时，只要按照想要更改的部分更改对应参数即可，不需要整体修改。坏处就是其他程序语言或数据库并不接受这种查询方式，所以在其他平台必须重新编写标准的 SQL 查询语法。

以实际例子来说明这两种方法的差别：

SQL 查询语法：SELECT name, address FROM sitesInfo WHERE name LIKE ?

- rawQuery(String sql, String[] selectionArgs)：

```
//单纯 SQL 查询语法
String sql = " SELECT name, address FROM sitesInfo WHERE name LIKE ? "
//以参数值替换"?"，因为"?"可能有多个，所以为字符串数组
String[] selectionArgs = {"%阳明山%"};
//仅需要传递两个参数
Cursor cursor = db.rawQuery(sql, args);
```

- query(String table, String[] columns, String selection, String[] selectionArgs, String groupBy, String having, String orderBy)：

```
//欲查询的字段
String[] columns = {"name", "address"};
//WHERE 条件式，所以字符串内不需要再添加 WHERE
String selection = "name LIKE ?";
//以参数值替换"?"，与 rawQuery()相同
String[] selectionArgs = {"%阳明山%"};
//需要传递许多参数，每个参数都代表 SQL 查询语法的一部分，null 代表略过该项条件
Cursor cursor = db.query("sitesInfo", columns , selection , selectionArgs, null, null, null);
```

255

由上述说明可知，无论调用 rawQuery()或 query()任何一种方法，都会返回 Cursor 对象[①]，Cursor 界面说明如表 8-4 所示：

表 8-4
Cursor 界面

方　　法
public abstract boolean moveToNext() 将指针移动至下一个数据，如果成功，移至下一个数据会返回 true
public abstract int getColumnCount() 取得字段总数量
public abstract String getString (int columnIndex) 取得指定字段的值。 columnIndex：代表字段索引，第 1 个字段的索引是 0 而不是 1(zero-based)

照理说，执行查询语法，应该会返回整个结果集(result set，查询结果数据的总集合)，但无论 JDBC 或是 Android 的做法都不是直接将结果集返回，这是因为担心数据量太大会超过内存能够负荷的大小。查询后实际上返回的是 Cursor 对象，而该对象其实是一个指针，初始位置在结果集第一个数据的前面，假设查询字段为 id 与 name，如图 8-2 所示。

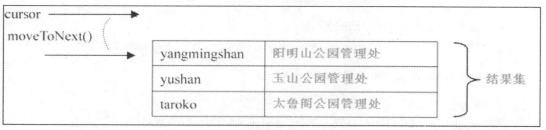

图 8-2

之后每调用一次 moveToNext()，cursor 就会向下一个数据移动。当 cursor 移到数据行上时，通过调用 getString()来取得指定字段的值，例如图 8-2；当 cursor 移到第 1 行时，调用 cursor.getString(0)会取得 "yangmingshan"。之后不断调用 moveToNext()即可按照顺序取得各个数据。

8.6.1　输入欲查询的数据

要让用户能够查询 SQLite 数据库内的数据，必须创建合适的 UI。最直接的方式就是让用户自行输入想要查询的数据。要求用户精准输入要查询的数据太强人所难，应该允许用户输入关键词即可查询，也就是所谓的模糊查询功能。在数据库上进行模糊查询并不难，只要 SQL 语法搭配 LIKE 关键词即可达成。

[①] Cursor 对象非常类似 java.sql.ResultSet 对象，所以通过 Cursor 对象来取得查询结果的方式也非常类似 ResultSet 对象。

范例 QueryData

范例(如图 8-3 所示)说明：
- 用户输入想去游玩的地点后按"查询"按钮。
- 输入的地名将与数据库存储的地名进行LIKE 比较，如果数据库有符合条件的数据，则取出完整地名与地址，并以 Toast 消息框呈现。
- 如果输入的地名无法找到对应的数据，则以 Toast 消息框呈现找不到指定地点。

图 8-3

QueryData/src/org/queryData/SitesDBHlp.java

```
37.     public void fillDB() {
38.         SQLiteDatabase db = getWritableDatabase();
39.         ContentValues[] values = new ContentValues[3];
40.         for(int i=0; i<values.length; i++)
41.             values[i] = new ContentValues();
42.
43.         values[0].put("id", "yangmingshan");
44.         values[0].put("name", "阳明山公园");
45.         values[0].put("phoneNo", "02-28613601");
46.         values[0].put("address", "台北市北投区竹子湖路 1 之 20 号");
47.
48.         values[1].put("id", "yushan");
49.         values[1].put("name", "玉山公园管理处");
50.         values[1].put("phoneNo", "049-2773121");
51.         values[1].put("address", "南投县水里乡中山路一段 300 号");
```

```
52.
53.        values[2].put("id", "taroko");
54.        values[2].put("name", "太鲁阁公园管理处");
55.        values[2].put("phoneNo", "03-8621100");
56.        values[2].put("address", "花莲县秀林乡 258 号");
57.
58.        for(ContentValues row : values){
59.            db.insert(TABLE_NAME, null, row);
60.        }
61.        db.close();
62.    }
```

37 行：调用 fillDB() 会预先新增 3 个数据到指定的数据表内，方便之后查询。

39~56 行：准备 ContentValues 数组存储 3 个旅游地点信息。

58~60 行：调用 insert() 按照顺序将每个 ContentValues 对象内的数据新增至指定数据表内。

```
64.    public ArrayList<String> getAddress(String name){
65.        SQLiteDatabase db = getReadableDatabase();
66.        String sql = "SELECT name, address FROM " + TABLE_NAME +
67.                    " WHERE name LIKE ?";
68.        String[] args = {"%" + name + "%"};
69.        Cursor cursor = db.rawQuery(sql, args);
70.        ArrayList<String> addresses = new ArrayList<String>();
71.        int columnCount = cursor.getColumnCount();
72.        while(cursor.moveToNext()){
73.            String name_addr = "";
74.            for(int i=0; i<columnCount; i++)
75.                name_addr += cursor.getString(i) + "\n    ";
76.            addresses.add(name_addr);
77.        }
78.        cursor.close();
79.        db.close();
80.        return addresses;
81.    }
```

64 行：当用户输入地名时，会调用 getAddress() 并将地名传递进来。之后便将输入的地名与数据库内所存储的地名进行比较，以便找出完整地名与地址后返回。

66~68 行：利用 SELECT 语法搭配 LIKE 与%来进行模糊查询，只要用户输入的地名和数据表的 name 字段所存的地名有局部相同，即可比较成功。

69 行：rawQuery() 专门用来执行单纯的 SQL 查询语法。执行完毕会返回 Cursor 对象。

71 行：调用 getColumnCount() 取得查询结果的字段数。

72 行：调用 moveToNext() 移动 cursor 至下一行。

75 行：调用 getString() 取得指定字段索引的值。

76 行：将取得的完整地名与地址串接成一个字符串后新增至 List 中，最后在 80 行将其返回。

QueryData/src/org/queryData/QueryData.java

```
47.     private void findViews() {
48.         etPlaceName = (EditText)findViewById(R.id.etPlaceName);
49.         btnSubmit = (Button)findViewById(R.id.btnSubmit);
50.         btnSubmit.setOnClickListener(new OnClickListener() {
51.             @Override
52.             public void onClick(View v) {
53.                 StringBuilder address = new StringBuilder("");
54.                 String placeName = etPlaceName.getText().toString().trim();
55.                 if(placeName.length() > 0){
56.                     ArrayList<String> addresses = dbHlp.getAddress(placeName);
57.                     if(addresses.size() > 0){
58.                         for(String addr: addresses)
59.                             address.append(addr + "\n");
60.                     }else{
61.                         address.append(getString(R.string.placeNotFound));
62.                     }
63.                 } else{
64.                     address.append(getString(R.string.inputPlaceName));
65.                 }
66.                 Toast.makeText(QueryData.this,
67.                         address, Toast.LENGTH_LONG).show();
68.             }
69.         });
70.     }
```

52~54 行：当按下查询按钮时，取得用户输入的地名。

55、63~65 行：检查用户是否输入地名，如果未输入，则要求重新输入。

56 行：调用 getAddress() 并传递用户输入的地名，会在数据库内取得对应数据。

57~62 行：如果 addresses 内有存储数据，就将数据一一取出并附加在 address(StringBuilder 对象)；如果 addresses 内没有任何数据，就附加"找不到指定地点"。

66~67 行：最后将 address 内存的信息以 Toast 消息框呈现。

8.6.2 数据浏览

前一个范例可以让用户自行输入想要查询的数据来取得对应的结果数据。但是用户也有可能想要仔细地浏览每一个数据。接下来的范例也会创建一个 Site 类，每一个 Site 对象都存储一个旅游地信息，并将 Site 对象存入 List 集合内，方便用户浏览数据。

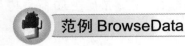 范例 BrowseData

范例(如图 8-4 所示)说明：
　　用户按下"下一个"或"上一个"按钮来浏览旅游地信息，而且会显示目前在第几个数据以及数据的总个数。

图 8-4

BrowseData/src/org/browseData/SitesDBHlp.java

```java
70.     public ArrayList<Site> getAllSites(){
71.         SQLiteDatabase db = getReadableDatabase();
72.         String[] columns = {COL_id, COL_name, COL_phoneNo, COL_address};
73.         Cursor cursor = db.query(TABLE_NAME, columns, null, null, null, null, null);
74.         ArrayList<Site> sites = new ArrayList<Site>();
75.         while(cursor.moveToNext()){
76.             String id = cursor.getString(0);
77.             String name = cursor.getString(1);
78.             String phoneNo = cursor.getString(2);
79.             String address = cursor.getString(3);
80.             Site site = new Site(id, name, phoneNo, address);
81.             sites.add(site);
82.         }
83.         cursor.close();
84.         db.close();
85.         return sites;
86.     }
```

72 行：将欲查询的域名放入字符串数组内。

73 行：调用 query()并按照顺序传入数据表名称(TABLE_NAME)与列名(columns)。各条件设为 null，代表不设置条件，所以会选择全部数据行。

75~82 行：将数据行的各个列值取出后当作参数并调用 Site 构造函数以创建 Site 对象，之后将 Site 对象按照顺序放至 List 集合内，并于 85 行返回该 List 集合。

BrowseData/src/org/browseData/BrowseData.java

```
11.     public class BrowseData extends Activity {
12.         private TextView tvRow; //数据个数信息
13.         private TextView tvId; //旅游地代号
14.         private TextView tvName; //旅游地名称
15.         private TextView tvPhoneNo; //联系电话
16.         private TextView tvAddress; //地址
17.         private Button btnNext; //下一个按钮
18.         private Button btnBack; //上一个按钮
19.         private SitesDBHlp dbHlp; //SQLiteOpenHelper 对象
20.         private ArrayList<Site> sites; //旅游地信息
21.         private int index; //按照 index 取得 sites 内数据
22.
23.         @Override
24.         public void onCreate(Bundle savedInstanceState) {
25.             super.onCreate(savedInstanceState);
26.             setContentView(R.layout.main);
27.             initDB();
28.             findViews();
29.             showSites(index);
30.         }
31.
32.         private void initDB() {
33.             if(dbHlp == null)
34.                 dbHlp = new SitesDBHlp(this);
35.             dbHlp.fillDB();
36.             sites = dbHlp.getAllSites();
37.         }
38.
39.         //显示指定旅游地的信息
40.         private void showSites(int index) {
41.             if(sites.size() > 0){
42.                 tvRow.setText((index + 1) + "/" + sites.size() +
43.                         getString(R.string.rowid_rowCount));
44.                 tvId.setText(sites.get(index).getId());
```

```
45.            tvName.setText(sites.get(index).getName());
46.            tvPhoneNo.setText(sites.get(index).getPhoneNo());
47.            tvAddress.setText(sites.get(index).getAddress());
48.        }else{
49.            tvRow.setText("0/0" + getString(R.string.noData));
50.            tvId.setText("");
51.            tvName.setText("");
52.            tvPhoneNo.setText("");
53.            tvAddress.setText("");
54.        }
55.    }
```

29 行：调用 showSites() 显示指定旅游地的信息。

35 行：调用 fillDB() 预先新增几个范例数据。

36 行：调用 SitesDBHlp 的 getAllSites() 取得全部旅游地信息。

40~55 行：调用 showSites(index)，显示该索引所代表的信息，详细说明如下：

- 41 行：首先检查数据表内是否有数据
- 42 行：如果有数据，则显示是第几个数据与数据总个数。
- 44~47 行：将该个数据的每个列值都通过 TextView 组件呈现出来。
- 49 行：如果没有数据，则将"第几个/总个数"的值设为 0/0。
- 50~53 行：将 TextView 组件上的数据清空。

```
66.        //按下"下一个"按钮
67.        btnNext.setOnClickListener(new OnClickListener() {
68.            @Override
69.            public void onClick(View v) {
70.                index++;
71.                if(index >= sites.size())
72.                    index = 0;
73.                showSites(index);
74.            }
75.        });
```

67~75 行：按下"下一个"按钮后会显示下一个旅游地信息。

70 行：因为指定显示下一个数据，所以必须将索引加 1(代表将取出 sites 内的下一个数据)。

71~72 行：index 持续升序可能会超过 sites 的最大索引，当超过时就将 index 归 0，代表回到第一个数据。

73 行：调用 40 行 showSites()，显示该索引所代表的信息。

```
77.        //按下"上一个"按钮
78.        btnBack.setOnClickListener(new OnClickListener() {
```

```
79.         @Override
80.         public void onClick(View v) {
81.             index--;
82.             if(index < 0)
83.                 index = sites.size()-1;
84.             showSites(index);
85.         }
86.     });
```

78~86 行：按下"上一个"按钮后会显示上一个旅游地信息。

- 81 行：因为指定显示上一个数据，所以必须将索引减 1(代表将取出 sites 内的上一个数据)。
- 82~83 行：index 持续降序可能会造成索引为负值，当索引为负值时就将 index 设为最大的索引值，代表回到最后一个数据。
- 84 行：调用 40 行 showSites()，显示该索引所代表的信息。

```
89.     @Override
90.     public void onResume() {
91.         super.onResume();
92.         if(dbHlp == null)
93.             dbHlp = new SitesDBHlp(this);
94.         sites = dbHlp.getAllSites();
95.         showSites(index);
96.     }
97.
98.     @Override
99.     public void onPause() {
100.        super.onPause();
101.        if(dbHlp != null){
102.            dbHlp.close(); //关闭数据库连接
103.            dbHlp = null;
104.        }
105.        sites.clear(); //清空旅游地信息
106.    }
```

90~96 行：调用 onResume()代表用户可以开始与程序画面交互，如果 dbHelp(SQLiteOpenHelper 对象)不为 null，应该通过该对象调用 getAllSites()以取得所有旅游地信息，并调用 showSites() 显示第一个数据在画面上。

99~106 行：调用 onPause()，代表本程序暂停，除了要调用 close()关闭数据库对象并将其设为 null，也应该调用 clear()清空 sites 内容，以节省内存空间。

8.7 SQLite 修改与删除功能

1. 修改数据

修改数据与新增数据十分类似，差别在修改数据要调用 SQLiteDatabase 的 update()，此方法已在前面表 8-2 说明过。修改数据的步骤如下：

STEP 1 创建、打开数据库并返回数据库对象。

```
SQLiteDatabase db = getWritableDatabase();
```

STEP 2 创建 ContentValues 对象存储欲修改的数据。

```
ContentValues values = new ContentValues();
```

STEP 3 将欲修改数据放到指定的字段内，并将 id 值当作修改的条件。

```
values.put("name", "玉山公园管理处"); //欲修改 name 字段
values.put("phoneNo", "049-2773121"); //欲修改 phoneNo 字段
values.put("address", "中山路一段 300 号"); //欲修改 address 字段
String whereClause = "id = ?"; //id 值当作修改的条件
String[] selectionArgs = {"yangmingshan"}; //"yangmingshan"将替换 whereClause 的 "?"
```

STEP 4 执行修改命令后会返回成功修改个数。

```
int count = db.update("sitesInfo", values, whereClause, selectionArgs);
```

2. 删除数据

删除数据需要调用 SQLiteDatabase 的 delete()，此方法也已在前面表 8-2 说明过。删除数据与修改数据最大的差别在于删除数据不需要提供 ContentValues 对象而只要提供删除的条件即可。删除数据的步骤如下：

STEP 1 创建、打开数据库并返回数据库对象。

```
SQLiteDatabase db = getWritableDatabase();
```

STEP 2 将数据的 id 值当作删除的条件。

```
String whereClause = "id = ?"; //id 值当作删除的条件
String[] selectionArgs = {"yangmingshan"}; //"yangmingshan"将替换 whereClause 的 "?"
```

STEP 3 执行删除命令后会返回成功删除个数。

```
int count = db.delete("sitesInfo", whereClause, selectionArgs);
```

移动数据库 SQLite 08

 范例 UpdateData

范例(如图 8-5 所示)说明:

- 用户按下"下一个"或"上一个"按钮来浏览旅游地信息,而且会显示目前在第几个数据以及数据总个数。
- 修改完毕后按下"修改"按钮会根据旅游地代号将修改结果存入对应的数据表内,并以 Toast 消息框呈现"1 个数据修改完毕"。
- 按下"删除"按钮会根据旅游地代号从对应的数据表内删除该个数据,并以 Toast 呈现"1 个数据删除完毕"。

图 8-5

UpdateData/src/org/updateData/SitesDBHlp.java

```
69.     //取得全部旅游地信息
70.     public ArrayList<Site> getAllSites(){
71.         SQLiteDatabase db = getReadableDatabase();
72.         String[] columns = {COL_id, COL_name, COL_phoneNo, COL_address};
73.         Cursor cursor = db.query(TABLE_NAME, columns, null, null, null, null, null);
74.         ArrayList<Site> sites = new ArrayList<Site>();
75.         while(cursor.moveToNext()){
76.             String id = cursor.getString(0);
77.             String name = cursor.getString(1);
78.             String phoneNo = cursor.getString(2);
79.             String address = cursor.getString(3);
80.             Site site = new Site(id, name, phoneNo, address);
81.             sites.add(site);
```

```
82.         }
83.         cursor.close();
84.         db.close();
85.         return sites;
86.     }
87.
88.     //修改旅游地信息
89.     public int updateDB(Site site){
90.         SQLiteDatabase db = getWritableDatabase();
91.         ContentValues values = new ContentValues();
92.         values.put(COL_name, site.getName());
93.         values.put(COL_phoneNo, site.getPhoneNo());
94.         values.put(COL_address, site.getAddress());
95.         String whereClause = COL_id + "=" + site.getId() + "";
96.         int count = db.update(TABLE_NAME, values, whereClause, null);
97.         db.close();
98.         return count;
99.     }
100.
101.     //删除旅游地信息
102.     public int deleteDB(String id){
103.         SQLiteDatabase db = getWritableDatabase();
104.         String whereClause = COL_id + "=" + id + "";
105.         int count = db.delete(TABLE_NAME, whereClause, null);
106.         db.close();
107.         return count;
108.     }
```

70~86 行：取得全部旅游地信息，与范例 BrowseData 的 SitesDBHlp.java 同名方法内容相同，不再赘述。

92~94 行：将欲修改的数据置入 ContentValues 对象内。

95 行：将数据的 id 值当作修改的条件。

96 行：调用 update() 会按照 ContentValues 数据与条件修改指定数据表内容并返回成功修改个数。

104 行：将数据的 id 值当作删除的条件。

105 行：调用 delete() 会根据条件删除指定数据表内容并返回成功删除个数。

UpdateData/src/org/updateData/UpdateData.java

```
43.     //呈现指定旅游地的信息
44.     private void showSites(int index) {
45.         if(sites.size() > 0){
46.             tvRow.setText((index + 1) + "/" + sites.size() +
47.                     getString(R.string.row));
```

```java
48.            tvId.setText(sites.get(index).getId());
49.            etName.setText(sites.get(index).getName());
50.            etPhoneNo.setText(sites.get(index).getPhoneNo());
51.            etAddress.setText(sites.get(index).getAddress());
52.        }else{
53.            tvRow.setText("0/0" + getString(R.string.noData));
54.            tvId.setText("");
55.            etName.setText("");
56.            etPhoneNo.setText("");
57.            etAddress.setText("");
58.        }
59.    }
60.
61.    private void findViews() {
62.        tvRow = (TextView)findViewById(R.id.tvRow);
63.        tvId = (TextView)findViewById(R.id.tvId);
64.        etName = (EditText)findViewById(R.id.etName);
65.        etPhoneNo = (EditText)findViewById(R.id.etPhoneNo);
66.        etAddress = (EditText)findViewById(R.id.etAddress);
67.        btnNext = (Button)findViewById(R.id.btnNext);
68.        btnBack = (Button)findViewById(R.id.btnBack);
69.        btnUpdate = (Button)findViewById(R.id.btnUpdate);
70.        btnDelete = (Button)findViewById(R.id.btnDelete);
71.
72.        //按下"下一个"按钮
73.        btnNext.setOnClickListener(new OnClickListener() {
74.            @Override
75.            public void onClick(View v) {
76.                index++;
77.                if(index >= sites.size())
78.                    index = 0;
79.                showSites(index);
80.            }
81.        });
82.
83.        //按下"上一个"按钮
84.        btnBack.setOnClickListener(new OnClickListener() {
85.            @Override
86.            public void onClick(View v) {
87.                index--;
88.                if(index < 0)
89.                    index = sites.size()-1;
90.                showSites(index);
```

```
91.            }
92.        });
93.
94.        //按下"修改"按钮
95.        btnUpdate.setOnClickListener(new OnClickListener() {
96.            @Override
97.            public void onClick(View v) {
98.                String id = tvId.getText().toString().trim();
99.                String name = etName.getText().toString().trim();
100.               String phoneNo = etPhoneNo.getText().toString().trim();
101.               String address = etAddress.getText().toString().trim();
102.               if(id.length() <= 0 || name.length() <= 0){
103.                   Toast.makeText(UpdateData.this, getString(R.string.blank),
104.                           Toast.LENGTH_SHORT).show();
105.                   return;
106.               }
107.               Site site = new Site(id, name, phoneNo, address);
108.               int count = dbHlp.updateDB(site);
109.               Toast.makeText(UpdateData.this,
110.                       count + getString(R.string.updated),
111.                       Toast.LENGTH_SHORT).show();
112.               sites = dbHlp.getAllSites();
113.               index = 0;
114.               showSites(index);
115.           }
116.       });
117.
118.       //按下"删除"按钮
119.       btnDelete.setOnClickListener(new OnClickListener() {
120.           @Override
121.           public void onClick(View v) {
122.               String id = tvId.getText().toString().trim();
123.               int count = dbHlp.deleteDB(id);
124.               Toast.makeText(UpdateData.this,
125.                       count + getString(R.string.deleted),
126.                       Toast.LENGTH_SHORT).show();
127.               sites = dbHlp.getAllSites();
128.               index = 0;
129.               showSites(index);
130.           }
131.       });
132.   }
```

98~101 行：取得 id 与用户输入的数据。

107~111 行：将取得的数据存入 Site 对象，并根据该对象修改指定数据表，最后以 Toast 消息框呈现修改成功个数。

112~114 行：修改完毕后调用 getAllSites()取得所有旅游地信息，并调用 showSites()显示第一个数据在画面上。

122 行：取得 id 数据。

123~126 行：根据 id 删除指定数据行，最后以 Toast 消息框呈现删除成功个数。

127~129 行：删除完毕后调用 getAllSites()取得所有旅游地信息，并调用 showSites()显示第一个数据在画面上。

第9章 Google地图

本章学习目标：

- 申请 Google 地图的 API 密钥
- 在 Google 地图上呈现自己位置
- 在 Google 地图上指定位置
- 标记的使用
- LocationListener 与 LocationManager
- 以地名/地址查询位置
- 导航功能

9.1 申请 Google 地图的 API 密钥

MapView 是一个专门用来呈现 Google 地图(Google Maps)的组件,用户浏览 Google 地图或将地图放大、缩小,都是在 MapView 上面操作。而 MapView 就如同其他 View 组件(例如 Button、TextView)一样在 main.xml 文件内设置其页面位置,如下所示。

```
1.  <?xml version="1.0" encoding="utf-8"?>
2.  <LinearLayout mlns:android="http://schemas.android.com/apk/res/android"
3.      android:orientation="vertical"
4.      android:layout_width="match_parent"
5.      android:layout_height="match_parent" >
6.    <com.google.android.maps.MapView
7.        android:id="@+id/mView"
8.        android:layout_width="match_parent"
9.        android:layout_height="match_parent"
10.       android:apiKey="0oNQD7b0k4dZPxksCRS3QPHJn44lQOt04A2deHg"
11.       android:clickable="true" />
12. </LinearLayout>
```

第 11 行 apiKey 指的就是 Android Maps API Key(以下简称 API 密钥),必须申请此密钥才能使用 MapView 功能。如果密钥正确,Google 地图可以正常显示,如图 9-1 所示;如果不正确,地图无法显示,而只会显示网格线,如图 9-2 所示。

图 9-1

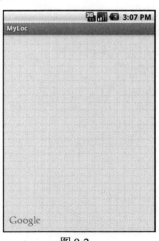

图 9-2

申请 API 密钥的步骤如下。

STEP 1 单击 Eclipse 菜单的 Windows │ Preferences │ Android │ Build,在右边窗格的 Default debug keystore 字段中找到默认的 debug.keystore[1]，如图 9-3 所示。

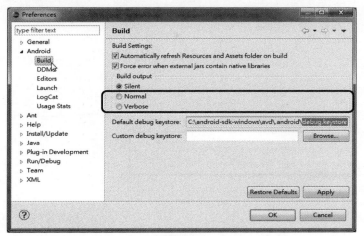

图 9-3

STEP 2 将 JDK 的 keytool 工具所在路径[2]添加到 path 环境变量中。

STEP 3 请按照 debug.keystore 文件(假设在 C:\Users\RON\.android 目录中)所在路径输入命令，例如：

C:\Users\RON\.android>keytool -list -keystore debug.keystore

STEP 4 接着会要求输入 keystore 密码，默认为 android，也可以不输入而直接按 Enter 键，会产生 MD5 认证指纹[3]，如下所示。

输入 keystore 密码：
Keystore 类型： JKS
Keystore 提供者： SUN
您的 keystore 包含 1 输入
androiddebugkey, 2011/12/12, PrivateKeyEntry,
认证指纹 (MD5)：C1:98:DE:1E:7F:21:DA:FE:59:C9:55:48:8F:67:E3:49

STEP 5 在 Google 搜索 Android Maps API Key 并前往对应网页；在 Sign Up for the Android Maps API 网页下方勾选 I have read and agree with the terms and conditions，并将 Step 2 产生的认证指纹粘贴至 My certificate's MD5 fingerprint 字段，然后按 Generate API Key 按钮，如图 9-4 所示。

---

① debug.keystore 密钥库内储存的是 debug 密钥(debug key)，是开发阶段使用的密钥，所以又称为开发者密钥。debug.keystore 是由 SDK 工具产生的。
② keytool 属于 JDK 的工具，所以会放在 JDK 的安装目录内(例如：C:\Program Files\Java\jdk1.6.0_21\bin)。
③ 在 Java 7 版(1.7.0)中，使用 keytool 工具产生的认证指纹可能是 SHA1，无法用于申请 API 密钥，要在命令最后加上 "-v" 即可显示 MD5 与 SHA1 的认证指纹：keytool -list -keystore debug.keystore -v。

273

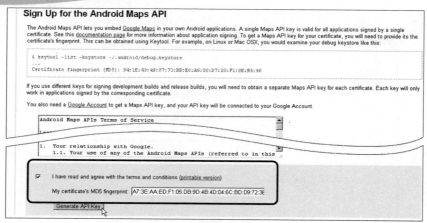

图 9-4

STEP 6 输入 Google 账号[①]、密码后按 Sign in 按钮,如图 9-5 所示。

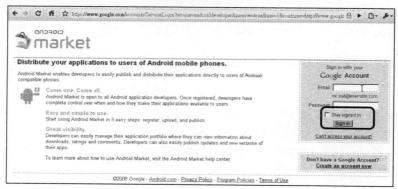

图 9-5

STEP 7 API 密钥产生完成,如图 9-6 所示。之后使用到 MapView 功能,需要输入此密钥号码。

图 9-6

---

① Google Gmail 邮箱账号就是 Google 账号,所以想要获得 Google 账号,申请一个 Gmail 电子邮箱即可。

Google 地图

不可不知

如果更换计算机开发 Google 地图功能，但却不想申请新的 API 密钥，可将原本用来申请 API 密钥的 debug.keystore 文件复制到欲开发的计算机上，接着单击 Eclipse 菜单的 Windows｜Preferences｜Android｜Build，然后在右边窗格的 Custom debug keystore 文本框的右侧，按下 Browse 按钮指定该 debug.keystore 文件的路径，即可使用原本已经申请好的密钥，如图 9-7 所示。

图 9-7

9.2 在 Google 地图上呈现自己位置

应用程序使用到 Google 地图功能时，就必须安装 Google API，安装步骤如下：

STEP 1 在 Eclipse 中单击菜单的 Windows｜Android SDK Manager。

STEP 2 勾选欲安装的 Google API 并按下右下角的 Install package 按钮开始安装，如图 9-8 所示。如果已经安装(最右边 Status 字段为 Installed)，则不需再次安装。

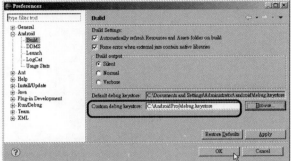

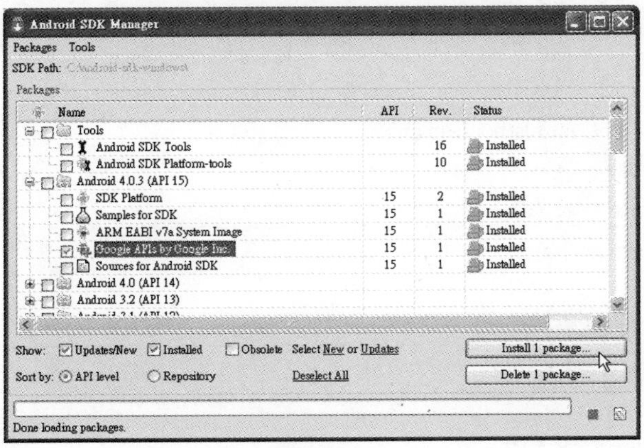

图 9-8

275

 **不可不知**

如果想在模拟器上呈现 Google 地图，必须创建可以支持 Google 地图的模拟器。如果已经按照前述步骤安装好 Google API，就可在创建模拟器时选择支持 Google 地图的 Google API，如图 9-9 所示。最后将欲运行的 Google 地图范例程序指定在该模拟器上运行即可。

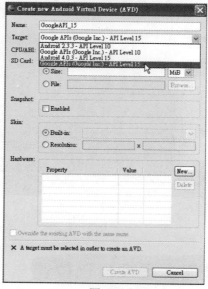

图 9-9

9.2.1 显示与缩放 Google 地图

MapView 是一个专门用来呈现 Google 地图的组件，要浏览 Google 地图，就必须先加载 MapView 组件。加载 MapView 后，如果想要操作地图(例如：缩放地图或移动指定地点至画面的正中间)，就必须使用 MapController。MapView 与 MapController 类的相关方法说明如表 9-1 所示。

表 9-1

MapView 类
public void setBuiltInZoomControls(boolean on)   是否打开缩放控件。 on：true 代表打开，false 代表关闭
public void setTraffic(boolean on)   是否打开一般地图   on：true 代表打开，false 代表关闭
public void setStreetView(boolean on)   是否打开街道地图   on：true 代表打开，false 代表关闭

(续表)

MapController 类
public void setSatellite(boolean on) 是否打开卫星地图 on：true 代表打开，false 代表关闭
public MapController getController() 取得 MapController 对象以便之后缩放或移动地图
public final List<Overlay> getOverlays() 取得地图上的所有 Overlay(图层)，并存入 List 后返回，该 List 内的 Overlay 会按 index 从小到大的顺序绘制在 MapView 上
public int setZoom(int zoomLevel) 设置地图的缩放等级(zoomlevel) zoomLevel：缩放等级①
public void animateTo(GeoPoint point) 移动地图上的指定位置至画面的正中间 point：指定的位置

第 6 章已经说明 Activity 的生命周期。如果不仅要符合 Activity 的生命周期管理，又要能够操作 Google 地图，就必须使用 Activity 子类——MapActivity。由此可知，要使用 Google 地图功能，就必须自行创建类继承 MapActivity。因为 MapActivity 的使用方式与 Activity 基本上一样，就不在此赘述。

9.2.2 呈现自己位置

1. 定位

要找到自己位置，就必须使用定位技术，现在最常用的定位技术有两种，说明如下。

(1) GPS(Global Positioning System)定位：使用卫星定位功能。优点为精确度高；缺点为室内无法接收到卫星信号，所以无法使用 GPS 定位。

(2) 基站三角定位：利用附近基站的信息来作定位。优点为在室内也可接收定位结果；缺点为不够精确。使用 WiFi 联机上网，定位的原理也相同。

2. 图层

如果只有一张地图，而且要在地图上标记自己现在的位置，大概不会有人直接将自己位置标在地图上。原因很简单，因为自己所在的位置会不断更新，一旦直接标在地图上，之后要将其清除，就可能会影响地图。最好的做法就是在地图上添加一个透明图层(overlay)，而将自己位置标记在该透明图层上，这样从上方往下看，即可看到地图与自己位置的综合信息，而且之后移除或是修改该标记都不会影响地图。

---

① 地图的缩放等级介于值 1~21 之间，值越大代表将地图放得越大，看得越清楚。但各地区可放大的程度不同，有些地区可放大到 21，有些地区却只能放大到 15，这是由各地区地图信息收集的精细程度不同所造成，与移动设备无关。

Google 地图的做法也是如此，在地图上可以新建许多不同的图层，如同透明图层般可以新建地图信息在上面。如果地图上有许多信息，例如火车站、停车场等，就可以加上火车站图层、停车场图层，以方便管理；如果用户不想看到停车场信息，只要移除该图层即可。

　　Google 地图的图层类为 Overlay，它有两个重要子类：MyLocationOverlay(自己位置图层)与 ItemizedOverlay(标记图层)。这一节要介绍的是 MyLocationOverlay 类，说明如表 9-2 所示。

表 9-2

构　造　函　数
public MyLocationOverlay(Context context, MapView mapView) 创建 MyLocationOverlay 对象 ● context：代表提供服务的对象，通常是指 MapActivity 对象 ● mapView：指定在哪一个 mapView 上面创建 MyLocationOverlay。简单地说，就是要在哪个地图上面创建自己位置图层
方　　　法
public boolean runOnFirstFix(Runnable runnable) 移动设备第一次进行定位时会调用此方法，同时会产生新的线程执行 run() runnable：指 runnable 对象，必须实现 run()

 范例 MyLoc

范例(如图 9-10 所示)说明：
- 在模拟器上一开始显示的自己位置是模拟器系统默认位置，而实机才是显示真实的自己位置。
- 可以使用 DDMS 的 Location Controls 窗格改变模拟器的自己位置[①]。

图 9-10

---

① 关于使用 DDMS 改变自己位置的方法，参见 3.3 节。

- 单击画面即可弹出缩放工具，方便用户缩放地图。
- 若使用实机测试，当移动设备移动时可以看到 Google 地图上的自己位置也会跟着移动。

MyLoc/AndroidManifest.xml

```
1.   <?xml version="1.0" encoding="utf-8"?>
2.   <manifest xmlns:android="http://schemas.android.com/apk/res/android"
3.     package="org.myLoc"
4.     android:versionCode="1"
5.     android:versionName="1.0">
6.     <application
7.       android:icon="@drawable/icon"
8.       android:label="@string/app_name">
9.       <activity
10.         android:name=".MyLoc"
11.         android:label="@string/app_name">
12.         <intent-filter>
13.           <action android:name="android.intent.action.MAIN" />
14.           <category android:name="android.intent.category.LAUNCHER" />
15.         </intent-filter>
16.       </activity>
17.       <uses-library android:name="com.google.android.maps" />
18.     </application>
19.     <uses-sdk android:minSdkVersion="8" />
20.     <uses-permission android:name="android.permission.ACCESS_FINE_LOCATION" />
21.     <uses-permission android:name="android.permission.ACCESS_COARSE_LOCATION"/>
22.     <uses-permission android:name="android.permission.INTERNET" />
23.   </manifest>
```

17 行：com.google.android.maps 不属于 Android 标准函数库的包[①]，若欲使用该包，必须在 AndroidManifest.xml 文件的<application>标签内加上<uses-library ...>，否则运行时会发生找不到对应类的错误。

20 行：允许应用程序使用 GPS 定位，ACCESS_FINE_LOCATION 的 FINE 代表细致的、质量好的。

21 行：允许应用程序使用基站或 WiFi 三角定位，ACCESS_COARSE_LOCATION 的 COARSE 代表粗糙的、质量较差的。

22 行：允许应用程序使用网络联机。要使用 Google 地图，必须通过网络访问地图信息。

不可不知

如果只允许 ACCESS_FINE_LOCATION 和 INTERNET，代表应用程序只能使用 GPS 定位，

---

① Google 地图的信息由 Google 向各地区专门收集地图信息的机构所购买，其知识产权属于 Google 所有。访问 Google 地图的 Map API，Google 也因为商业考虑而不将其源代码开放，因此有别于 Android API。

而无法使用基站或 WiFi 三角定位。

MyLoc/src/org/myLoc/MyLoc.java

```
4.    import com.google.android.maps.MapActivity;
5.    import com.google.android.maps.MapController;
6.    import com.google.android.maps.MapView;
7.    import com.google.android.maps.MyLocationOverlay;
```

4~7 行：这个范例使用到的组件大多属于 com.google.android.maps 包。

```
9.    public class MyLoc extends MapActivity {
10.       private MapView mView;
11.       private MapController mControl;
12.       private MyLocationOverlay mlOverlay;
```

9 行：MapActivity 主要用来放置 MapView 组件；它同时也是 Activity 的子类，所以生命周期与 Activity 相同。

10 行：MapView 专门用来呈现 Google 地图。

11 行：MapController 可以控制地图，例如缩放与移动指定地点至画面的正中间。

12 行：MyLocationOverlay 是标示自己位置的图层。

```
22.       private void findViews() {
23.           mView = (MapView) findViewById(R.id.mView);
24.       }
```

23 行：取得定义在 layout 文件的 MapView 对象。

```
26.       private void initMapView() {
27.           mView.setBuiltInZoomControls(true);
28.           mView.setTraffic(true);
29.           mlOverlay = new MyLocationOverlay(this, mView);
30.           mControl = mView.getController();
31.           mlOverlay.runOnFirstFix(new Runnable() {
32.               public void run() {
33.                   mControl.setZoom(15);
34.                   mControl.animateTo(mlOverlay.getMyLocation());
35.               }
36.           });
37.           mView.getOverlays().add(mlOverlay);
38.       }
```

26 行：调用 initMapView() 以初始化 Google 地图。

27 行：是否打开缩放控件，true 代表打开。

28 行：是否打开一般地图，true 代表打开。

29 行：创建自己位置图层，this 在本范例中指的是 MapActivity 对象。

30 行：取得 MapController 对象以便之后缩放或移动地图。

31~32 行：runOnFirstFix()在移动设备第一次进行定位时被调用，并产生新的线程执行 run()内容。

33 行：设置地图的缩放等级为 15。

34 行：getMyLocation()可以取得自己位置[①]；animateTo()可以移动地图上的指定位置至画面的正中间。所以 34 行是指将自己位置移至画面的正中间。

37 行：getOverlays()返回 List<Overlay>，List 内的 Overlay 都会被绘制，所以调用 add()加上自己位置图层即可在地图上显示自己位置。

```
40.        @Override
41.        protected void onResume() {
42.            super.onResume();
43.            mlOverlay.enableMyLocation();
44.        }
```

41~44 行：调用 onResume()代表用户可以开始与应用程序的画面交互，此时应该调用 enableMyLocation() 随时监听自己位置是否被更新。

```
46.        @Override
47.        protected void onPause() {
48.            super.onPause();
49.            mlOverlay.disableMyLocation();
50.        }
```

47~50 行：调用 onPause()，代表本程序暂停，而改为运行其他程序；此时应该调用 disableMyLocation()暂停监听本身位置的更新，避免浪费资源在监听更新上。

```
52.        @Override
53.        protected boolean isRouteDisplayed() {
54.            return false;
55.        }
```

53~55 行：为了计算，服务器需要知道是否显示任何路线信息，例如开车的方向等。若要实现此方法，必须如实地报告路线信息。返回 false 代表不显示。

9.3 在 Google 地图上指定位置

由前面范例 MyLoc 可知，要取得自己位置非常简单，只要调用 MyLocationOverlay.getMy-

[①] 返回的是 GeoPoint 对象，此对象存储着两个重要的信息：经度与纬度。

Location()即可,而且调用此方法返回的是 GeoPoint 对象,该对象存储着两个重要的信息:经度与纬度。原来 Google 地图以经纬度当作坐标[①],以方便定位,这样的观念非常符合一般人以经纬度定位的想法。所以若要移动 Google 地图至指定的位置(例如:玉山国家公园入口处),就必须先取得该位置的经纬度并通过构造函数以创建 GeoPoint 对象,GeoPoint 类的说明如表 9-3 所示。

表 9-3

构 造 函 数
public GeoPoint(int latitudeE6, int longitudeE6) 利用纬度和经度创建 GeoPoint 对象 ● latitudeE6:纬度(正数代表北半球纬度;负数代表南半球纬度),一般纬度都有浮点数(例如:23.791952),但 latitudeE6 参数数据类型为 int,为了兼顾精确度,所以一般先将纬度值乘以 1000000(也就是 1E6 = 1 * 10^6),然后再强制转换成 int 类型 ● longitudeE6:经度(正数代表东经;负数代表西经)。值转换方式与纬度相同,不再赘述

 范例 ManySites

范例(如图 9-11 所示)说明:

- 按移动设备上的 Menu 按键会弹出地点选项。
- 选择地点会移动该地点至画面的正中间。

图 9-11

---

① Google 所发展的 KML 就以经纬度当作坐标信息,参见 http://en.wikipedia.org/wiki/Keyhole_Markup_Language。

ManySites/src/org/manySites/ManySites.java

```
34.     private void initGeoPoint() {
35.         yangmingshan = new GeoPoint(
36.             (int)(25.091075 * 1E6), (int)(121.559834 * 1E6));
37.         yushan = new GeoPoint(
38.             (int)(23.791952 * 1E6), (int)(120.861379 * 1E6));
39.         taroko = new GeoPoint(
40.             (int)(24.151287 * 1E6), (int)(121.625537 * 1E6));
41.     }
```

initGeoPoint()主要用于初始化阳明山、玉山、太鲁阁公园所代表的 GeoPoint 对象。

```
57.     @Override
58.     public boolean onCreateOptionsMenu(Menu menu) {
59.         MenuInflater inflater = getMenuInflater();
60.         inflater.inflate(R.menu.mymenu, menu);
61.         return true;
62.     }
```

58~62 行：显示 mymenu.xml 所设置的地点选项。

```
64.     @Override
65.     public boolean onOptionsItemSelected(MenuItem item) {
66.         switch (item.getItemId()) {
67.         case R.id.yangmingshan:
68.             mControl.animateTo(yangmingshan);
69.             return true;
70.         case R.id.yushan:
71.             mControl.animateTo(yushan);
72.             return true;
73.         case R.id.taroko:
74.             mControl.animateTo(taroko);
75.             return true;
76.         case R.id.myloc:
77.             mControl.animateTo(mlOverlay.getMyLocation());
78.             return true;
79.         case R.id.exit:
80.             finish();
81.             return true;
82.         default:
83.             return super.onOptionsItemSelected(item);
84.         }
85.     }
```

65 行：当用户单击 Menu 的任一选项时会自动调用 onOptionsItemSelected()。

66 行：调用 getItemId()会取得该项目的 ID，并与下面各个 case 所代表的 ID 作比较。

67~75 行：若单击的是景点(例如：阳明山、玉山、太鲁阁公园)，则调用 animateTo()将地图移转至该景点。

76~78 行：若单击的是"我的位置"，则调用 animateTo()将地图移转至"我的位置"。

79~81 行：若单击的是"结束"，则结束程序。

9.4 标记的使用

前一个范例 ManySites 虽然可以将地图的指定地点移至画面的正中间，但该地点上面没有任何标记(marker)，所以用户会不清楚该地点的确切位置在哪里，因此需要在地图上作标记。之前曾提及可以在 Google 地图上放置图层，对应类为 Overlay，MyLocationOverlay 属于其子类，而 ItemizedOverlay 也属于其子类。ItemizedOverlay 专门用来新建标记，所以又称为标记图层，ItemizedOverlay<Item extends OverlayItem>类的说明如表 9-4 所示。该标记图层内存储着标记(数据类型为 OverlayItem)，OverlayItem 类的说明如表 9-5 所示。

表 9-4

构 造 函 数
public ItemizedOverlay(Drawable defaultMarker) 创建 ItemizedOverlay 对象 defaultMarker：标记在地图上的图标，基本上是一个图形文件
方　　法
protected abstract Item createItem(int i) 创建标记并放在地图上。调用 populate()后才会执行此方法 i：欲创建标记的索引
public abstract int size() 返回标记图层上面有几个标记，也就是标记总数量。调用 populate()后会执行此方法
protected boolean onTap(int index) 当用户单击标记时，此方法会被调用 index：被单击标记的索引

表 9-5

构 造 函 数
public OverlayItem(GeoPoint point, String title, String snippet) 创建 OverlayItem 对象 point：地点的坐标 ● 　title：标题 ● 　snippet：简单说明

Google 地图

(续表)

方 法
public java.lang.String getSnippet() 取得标记的简短说明

 范例 Marker

范例(如图 9-12 所示)说明：
- 按移动设备上的 Menu 按钮会弹出选项。
- 选择地点会将该地点移至画面的正中间，并在该点加上标记。
- 单击该标记会弹出地名。

图 9-12

Marker/src/org/marker/Marker.java

```
65.     private class MarkerOverlay extends ItemizedOverlay<OverlayItem> {
66.         private List<OverlayItem> olItems = new ArrayList<OverlayItem>();
67.         public MarkerOverlay(Drawable defaultMarker) {
68.             super(boundCenterBottom(defaultMarker));
69.             olItems.add(new OverlayItem(yangmingshan, "阳明山", "阳明山公园"));
70.             olItems.add(new OverlayItem(yushan, "玉山", "玉山公园"));
71.             olItems.add(new OverlayItem(taroko, "太鲁阁", "太鲁阁公园"));
```

285

```
72.            populate();
73.        }
74.
75.        @Override
76.        protected OverlayItem createItem(int i) {
77.            return olItems.get(i);
78.        }
79.
80.        @Override
81.        public int size() {
82.            return olItems.size();
83.        }
84.
85.        @Override
86.        protected boolean onTap(int index) {
87.            Toast.makeText(
88.                    Marker.this,
89.                    olItems.get(index).getSnippet(),
90.                    Toast.LENGTH_SHORT)
91.                    .show();
92.            return true;
93.        }
94.    }
```

65 行：ItemizedOverlay 类的主要目的就是产生标记图层，一旦 MarkerOverlay 类继承该类，就可以产生出想要的标记图层。

66 行：先准备 List 集合，方便之后新建标记。

68 行：defaultMarker 用来标记在地图上的图标。当用户指定地点时，该地点会放在整个画面的正中间，而 boundCenterBottom()将该标记图标底部中点位置对齐该地点(因为地点在画面的正中间，所以图标的底部中点就在画面的正中间)。

69~71 行：调用 add() 将标记一一添加到标记图层。为了让读者更明了 add()的意义，所以直接将文字("阳明山"、"阳明山公园" 等)放在程序内。正确做法应该将文字放在文本文件内，方便之后支持多国语言。

72 行：调用 populate()后会按照顺序自动调用(可以用 debug 模式自行观察)。

- 81 行 size()：取得 olItems 内已经存储好的标记总个数(因为第 69~71 行已经添加 3 个标记，所以标记总个数为 3)。
- 76 行 createItem()：按照 82 行返回的标记总个数来决定调用的次数(因为标记总个数为 3，所以会调用 3 次)。第一次调用 createItem()会传入 0(索引)，77 行就会返回 69 行的阳明山标记，并将该标记放在地图上；之后按照顺序将玉山、太鲁阁标记一一放在地图上。

76~78 行：改写 ItemizedOverlay.createItem()以返回各个标记，并参见 72 行说明。

Google 地图

81~83 行：改写 ItemizedOverlay.size() 以返回标记总个数，并参见 72 行说明。

86~93 行：当用户单击标记时，ItemizedOverlay.onTap(int index) 会被调用——index 代表标记的索引，例如用户单击阳明山标记，index 会为 0。89 行调用 getSnippet() 会取得标记的简短说明。简单地说，当用户单击地点标记时会弹出该地点的说明。

```
96.     private void initMarkerOverlay() {
97.         Drawable marker = getResources().getDrawable(
98.             android.R.drawable.ic_menu_myplaces);
99.         markOverlay = new MarkerOverlay(marker);
100.        mView.getOverlays().add(markOverlay);
101.    }
```

96 行：调用 initMarkerOverlay() 以初始化标记图层的内容。

97~98 行：取得系统内置图标"＊"。

99 行：调用前述 67 行中的 MarkerOverlay 构造函数以创建标记图层，并将指定好的图标传入。

100 行：mView.getOverlays() 会取得 Google 地图上面的图层，调用 add() 将标记图层添加到地图的图层，即可将标记图层显示在地图上。

9.5 LocationListener 与 LocationManager

如果只想单纯取得移动设备所在位置而不想使用 Google Map(例如：汽车导航公司有自己的地图)，可以只使用 LocationManager 并搭配 LocationListener 功能来监听移动设备的位置是否有改变，而无须使用 Google Map API。如果希望位置一改变就会通知，以方便自动运行某些程序，就必须实现 LocationListener 界面，其方法说明如表 9-6 所示。

表 9-6

public abstract void onLocationChanged (Location location)
当位置改变时会自动调用此方法并传入新的位置
location：新的位置
public abstract void onProviderDisabled (String provider)
当用户将定位功能(GPS 或网络定位)关闭时会自动调用此方法
provider：与此变动有关的位置信息提供者①(例如："gps"或"network")
public abstract void onProviderEnabled (String provider)
当用户将定位功能打开时会自动调用此方法
provider：与此变动有关的位置信息提供者(例如："gps"或"network")

---

① 其实选择 provider(位置信息提供者)就是指想要何种定位功能；如果 provider 的值为 "gps"，就代表设置为 GPS 定位。

(续表)

public abstract void onStatusChanged (String provider, int status, Bundle extras)

当定位状态改变时(例如本来由 GPS 定位,但 GPS 失效,则改由网络定位)会自动调用此方法
(1) provider:与此变动有关的位置信息提供者(例如:"gps"或"network")
(2) status:可能的值有 3 种,以 LocationProvider 对应的常量作说明
  ● LocationProvider.AVAILABLE:可以使用
  ● LocationProvider.OUT_OF_SERVICE:无法提供服务,而且短时间内无法回复
  ● LocationProvider.TEMPORARILY_UNAVAILABLE:暂时无法使用,但应该很快就可回复
(3) extras:内容包含位置信息提供者的状态

如果想要指定由 GPS 来定位,当 GPS 无法提供服务时才改成其他定位功能,可以利用 LocationManager 来达成。如果想要每隔一段时间(例如 1 分钟)就更新自己位置信息,也可通过 LocationManager 并且搭配 LocationListener 来达成。LocationManager 类的方法说明如表 9-7 所示。

表 9-7

public boolean isProviderEnabled (String provider) 返回指定 provider 的定位功能状态,如果可以使用就返回 true,否则返回 false provider:位置信息提供者(例如:"gps" 或 "network")

public String getBestProvider (Criteria criteria, boolean enabledOnly)
按照给定的 criteria(规则)返回最佳的 provider,规则可设置下列条件[①]:
(1) power requirement(电量需求):可设置成 Criteria.POWER_HIGH(高电量)、Criteria.POWER_LOW(低电量)、Criteria.POWER_MEDIUM(中电量)
(2) accuracy(精确度):可设置成 Criteria.ACCURACY_COARSE(精确度差)、Criteria.ACCURACY_FINE(精确度好)
(3) speed(速度):是否提供速度信息
(4) altitude(高度):是否提供高度信息
  ● criteria:可通过 criteria 条件的设置来取得最佳 provider
  ● enabledOnly:如果设置为 true,代表只返回目前可提供定位功能的 provider

public Location getLastKnownLocation (String provider) 返回移动装置最新位置 provider:位置信息提供者(例如:"gps" 或 "network")

public void requestLocationUpdates (String provider, long minTime, float minDistance, LocationListener listener)
要求更新位置。如果实现 LocationListener 接口,每当有更新位置就会自动调用 LocationListener.onLocationChanged()
provider:位置信息提供者
  ● minTime:要求在多少时间内更新位置,单位是毫秒(milliseconds,1 秒=1000 毫秒)。更新频率越高越耗电力
  ● minDistance:要求在多短的移动距离内更新位置,单位是公尺(meters)
  ● listener:实现 LocationListener 的对象,必须实现 onLocationChanged()

public void removeUpdates (LocationListener listener) 取消更新位置的要求,之后当位置更新时就不再调用 LocationListener 的 onLocationChanged() listener:实现 LocationListener 的对象。取消调用指定 LocationListener 对象的 onLocationChanged()

[①] 也可参照 API 文件中关于 Criteria 类的说明。

Google 地图

 范例 MyLocTrack

范例(如图 9-13 所示)说明：
- 每隔一段时间会自动取得最新位置，并将该位置移至画面的正中间。
- 为该位置加上标记。
- 单击该标记会弹出 provider 名称。

图 9-13

MyLocTrack/src/org/myLocTrack/MyLocTrack.java

```
38.     private void initLocMgr() {
39.         locMgr = (LocationManager) getSystemService(
40.             Context.LOCATION_SERVICE);
41.         Criteria criteria = new Criteria();
42.         criteria.setAccuracy(Criteria.ACCURACY_FINE);
43.         best = locMgr.getBestProvider(criteria, true);
44.     }
```

39~40 行：取得 LocationManager 的系统服务对象以定时更新移动设备的位置。

41 行：调用 Criteria 的默认构造函数(default constructor)代表没有设置任何条件，而交由系统自行决定定位功能。

42 行：调用 setAccuracy()并传入 Criteria.ACCURACY_FINE，代表要系统选择较精确的定位功能。

43 行：getBestProvider()的第 1 个参数为 criteria 代表会按照 42 行所制订规则(精确定位)来选定 provider，并返回 provider 名称。因为选择精确定位，应该会返回"gps"。第 2 个参数为 true 代表只返回目前可提供定位功能的 provider。

```
50.    private void initMapView() {
51.        mView.setBuiltInZoomControls(true);
52.        mView.setTraffic(true);
53.        mControl = mView.getController();
54.        mControl.setZoom(15);
55.        Location myLoc = locMgr.getLastKnownLocation(best);
56.        if(myLoc != null){
57.            myGP = locToGP(myLoc);
58.            mControl.animateTo(myGP);
59.            addMyLocMarker(myGP);
60.        }
61.    }
62.
63.    private GeoPoint locToGP(Location location) {
64.        int latitudeE6 = (int) (location.getLatitude() * 1E6);
65.        int longitudeE6 = (int) (location.getLongitude() * 1E6);
66.        GeoPoint gp = new GeoPoint(latitudeE6, longitudeE6);
67.        return gp;
68.    }
69.
70.    private void addMyLocMarker(GeoPoint myGP) {
71.        myLocMarker = new MarkerOverlay(myGP);
72.        List<Overlay> overlays = mView.getOverlays();
73.        if(overlays.size() != 0){
74.            for(int i=0; i<overlays.size(); i++){
75.                if(overlays.get(i) instanceof MarkerOverlay)
76.                    overlays.remove(i);
77.            }
78.        }
79.        overlays.add(myLocMarker);
80.    }
```

55 行：如果位置信息有更新，调用 getLastKnownLocation()会取得最新位置(数据类型为 Location)；如果位置无法更新(可能因为 GPS 失效或网络不通)，取得的位置就不一定是最新位置，而仅是上次成功定位所取得的位置。当然也有可能无法取得最新位置，这个时候会返回 null。

56 行：如果 myLoc 为 null，57 行就会产生 NullPointerException，为了避免执行产生的例外事件，所以使用 if 语句判断。

57~58 行：要想将最新位置放在 Google 地图的正中间，需要调用 animateTo()，但是 animateTo()只接受 GeoPoint 类型的参数，无法接受 55 行返回的 Location 对象，所以需要将 Location 对象转换成 GeoPoint 对象，因此在 63 行自定义转换方法——locToGP()[①]。

59 行：调用 70 行自定义方法——addMyLocMarker()，可以在最新位置上添加标记。

---

① 如果只是单纯取得位置而不想使用 Google 地图，就无须将 Location 对象转换成 GeoPoint 对象。因为本范例最终会将取得位置呈现在 Google 地图上，所以才需要转换。

63~68 行：locToGP()的主要功能就是接收 Location 对象，并将其转换成 GeoPoint 对象。Location.getLatitude()会取得纬度；Location.getLongitude()会取得经度；再将纬度与经度当作参数调用 GeoPoint 构造函数即可创建 GeoPoint 对象。

70 行：addMyLocMarker()的主要功能就是接收 GeoPoint 对象，并在该对象所代表的位置上添加标记。

71 行：内部类 MarkerOverlay 与前面 Marker 范例的同名类功能一样，虽然做了一些小改变，但只要熟悉 Marker 范例即可了解，因此不在此重复讲解该类。

73~79 行：先检查地图上是否已经有之前插入的 MarkerOverlay(自行定义的标记图层)，如果有，必须先移除。否则当位置有更新时，旧位置上的标记没有被移除，新位置又加上标记，就会呈现如图 9-14 所示(更新两次，造成 3 个标记)的情况，除非希望标记走过的路程，否则应该移除旧位置的标记。

图 9-14

73 行先判断地图上有没有图层，如果有，就需要执行 74~77 行的循环，以 instanceof 检查是否有属于 MarkerOverlay 的标记图层。如果有，就代表是之前已经插入的图层，执行 76 行中的 remove()，移除该图层。最后执行 79 行添加新位置的标记图层。

```
112.      LocationListener listener = new LocationListener() {
113.          @Override
114.          public void onLocationChanged(Location location) {
115.              myGP = locToGP(location);
116.              mControl.animateTo(myGP);
117.              addMyLocMarker(myGP);
118.          }
119.
120.          @Override
121.          public void onProviderDisabled(String provider) {
122.              Toast.makeText(MyLocTrack.this,
123.                  " onProviderDisabled ", Toast.LENGTH_LONG).show();
```

```
124.            }
125.
126.            @Override
127.            public void onProviderEnabled(String provider) {
128.                Toast.makeText(MyLocTrack.this,
129.                        " onProviderEnabled ", Toast.LENGTH_LONG).show();
130.            }
131.
132.            @Override
133.            public void onStatusChanged(String provider, int status, Bundle extras) {
134.                Toast.makeText(MyLocTrack.this,
135.                        " onStatusChanged ", Toast.LENGTH_LONG).show();
136.            }
137.        };
```

112 行：listener 对象是由匿名内部类实现 LocationListener 接口后所产生的对象。

114 行：当移动设备的位置改变时会自动调用此方法并传入新的位置(Location 对象)。

115 行：locToGP()可以将 Location 对象转换成 GeoPoint 对象。

116 行：将接收到的新位置移至画面的正中间。

117 行：将接收到的新位置加上标记。

121、127、133 行的方法参见前面表 9-6 的说明，不再赘述。

```
139.        @Override
140.        protected void onResume() {
141.            super.onResume();
142.            if(locMgr.isProviderEnabled(LocationManager.GPS_PROVIDER))
143.                best = LocationManager.GPS_PROVIDER;
144.            locMgr.requestLocationUpdates(best, 10000, 1, listener);
145.        }
146.
147.        @Override
148.        protected void onPause() {
149.            super.onPause();
150.            locMgr.removeUpdates(listener);
151.        }
```

142 行：调用 onResume()代表用户开始与应用程序的画面交互，这时应该重新检查并指定最适当的 provider。欲使用 GPS 定位功能，应先调用 isProviderEnabled()检查 GPS 定位是否可以运作，如果可以，就使用 GPS 定位功能(当然也可以继续使用 42 行定义的 criteria 来决定使用何种定位功能)。LocationManager.GPS_PROVIDER 是常量，值为 "gps"。

144 行：在 onResume()状态下应该调用 requestLocationUpdates()要求更新位置，每当有更新位置时，就会自动调用 114 行的 onLocationChanged()。best 代表 provider；10000(毫秒)代表要求在 10 秒内更新位置；1 代表要求最短在 1 公尺的移动距离内更新位置；listener 代表

LocationListener 对象。

148~151 行：调用 onPause()，代表本程序暂停，而改为运行其他程序；此时应该调用 removeUpdates() 以暂时取消更新位置的要求，避免在监听更新上浪费资源。

9.6 以地名/地址查询位置

Geocoder 类可以将地名/地址转换成经纬度；也可以将经纬度转换成地名/地址。但更精确地说，应该是 Geocoder 类可以将地名/地址转换成 Address 对象，然后再由 Addrees 对象转换成经纬度；也可以将经纬度转换成 Address 对象，然后再将 Address 对象转换成地名/地址。一般而言，不论何种转换，都有可能得到多个 Address 对象，并且存放在 List 内。这就像使用 Web 版的 Google 地图，输入"阳明山公园"，会找到多个数据，如图 9-15 所示。而每个数据在 Android 中是用一个 Address 对象存储的。

图 9-15　输入后找到多个符合的数据

Geocoder 类说明如表 9-8 所示，Address 类说明如表 9-9 所示。

表 9-8

构　造　函　数
public Geocoder (Context context) 创建 Geocoder 对象 context：现行的 Activity 对象

(续表)

方 法
public List<Address> getFromLocationName(String locationName, int maxResults) 将地名/地址转换成经纬度 ● locationName：代表欲查询的地名/地址(例如：阳明山公园就是地名，而台北市阳明山竹子湖路 1-20 号就是地址) ● maxResults：因为符合输入的地名/地址的数据可能不止一个，一般都会指定参数 maxResults 的值来限定返回的数据数，API 文件建议返回的数据数限定在 1~5 个以方便处理(而且列在前面的数据应该更加符合欲搜索的地名/地址)
public List<Address> getFromLocation (double latitude, double longitude, int maxResults) 将经纬度转换成地名/地址 ● latitude：纬度 ● longitude：经度 ● maxResults：限定返回的数据数

表 9-9

方 法
public double getLatitude() 返回纬度
public double getLongitude() 返回经度
public String getFeatureName() 返回地名，例如"阳明山公园"，如果没有地名，则返回 null
public int getMaxAddressLineIndex () 返回最后一个地址列的索引(一个完整地址可能由多个地址列组成)
public String getAddressLine (int index) 按照指定的索引返回对应的地址行 index：索引

 不可不知

在模拟器上，无论输入任何地址，调用 getFromLocationName()只会返回 null 而不会返回任何数据，换句话说，模拟器没有地址解析功能。类似问题早已有人提出，不过这个问题仍然未被解决[①]。还好以实机测试不会发生这种问题，所以下列执行结果(如图 9-16 所示)是获取实机画面而非模拟器的画面。

以上问题在 Android 4.0 以后版本的模拟器中已经获得解决。

---

① 相关讨论参见 http://code.google.com/p/android/issues/detail?id=1537。

Google 地图 09

范例 AddrToGP

范例(如图 9-16 所示)说明：
- 一开始先显示自己位置。
- 如果不输入地址就按下"查询"按钮，会弹出"未输入地址"信息。
- 如果输入的地址找不到对应的位置，会弹出"找不到该地址"信息。
- 输入地名/地址后若找到对应位置，会将该位置加上标记，并将其移至画面的正中间。
- 单击该位置会显示地名/地址。

图 9-16

AddrToGP/src/org/addrToGP/AddrToGP.java

```
41.     OnClickListener listener = new OnClickListener() {
42.         @Override
43.         public void onClick(View v) {
44.             input = etAddress.getText().toString().trim();
45.             if (input.length() > 0) {
46.                 Geocoder geocoder = new Geocoder(AddrToGP.this);
47.                 List<Address> addresses = null;
48.                 Address address = null;
49.                 try {
50.                     addresses = geocoder.getFromLocationName(input, 1);
51.                 } catch (IOException e) {
52.                     Log.e("AddrToGP", e.toString());
53.                 }
54.                 if(addresses == null || addresses.isEmpty()){
55.                     Toast.makeText(AddrToGP.this, R.string.addressNotFound,
56.                         Toast.LENGTH_SHORT).show();
57.                 }else{
58.                     address = addresses.get(0);
```

```
59.            double geoLatitude = address.getLatitude() * 1E6;
60.            double geoLongitude = address.getLongitude() * 1E6;
61.            gp = new GeoPoint((int) geoLatitude, (int) geoLongitude);
62.            mControl.animateTo(gp);
63.            addMyLocMarker(gp);
64.        }
65.    }else {
66.        Toast.makeText(AddrToGP.this, R.string.addressIsEmpty,
67.            Toast.LENGTH_SHORT).show();
68.        }
69.    }
70. };
```

43~44 行：按下"查询"按钮会取得用户输入的文本。

45 行：检查用户是否输入，如果未输入则执行 65 行，弹出"未输入地址"的警示信息。

49~53 行：按照输入的地名/地址来取得对应的数据，并且限定仅返回一个数据(取得的对应数据属于 Address 对象，系统将其存储在 List<Address>集合内后返回)。调用 getFromLocation-Name()可能会产生 IOException 例外事件(例如：无法连接网络)，使用 try-catch 块处理。

54~56 行：如果无法从输入的地名/地址找到对应的数据，就弹出"找不到该地址"的警示信息。

58 行：既然限定仅返回一个数据，所以只要调用 get(0)，取得第一个数据(Address 对象)即可。

59~61 行：取得 Address 对象后只要调用 getLatitude()、getLongitude()，即可取得对应的纬度与经度以创建 GeoPoint 对象。

62~63 行：一旦有了 GeoPoint 对象，即可在地图上显示其对应位置，并加上标记。

9.7 导航功能

Google 地图服务器提供导航功能，只要通过 HTTP 协议传送出发地与目的地的经纬度，该服务器就可以计算并显示从出发地至目的地应走的道路、方向和距离，并且会在地图上标示建议路线。该服务器虽然支持实时导航，但功能过于简单，既无法按照用户行进的方向调整地图，也无语音告知功能；希望该导航功能未来能够像 Garmin 或 Mio 等汽车导航般功能齐全。欲开发导航功能，步骤如下：

STEP 1 取得出发地(自己位置)与目的地的经纬度：必须使用浮点数，对于 GeoPoint 对象，在取出经纬度后要除以 10^6，并转换成字符串。

```
//出发地位置
String fromGPStr = String.valueOf(fromGP.getLatitudeE6() / 1E6) + ","
    + String.valueOf(fromGP.getLongitudeE6() / 1E6);
//目的地位置
String destGPStr = String.valueOf(destGP.getLatitudeE6() / 1E6) + ","
```

Google 地图

   + String.valueOf(destGP.getLongitudeE6() / 1E6);

STEP 2 设置欲前往的 URI(必须包含出发地与目的地位置信息)。

```
Uri uri = Uri.parse("http://maps.google.com/maps?f=d&saddr="
    + fromGPStr + "&daddr=" + destGPStr + "&hl=tw");
```

STEP 3 创建 Intent 对象并设置 action 为 ACTION_VIEW(呈现数据给用户);调用 setData() 将 URI 信息附加到 Intent 对象上,最后调用 startActivity() 打开 Activity 显示导航结果。

```
Intent intent = new Intent();
intent.setAction(android.content.Intent.ACTION_VIEW);
intent.setData(uri);
startActivity(intent);
```

STEP 4 因为会使用到定位功能,所以必须在 manifest 文件中设置<uses-permission>。

```
//允许应用程序使用 GPS 定位
<uses-permission android:name="android.permission.ACCESS_FINE_LOCATION" />
//允许应用程序使用基站或 WiFi 三角定位
<uses-permission android:name="android.permission.ACCESS_COARSE_LOCATION" />
```

 范例 RouteEx

范例说明:

- 输入地名(例如:阳明山公园)或地址,按下"导航"按钮后选择"地图"会开始导航,如图 9-17 所示。
- 以移动设备的位置当作出发地,输入的地点当作目的地,列出从出发地至目的地应走的道路、方向和距离,如图 9-18 所示。
- 单击图 9-18 右上角的按钮会显示规划好的路线(路线会以青色标示),如图 9-19 所示。

图 9-17

图 9-18

图 9-19

RouteEx/src/org/routeEx/RouteEx.java

```
23.     public class RouteEx extends MapActivity{
24.         private Button btnRoute;
25.         private EditText etAddress;
26.         private LocationManager locMgr;
27.         private String best;
28.
29.         @Override
30.         public void onCreate(Bundle savedInstanceState) {
31.             super.onCreate(savedInstanceState);
32.             initLocMgr();
33.             setContentView(R.layout.main);
34.             findViews();
35.         }
36.
37.         private void initLocMgr() {
38.             locMgr = (LocationManager) getSystemService(Context.LOCATION_SERVICE);
39.             Criteria criteria = new Criteria();
40.             best = locMgr.getBestProvider(criteria, true);
41.         }
```

32、37~41 行：调用 initLocMgr()，初始化 LocationManager 以方便之后取得自己位置。

```
43.         private void findViews() {
44.             etAddress = (EditText) findViewById(R.id.etAddress);
45.             btnRoute = (Button) findViewById(R.id.btnRoute);
46.             btnRoute.setOnClickListener(new OnClickListener() {
47.                 @Override //按下"导航"按钮
48.                 public void onClick(View v) {
49.                     GeoPoint myGP = null;
50.                     GeoPoint destGP = null;
51.                     Location myLoc = locMgr.getLastKnownLocation(best);
52.                     if (myLoc != null) {
53.                         myGP = locToGP(myLoc);
54.                     } else{
55.                         Toast.makeText(RouteEx.this,
56.                             R.string.lastLocNotFound,
57.                             Toast.LENGTH_SHORT).show();
58.                         return;
59.                     }
60.
61.                     String addr = etAddress.getText().toString().trim();
62.                     if(addr.length() > 0){
63.                         destGP = addrToGP(addr);
64.                     } else{
```

```
65.                    Toast.makeText(RouteEx.this,
66.                            R.string.addressIsEmpty,
67.                            Toast.LENGTH_SHORT).show();
68.                    return;
69.                }
70.
71.                route(myGP, destGP);
72.            }
73.        });
74.    }
```

51 行：取得移动设备的最新位置(出发地)。
53 行：调用 77 行中的 locToGP()将 Location 转换成 GeoPoint。
61 行：取得用户输入的目的地地名/地址。
63 行：调用 85 行中的 addrToGP()将地名/地址转换成 GeoPoint。
71 行：调用 108 行中的 route()，并将出发地与目的地的 GeoPoint 位置传入，以打开导航功能。

```
76.        //将传入的 Location 解析成 GeoPoint 对象
77.        private GeoPoint locToGP(Location location) {
78.            int latitudeE6 = (int) (location.getLatitude() * 1E6);
79.            int longitudeE6 = (int) (location.getLongitude() * 1E6);
80.            GeoPoint gp = new GeoPoint(latitudeE6, longitudeE6);
81.            return gp;
82.        }
83.
84.        //将传入的地址解析成 GeoPoint 对象
85.        private GeoPoint addrToGP(String addr) {
86.            Geocoder geocoder = new Geocoder(this);
87.            List<Address> addresses = null;
88.            Address address = null;
89.            GeoPoint gp = null;
90.            try {
91.                addresses = geocoder.getFromLocationName(addr, 1);
92.            } catch (IOException e) {
93.                Log.e("RouteEx", e.toString());
94.            }
95.            if (addresses == null || addresses.isEmpty()) {
96.                Toast.makeText(this, R.string.addressNotFound,
97.                        Toast.LENGTH_SHORT).show();
98.            } else {
99.                address = addresses.get(0);
100.               double geoLatitude = address.getLatitude() * 1E6;
101.               double geoLongitude = address.getLongitude() * 1E6;
102.               gp = new GeoPoint((int) geoLatitude, (int) geoLongitude);
103.           }
```

```
104.            return gp;
105.        }
106.
107.        //打开 Google 地图服务器的导航功能
108.        public void route(GeoPoint fromGP, GeoPoint destGP) {
109.            String fromGPStr = String.valueOf(fromGP.getLatitudeE6() / 1E6) + ","
110.                    + String.valueOf(fromGP.getLongitudeE6() / 1E6);
111.            String destGPStr = String.valueOf(destGP.getLatitudeE6() / 1E6) + ","
112.                    + String.valueOf(destGP.getLongitudeE6() / 1E6);
113.            Uri uri = Uri.parse("http://maps.google.com/maps?f=d&saddr="
114.                    + fromGPStr + "&daddr=" + destGPStr + "&hl=tw");
115.            Intent intent = new Intent();
116.            intent.setAction(android.content.Intent.ACTION_VIEW);
117.            intent.setData(uri);
118.            startActivity(intent);
119.        }
120.
121.        @Override
122.        protected boolean isRouteDisplayed() {
123.            return false;
124.        }
125.    }
```

77~82 行：调用 locToGP()将传入的 Location 对象解析成 GeoPoint 对象后传回。内容已在之前范例 MyLocTrack 中说明，不再赘述。

85~105 行：调用 addrToGP()将传入的地址(String 类型)解析成 GeoPoint 对象后返回。内容已在之前范例 AddrToGP 中说明，不再赘述。

108 行：调用 route()并根据传入的出发地与目的地打开 Google 地图服务器的导航功能。

109~110 行：将出发地的 GeoPoint 位置转换成 String 类型。

111~112 行：将目的地的 GeoPoint 位置转换成 String 类型。

113~114 行：设置前往 Google 地图服务器的 URI(必须包含出发地与目的地位置信息)。

116 行：设置 action 为 ACTION_VIEW，代表要呈现数据给用户。

117 行：将 URI 信息附加到 Intent 对象上。

118 行：打开 Activity 显示导航结果。

第 10 章

传感器应用

 本章学习目标:

- 传感器介绍
- 加速度传感器
- 方位传感器
- 接近传感器
- 光线传感器

10.1 传感器介绍

传感器(sensor)就是专门感应外界事物变化并将其变化转为数值的一种接收器。日常生活中常见的传感器有：温度计(感应外界温度变化)、指南针(感应南北极磁场)。另外对于受欢迎的游戏主机 Wii，其游戏杆内藏有加速度传感器，可以让 Wii 通过该传感器知道游戏杆倾斜的状况来作出适当的响应。

与 Android 传感器有关的函数库都放在 android.hardware 包内，目前所支持的传感器如表 10-1 所示①。

表 10-1

传　感　器	对　应　的　值
加速度传感器	Sensor.TYPE_ACCELEROMETER
重力传感器	Sensor.TYPE_GRAVITY
陀螺仪传感器	Sensor.TYPE_GYROSCOPE
光线传感器	Sensor.TYPE_LIGHT
线性加速度传感器	Sensor.TYPE_LINEAR_ACCELERATION
磁场传感器	Sensor.TYPE_MAGNETIC_FIELD
方位传感器	SensorManager.getOrientation()已替换 Sensor.TYPE_ORIENTATION
压力传感器	Sensor.TYPE_PRESSURE
接近传感器	Sensor.TYPE_PROXIMITY
旋转向量传感器	Sensor.TYPE_ROTATION_VECTOR
温度传感器	Sensor.AMBIENT_TEMPERATURE 已替换 Sensor.TYPE_TEMPERATURE
相对湿度传感器	Sensor.TYPE_RELATIVE_HUMIDITY

 不可不知

(1) 不是每一台 Android 移动设备都有表 10-1 所列的传感器。如果没有对应的传感器，即使写程序也无法取得对应的数据。

(2) Android 模拟器无法模拟传感器功能，虽然网络上有模拟传感器的软件②，但还是强烈建议直接在实机上测试。

不论是何种传感器，最重要的就是取得其对外界感应后所收集到的数值。数值以一个 float 数组的形式存储，通常采用 values[i]来代表(values 代表数组名，i 代表索引)，按照不同的传感

---

① 参考 http://developer.android.com/reference/android/hardware/Sensor.html。

② 参考 http://www.openintents.org/en/node/23。

器，数组的元素个数也会有所不同。例如，加速度与方位传感器都有 X 轴、Y 轴、Z 轴观念，所以有 3 个数值：values[0]、values[1]、values[2]，以存储对应信息；而接近传感器只有距离一个数值，所以只使用到 values[0]。每个数值代表的意义将于下列各个传感器小节再做详细说明。一般 Android 移动设备大部分都有加速度传感器、方位传感器、接近传感器与光线传感器的功能，所以本章专门探讨这 4 个传感器。虽然未提及其他传感器，但取得数值的方式皆相同。

10.2 加速度传感器

在说明加速度传感器之前，先说明 X 轴、Y 轴、Z 轴所代表的位置。图 10-1 属于 3D 坐标图，Android 采用 OpenGL ES 的坐标系统，说明如下：

(1) 屏幕左下角顶点为原点(x=0, y=0, z=0)，它与一般 2D 坐标系统原点在屏幕左上角不同。
(2) X 轴是由左向右的水平方向，所以向右 X 值增加，向左 X 值减少。
(3) Y 轴是由下向上的垂直方向，所以向上 Y 值增加，向下 Y 值减少。
(4) Z 轴是由后向前的方向，所以向前 Z 值增加，向后 Z 值减少。

接下来说明加速度传感器。加速度的单位是 m/sec^2(公尺/秒的平方)，而加速度传感器则是反映 X 轴、Y 轴、Z 轴受到地心引力的影响情况，重力方向恰与坐标方向相反，所以若符合重力方向与坐标方向相反，则会得到正的值，反之会得到负的值，如图 10-2 所示。

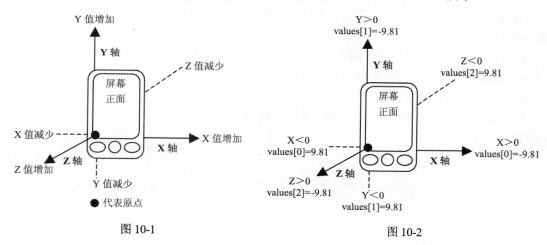

图 10-1   图 10-2

各种状态说明如下。

(1) 移动设备平躺(屏幕正面朝上)，如图 10-3 所示，此时 Z 轴受重力影响，values 值如下：
- values[0] = 0.0，代表 X 轴未受重力影响。
- values[1] = 0.0，代表 Y 轴未受重力影响。
- values[2] = 9.81，值为正代表 Z 轴后面方向(Z＜0)受重力影响。

若移动设备平躺但屏幕正面朝下，背盖朝上，如图 10-4 所示，则 values[2] = -9.81，代表 Z 轴前面方向受重力影响。

图 10-3　　　　　　　　　图 10-4

(2) 移动设备呈现如图 10-5 所示的垂直状态(称为 portrait)，此时 Y 轴受重力影响，values 值如下：
- values[0] = 0.0，代表 X 轴未受重力影响。
- values[1] = 9.81，值为正代表 Y 轴下面方向(Y < 0)受重力影响。
- values[2] = 0.0，代表 Z 轴未受重力影响。

若移动设备垂直方式上下颠倒，如图 10-6 所示，则 values[1] = -9.81，代表 Y 轴上面方向受重力影响。

 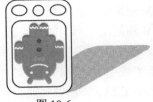

图 10-5　　　　　　　　　图 10-6

(3) 移动设备呈现如图 10-7 所示的横向直立状态(称为 landscape)，此时 X 轴受重力影响，values 值如下：
- values[0] = 9.81，代表 X 轴左面方向(X < 0)受重力影响。
- values[1] = 0.0，代表 Y 轴未受重力影响。
- values[2] = 0.0，代表 Z 轴未受重力影响。

若移动设备横向直立方式左右颠倒，如图 10-8 所示，则 values[0] = -9.81，代表 X 轴右面方向受重力影响。

 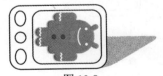

图 10-7　　　　　　　　　图 10-8

按照下列步骤可以取得传感器的相关信息与该传感器对外界感应后所得到的对应数值。

STEP 1 取得 SensorManager 对象：通过 SensorManager 对象才能取得各种传感器的信息，而要取得该对象，就必须调用 Context 的 getSystemService()，并指定欲取得的系统服务名称，这一点跟取得 NotificationManager、LocationManager 对象方式相同[①]。

```
SensorManager sensorMgr = (SensorManager)getSystemService(SENSOR_SERVICE);
```

---

① 欲知目前有哪些系统服务，请参考 API 文件 Context 的 getSystemService()的说明。

STEP 2 实现 SensorEventListener：实现 SensorEventListener 的 onSensorChanged()，当传感器的值改变时会自动调用此方法，并传入 SensorEvent 对象，通过该对象可以取得产生事件的传感器。

```
class MySensorEventListener implements SensorEventListener{
    public void onSensorChanged(SensorEvent event) {
        float[] sensorsValues = event.values; //传感器对外界感应后所收集到的数值
        Sensor sensor = event.sensor; //取得产生此事件的传感器
        String sensorName = sensor.getName(); //取得传感器名称
        int sensorType = sensor.getType(); //取得传感器种类
        float sensorPower = sensor.getPower(); //取得传感器的耗电量
    }

    public void onAccuracyChanged(Sensor sensor, int accuracy) {
        //当传感器的精确度改变时会调用此方法
    }
}
```

STEP 3 为指定的传感器注册 SensorEventListener：调用 SensorManager 的 registerListener() 为指定的传感器注册 SensorEventListener，当传感器的值变化时，SensorEventListener 的 onSensorChanged() 才会自动被调用。

registerListener(listener, sensor, int rate)

取得传感器相关信息所需使用的相关方法如表 10-2 所示。

表 10-2

Context 类
public Object getSystemService (String name) 根据指定名称取得对应系统服务的管理对象 name：欲取得的系统服务名称
SensorEventListener 界面
public abstract void onSensorChanged (SensorEvent event) 当传感器的值改变时会调用此方法 event：SensorEvent 对象，通过该对象可以取得传感器相关信息
public abstract void onAccuracyChanged (Sensor sensor, int accuracy) 当传感器的精确度改变时会调用此方法 ● sensor：产生此事件的传感器 ● accuracy：传感器新的精确度
SensorEvent 类
public Sensor sensor 产生事件的对应传感器
public final float[] values 传感器感知外界环境而收集的数值，不同传感器的数值与其代表的意义会不同

(续表)

Sensor 类
public String getName () 取得传感器名称
public int getType () 取得传感器种类
public float getPower () 取得传感器的耗电量
public float getMaximumRange () 取得传感器可检测的最大范围

SensorManager 类
public Sensor getDefaultSensor (int type) 根据指定的传感器种类返回对应的 Sensor 对象 type：Sensor 种类
public boolean registerListener (SensorEventListener listener, Sensor sensor, int rate) 为指定的传感器注册 SensorEventListener。如果移动设备有对应的传感器而且可以正常运作，则返回 true；否则返回 false listener：实现 SensorEventListener 的对象 sensor：欲注册的传感器 rate：设置事件发生后传送数值的频率，有下列 4 种(按照频率从低到高排列)： ● SENSOR_DELAY_NORMAL——适合屏幕的频率 ● SENSOR_DELAY_UI——适合用户接口的频率 ● SENSOR_DELAY_GAME——适合游戏的频率 ● SENSOR_DELAY_FASTEST——频率最高
public void unregisterListener (SensorEventListener listener) 解除所有对 SensorEventListener 注册的传感器 listener：实现 SensorEventListener 的对象
public void unregisterListener (SensorEventListener listener, Sensor sensor) 解除对 SensorEventListener 注册的特定传感器 ● listener：实现 SensorEventListener 的对象 ● sensor：欲解除注册的传感器

 范例 AccelerometerEx

范例(如图 10-9 所示)说明：
- 显示传感器名称、种类与耗电量。
- 将加速度传感器对外界感应后所得到的数值显示在画面上。
- values[0]、values[1]、values[2]分别反映移动设备 X 轴、Y 轴、Z 轴受到地心引力影响的情况；值都介于-9.81～+9.81 之间。

图 10-9

AccelerometerEx/src/org/accelerometerEx/AccelerometerEx.java

```
15.     @Override
16.     public void onCreate(Bundle savedInstanceState) {
17.         super.onCreate(savedInstanceState);
18.         setContentView(R.layout.main);
19.         sensorMgr = (SensorManager)getSystemService(SENSOR_SERVICE);
20.         findViews();
21.     }
22.
23.     private void findViews() {
24.         tvMsg = (TextView)findViewById(R.id.tvMsg);
25.     }
26.
27.     SensorEventListener listener = new SensorEventListener() {
28.         public void onSensorChanged(SensorEvent event) {
29.             Sensor sensor = event.sensor;
30.             StringBuilder sensorInfo = new StringBuilder();
31.             sensorInfo.append("sensor Name: " + sensor.getName() + "\n");
32.             sensorInfo.append("sensor Type: " + sensor.getType() + "\n");
33.             sensorInfo.append("used power: " + sensor.getPower() + " mA\n");
34.             sensorInfo.append("values: \n");
35.             float[] values = event.values;
36.             for (int i = 0; i < values.length; i++)
37.                 sensorInfo.append("-values[" + i + "] = " + values[i] + "\n");
38.             tvMsg.setText(sensorInfo);
39.         }
40.
```

```
41.        public void onAccuracyChanged(Sensor sensor, int accuracy) {
42.            //当传感器的精确度改变时会调用此方法
43.        }
44.    };
45.
46.    @Override
47.    protected void onResume() {
48.        super.onResume();
49.        sensorMgr.registerListener(listener,
50.                sensorMgr.getDefaultSensor(Sensor.TYPE_ACCELEROMETER),
51.                SensorManager.SENSOR_DELAY_UI);
52.    }
53.
54.    @Override
55.    protected void onPause() {
56.        super.onPause();
57.        sensorMgr.unregisterListener(listener);
58.    }
```

19 行：调用 getSystemService()，并指定 SENSOR_SERVICE 系统服务名称即可取得 SensorManager 对象。

27~28 行：以匿名内部类实现 SensorEventListener 的 onSensorChanged()，当传感器的值改变时会自动调用此方法。

29 行：取得传感器对象。

31 行：取得传感器名称。

32 行：取得传感器种类。

33 行：取得传感器耗电量。

35~37 行：取得传感器对外界感应后所搜集到的数值，并以 for-each 循环将对应的值取出。

47~52 行：在操作画面显示之前，先将加速度传感器注册到对应的 SensorEventListener，并设置传感器的传送频率为 SENSOR_DELAY_UI。

57 行：在操作画面消失之前，解除所有对 SensorEventListener 注册的传感器，以节省电力。

10.3 方位传感器

根据加速度传感器的数值只能判断受重力影响的方向，无法精确判断是如何旋转的，如图 10-10 所示。对于平躺且屏幕正面朝天的移动设备，无论顺时针或逆时针翻转移动设备，values[2] 值的变化都是 9.81→ 0，所以无法从加速度传感器的数值分析出是以哪种方式翻转移动装置，当然也就无法更精确反映用户的操作。如果有方位信息，就能得知移动设备目前的方位，更进一步可以判断出翻转方式。

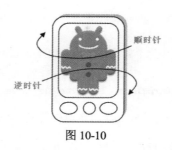

图 10-10

早期使用方位传感器只要将传感器指定为 Sensor.TYPE_ORIENTATION 即可。之后 Android 将 Sensor.TYPE_ORIENTATION 常量设置为 deprecated,而建议改为以调用 SensorManager.getOrientation()方式取得方位传感器的信息。因为前者观念上比较容易,而后者则较为复杂难懂,所以不仅市场上有许多学习教材仍然只介绍前者的概念,就连许多开发人员也只以前者的方式开发应用程序。本文会将这两种取值、解读数值方式加以比较与说明,让读者更容易了解它们的差异。因为 Android 的 API 文件已不建议采用前者,所以本文会先介绍后者,然后再说明前者的概念。

10.3.1 调用 getOrientation()取得方位信息

使用此种方式是通过加速度传感器的数值计算出移动设备的方位信息,所以必须先取得加速度传感器的数值。按照下列步骤可以取得方位数值:

STEP 1 取得加速度传感器与磁场传感器的数值(所以必须将加速度传感器与磁场传感器注册到对应的 SensorEventListener)。

```
public void onSensorChanged(SensorEvent event) {
    switch (event.sensor.getType()) {
        case Sensor.TYPE_ACCELEROMETER:
            //取得加速度传感器的数值
            accelerometer_values = (float[]) event.values.clone();
            break;
        case Sensor.TYPE_MAGNETIC_FIELD:
            //取得磁场传感器的数值
            magnitude_values = (float[]) event.values.clone();
            break;
        default:
            break;
    }
    //其他程序代码
}
```

STEP 2 调用 SensorManager.getRotationMatrix()并根据加速度传感器的数值来计算旋转矩阵

(rotation matrix)。

> /* 用来存储下面 accelerometer_values 参数(第 3 个参数)计算出来的旋转矩阵 */
> float[] R = new float[9];
> /* 第 2 个参数设置为 null 是因为不需要地磁倾斜度的信息，但第 4 个参数 magnitude_values 不可为 null，否则会产生 Exception */
> SensorManager.***getRotationMatrix(R***, null, ***accelerometer_values***, magnitude_values*)*;

STEP 3 调用 SensorManager.getOrientation()并根据旋转矩阵计算出移动装置的方位。

> float[] values = new float[3]; //存储由 R 计算出来的方位信息
> SensorManager.getOrientation(R, values);

取得方位信息所需 SensorManager 类的相关方法说明如表 10-3 所示。

表 10-3

SensorManager 类
public static boolean getRotationMatrix (float[] R, float[] I, float[] gravity, float[] geomagnetic) 将参数 gravity 数值转换成对应的旋转矩阵并存储在 R 参数内；将参数 geomagnetic 数值转换成对应的地磁倾斜矩阵并存储在 I 参数内。换句话说，就是根据加速度传感器的数值来计算旋转矩阵；根据磁场传感器的数值来计算地磁倾斜矩阵。如果成功，则返回 true，失败返回 false(例如自由落体) ● R：float 数组，内含 9 个 float 数字，用来存储旋转矩阵 ● I：float 数组，内含 9 个 float 数字，用来存储地磁倾斜矩阵 ● gravity：float 数组，内含 3 个 float 数字，必须指定加速度传感器(TYPE_ACCELEROMETER)的数值 ● geomagnetic：float 数组，内含 3 个 float 数字，必须指定磁场传感器(TYPE_MAGNETIC_FIELD)的数值
public static float[] getOrientation (float[] R, float[] values) 根据旋转矩阵计算出移动设备的方位。旋转矩阵的坐标系统属于世界坐标系统(the world coordinate system)，与移动设备的方位坐标系统不同，所以必须加以转换。返回值与 values 参数的值相同 ● R：旋转矩阵，其数值来自于 getRotationMatrix (float[] R, float[] I, float[] gravity, float[] geomagnetic)的第 1 个参数 R ● values：移动设备的方位数值，是一个内含 3 个 float 数字的数组，用来存储 R 计算完毕的结果

调用 getOrientation()会返回 float 数组，如前所述，通常以 values[i]来代表各种旋转情况，单位是弧度(radians)，其意义说明如下。

- values[0]：方位角(azimuth)，移动设备以罗盘方式旋转(沿着 Z 轴旋转)，这会改变方位角的值。如果符合如图 10-11 所示的 azimuth 箭头方向旋转，值会变大$(0 \rightarrow \pi)$[①]；反向则会变小$(0 \rightarrow -\pi)$。
- values[1]：投掷角(pitch)，移动设备以投掷方式旋转(沿着 X 轴旋转)，这会改变投掷角的值。如果符合如图 10-11 所示的 pitch 箭头方向旋转，值会变大$(0 \rightarrow \pi/2)$；反向则会变小$(0 \rightarrow -\pi/2)$。

---

① values 内的值都是弧度(radians)，而非角度(degrees)；$0 \rightarrow \pi$ 其实就是角度的变化由 0°→180°。而 π 是圆周率，值近似于 3.14159。

- values[2]：滚动角(roll)，移动设备以滚动方式旋转(沿着 Y 轴旋转)，这会改变滚动角的值。如果符合如图 10-11 所示的 roll 箭头方向旋转，值会变大(0 → π)；反向则会变小(0 → -π)。

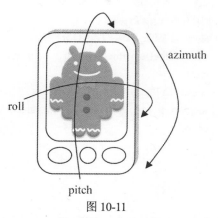

图 10-11

 范例 OrientationEx

范例(如图 10-12 所示)说明：
- 根据加速度传感器的数值计算出移动设备目前的方位数值并显示在画面上。
- values[0]、values[1]、values[2] 分别代表移动设备方位角、投掷角、滚动角的弧度，值都介于 -3.14159～+3.14159 之间。

图 10-12

OrientationEx/src/org/orientationEx/OrientationEx.java

```java
30.    SensorEventListener listener = new SensorEventListener() {
31.        public void onSensorChanged(SensorEvent event) {
32.            switch (event.sensor.getType()) {
33.                case Sensor.TYPE_ACCELEROMETER:
34.                    accelerometer_values = (float[]) event.values.clone();
35.                    break;
36.                case Sensor.TYPE_MAGNETIC_FIELD:
37.                    magnitude_values = (float[]) event.values.clone();
38.                    break;
39.                default:
40.                    break;
41.            }
42.
43.            if (magnitude_values != null && accelerometer_values != null) {
44.                float[] R = new float[9];
45.                float[] values = new float[3];
46.                SensorManager.getRotationMatrix(R, null,
47.                        accelerometer_values, magnitude_values);
48.                SensorManager.getOrientation(R, values);
49.                StringBuilder sensorInfo = new StringBuilder();
50.                for (int i = 0; i < values.length; i++)
51.                    sensorInfo.append("-values[" + i + "] = " + values[i] + "\n");
52.                tvMsg.setText(sensorInfo);
53.            }
54.        }
55.
56.        public void onAccuracyChanged(Sensor sensor, int accuracy) {}
57.    };
58.
59.    @Override
60.    protected void onResume() {
61.        super.onResume();
62.        if(!(sensorMgr.registerListener(listener, sensorMgr
63.                .getDefaultSensor(Sensor.TYPE_ACCELEROMETER),
64.                SensorManager.SENSOR_DELAY_UI) &&
65.            sensorMgr.registerListener(listener, sensorMgr
66.                .getDefaultSensor(Sensor.TYPE_MAGNETIC_FIELD),
67.                SensorManager.SENSOR_DELAY_UI))){
68.            Log.w("OrientationEx", "sensor not found!");
69.            sensorMgr.unregisterListener(listener);
70.        }
```

```
71.        }
72.
73.        @Override
74.        protected void onPause() {
75.            super.onPause();
76.            sensorMgr.unregisterListener(listener);
77.        }
```

32~41 行：取得传感器的种类后，判断如果属于加速度传感器，就将传感器数值存入 accelerometer_values 变量；如果属于磁场传感器，就将传感器数值存入 magnitude_values 变量。因为 event.values 取得的数组内容会随时变动，所以必须调用 clone() 将数组的值复制一份，而不仅是传送地址。

43 行：当 accelerometer_values 与 magnitude_values 不为 null 时执行后续语句；换句话说，就是取得加速度传感器与磁场传感器的数值时执行后续语句。

46~47 行：调用 getRotationMatrix()，根据加速度传感器的数值来计算旋转矩阵，并将结果存入 R 变量。第 2 个参数设置为 null 是因为不需要地磁倾斜度的信息。

48~52 行：调用 getOrientation()，根据 R 变量计算出移动装置的方位数值，并将结果存入 values 变量。之后将 values 内的数值一一显示在 TextView 组件上。

62~70 行：如果加速度传感器与磁场传感器中有任何一个无法运作，以日志文件记录并解除传感器的注册。

10.3.2 通过 Sensor.TYPE_ORIENTATION 取得方位信息

使用此方法取得方位传感器数值的方式与加速度传感器完全相同，所以不再赘述。values 数组内的 3 个元素值的单位是角度(degrees)而非弧度，所代表的意义说明如下。

(1) values[0]：方位角(azimuth)，移动设备以罗盘方式旋转(沿着 Z 轴旋转)，这会改变方位角的值。如果符合如图 10-13 所示的 azimuth 箭头方向旋转，值会变大(0 → 359)；反向则会变小。

- 0 代表移动设备的 Y 轴正向指着北方。
- 90 代表移动设备的 Y 轴正向指着东方。
- 180 代表移动设备的 Y 轴正向指着南方。
- 270 代表移动设备的 Y 轴正向指着西方。

(2) values[1]：投掷角(pitch)，移动设备以投掷方式旋转(沿着 X 轴旋转)，这会改变投掷角的值。如果符合如图 10-13 所示的 pitch 箭头方向旋转，值会变大(0 → 180)；反向则会变小(0 → -180)。

(3) values[2]：滚动角(roll)，移动设备以滚动方式旋转(沿着 Y 轴旋转)，这会改变滚动角的值。如果符合如图 10-13 所示的 roll 箭头方向旋转，值会变小(0 → -90)；反向则会变大(0 → 90)。[①]

[①] 这种滚动时角度的变化情况正好与数学上的概念相反，但是因为一开始 Android 函数库就已经如此设计，所以无法随意更改，否则之前已经开发好的应用程序会有兼容性问题。请参考 API 文件 SensorEvent 类的 Sensor.TYPE_ORIENTATION 说明。

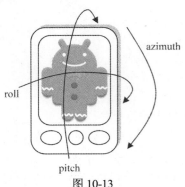

图 10-13

 范例 OrientationLegacyEx

范例(如图 10-14 所示)说明。
- 显示传感器名称、种类与耗电量。
- 将方位传感器对外界感应后所得到的数值显示在画面上。
- azimuth：方位角，值介于 0～359 之间。
- pitch：投掷角，值介于 -180～+180 之间。
- roll：滚动角，值介于 -90～+90 之间。

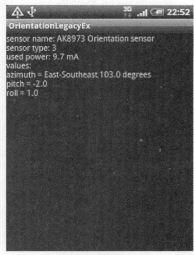

图 10-14

OrientationLegacyEx/src/org/orientationLegacyEx/OrientationLegacyEx.java

```
27.    SensorEventListener listener = new SensorEventListener() {
28.        public void onSensorChanged(SensorEvent event) {
29.            Sensor sensor = event.sensor;
30.            StringBuilder sensorInfo = new StringBuilder();
```

```
31.         sensorInfo.append("sensor name: " + sensor.getName() + "\n");
32.         sensorInfo.append("sensor type: " + sensor.getType() + "\n");
33.         sensorInfo.append("used power: " + sensor.getPower() + " mA\n");
34.         sensorInfo.append("values: \n");
35.         float[] values = event.values;
36.         float azimuth = values[0];
37.         String compassStr = "";
38.         if(azimuth >= 315)
39.             compassStr = "North-Northwest " + azimuth + " degrees";
40.         else if(azimuth >= 270)
41.             compassStr = "West-Northwest " + azimuth + " degrees";
42.         else if(azimuth >= 225)
43.             compassStr = "West-Southwest " + azimuth + " degrees";
44.         else if(azimuth >= 180)
45.             compassStr = "South-Southwest " + azimuth + " degrees";
46.         else if(azimuth >= 135)
47.             compassStr = "South-Southeast " + azimuth + " degrees";
48.         else if(azimuth >= 90)
49.             compassStr = "East-Southeast " + azimuth + " degrees";
50.         else if(azimuth >= 45)
51.             compassStr = "East-Northeast " + azimuth + " degrees";
52.         else
53.             compassStr = "North-Northeast " + azimuth + " degrees";
54.
55.         float pitch = values[1];
56.         float roll = values[2];
57.         sensorInfo.append("azimuth = " + compassStr + "\n");
58.         sensorInfo.append("pitch = " + pitch + "\n");
59.         sensorInfo.append("roll = " + roll + "\n");
60.         tvMsg.setText(sensorInfo);
61.     }
62.
63.     public void onAccuracyChanged(Sensor sensor, int accuracy) {}
64. };
```

38~53 行：根据方位角度判断属于哪一个方位。

10.4 接近传感器

取得接近传感器数值的方式与加速度传感器完全相同，所以不再赘述。一般而言，接近传感器的数值只有一个，就是接近传感器与物体之间的距离，而单位是公分。不过，也有接近传感器的数值仅有两个(例如：0.0 与 1.0 共两种变化)以代表接近或远离的意思；如果是这种情况，传感器应该要提供可感应的最大范围——通过调用 Sensor 的 getMaximumRange()取得。

 范例 ProximityEx

范例(如图 10-15 所示)说明。
- 显示传感器名称、种类、耗电量以及传感器能感应的最大范围。
- 将接近传感器对外界感应后所得到的数值显示在画面上。
- values[0]：不同的传感器所取得的值会有不同。可能是接近传感器与物体之间的距离；也可能仅有两个数值，分别代表接近或远离的意思。
- 画面背景原本为深红色，当接近传感器接近物体时会将画面的背景改为深蓝色。

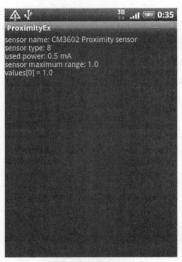

图 10-15

ProximityEx/src/org/proximityEx/ProximityEx.java

```
31.     SensorEventListener listener = new SensorEventListener() {
32.         public void onSensorChanged(SensorEvent event) {
33.             Sensor sensor = event.sensor;
34.             float[] values = event.values;
35.             StringBuilder sensorInfo = new StringBuilder();
36.             sensorInfo.append("sensor name: " + sensor.getName() + "\n");
37.             sensorInfo.append("sensor type: " + sensor.getType() + "\n");
38.             sensorInfo.append("used power: " + sensor.getPower() + " mA\n");
39.             sensorInfo.append("sensor maximum range: " +
40.                 sensor.getMaximumRange() + "\n");
41.             sensorInfo.append("values[0] = " + values[0] + "\n");
```

```
42.             tvMsg.setText(sensorInfo);
43.
44.             if(values[0] < 1.0)
45.                 linear.setBackgroundColor(Color.rgb(0, 0, 68));
46.             else
47.                 linear.setBackgroundColor(Color.rgb(68, 0, 0));
48.         }
49.
50.         public void onAccuracyChanged(Sensor sensor, int accuracy) {}
51.     };
```

40 行：调用 getMaximumRange()取得接近传感器能感应的最大范围(单位不一定是公分)。

44~47 行：当接近传感器接近物体时会将画面的背景改为深蓝色；远离时改为深红色。对于不同的接近传感器，1.0 所代表的意义不同；可能是指 1 公分，也可能是代表接近或远离[①]的意思。

10.5 光线传感器

光线传感器主要是感应移动装置环境四周的光线强弱程度，一般称为亮度(illuminance)，单位是 lux(流明[②]/平方公尺)[③]；光线传感器的数值只有一个，代表的就是亮度。家庭中的一般亮度建议在 100~300 lux 之间。一些日常的代表性亮度如表 10-4 所示。

表 10-4

环境或活动	亮度(或所需亮度)
星光	0.0003 lux
满月	0.2 lux
路灯	5 lux
看电视	30 lux
生活起居	100~300 lux
办公室、教室	300 lux
夜间棒球场	400 lux
阅读	500 lux
绘图	600 lux
阴天	8000 lux
手术	7000~10 000 lux
烈日	100 000 lux

① 以 HTC Desire 手机测试，0.0 代表接近，1.0 代表远离。
② 流明(Lumen)是人眼感知光能的量度，可参考 http://en.wikipedia.org/wiki/Lumen_(unit)。
③ 1 lux 大约等于 1 烛光在 1 公尺距离内的照度。

 范例 LightEx

范例(如图 10-16 所示)说明[1]：
- 显示传感器名称、种类、耗电量以及传感器能感应的最大范围。
- 将光线传感器对外界感应后所得到的数值显示在画面上。
- values[0]代表亮度，单位为 lux。
- 根据亮度来判断适合从事的活动。

图 10-16

LightEx/src/org/lightEx/LightEx.java

```
27.     SensorEventListener listener = new SensorEventListener() {
28.         public void onSensorChanged(SensorEvent event) {
29.             Sensor sensor = event.sensor;
30.             float[] values = event.values;
31.             StringBuilder sensorInfo = new StringBuilder();
32.             sensorInfo.append("sensor name: " + sensor.getName() + "\n");
33.             sensorInfo.append("sensor type: " + sensor.getType() + "\n");
34.             sensorInfo.append("used power: " + sensor.getPower() + " mA\n");
35.             sensorInfo.append(getString(R.string.maxRange) +
36.                     sensor.getMaximumRange() + "\n");
37.             sensorInfo.append("values[0] = " + values[0] + "\n");
38.
39.             if(values[0] >= 10000)
40.                 sensorInfo.append(getString(R.string.anyThing));
41.             else if(values[0] >= 7000)
```

---

[1] 以 HTC Desire 手机测试，如果亮度没有在 40 lux 以上，该款手机的光线传感器没有作用，所以画面上不会显示任何数据。如果读者有相同情形，建议以较强的灯光照射，即可显示数据。

```
42.                sensorInfo.append(getString(R.string.operation));
43.            else if(values[0] >= 500)
44.                sensorInfo.append(getString(R.string.read));
45.            else if(values[0] >= 100)
46.                sensorInfo.append(getString(R.string.dailyLife));
47.            else if(values[0] >= 30)
48.                sensorInfo.append(getString(R.string.watchTV));
49.            else if(values[0] >= 5)
50.                sensorInfo.append(getString(R.string.walk));
51.            else
52.                sensorInfo.append(getString(R.string.sleep));
53.
54.            tvMsg.setText(sensorInfo);
55.        }
56.
57.        public void onAccuracyChanged(Sensor sensor, int accuracy) {}
58.    };
59.
60.    @Override
61.    protected void onResume() {
62.        super.onResume();
63.        sensorMgr.registerListener(listener,
64.                sensorMgr.getDefaultSensor(Sensor.TYPE_LIGHT),
65.                SensorManager.SENSOR_DELAY_UI);
66.    }
```

29~37 行：取得传感器相关信息与感应的数值 values[0](代表的是亮度，单位为 lux)。
39~52 行：根据亮度强弱程度来判断适合从事的活动。
63~65 行：将光线传感器注册到对应的 SensorEventListener。

第 11 章 多媒体与相机功能

本章学习目标：

- Android 多媒体功能介绍
- 播放 Audio 文件
- 播放视频文件
- 录制音频文件
- 拍照功能
- 录像功能

11.1 Android 多媒体功能介绍

Android 系统支持许多常见的多媒体格式,让开发人员可以很简单地通过音频、视频等相关 Android API,就可以整合影音录制与播放的功能到应用程序内。Android 系统主要支持的多媒体格式如表 11-1[①]所示,其中编码器代表可以制作该格式的文件;解码器代表可以播放该格式的文件。例如 Android 系统同时支持 AAC 编码器与解码器,代表可以录制 AAC 的音频文件,也可以播放该类型的音频文件。但是真正的移动设备可能因为制造商的关系而支持更多其他种类的多媒体格式,并不局限于表 11-1 所列的格式。

表 11-1

类型	格式	编码器	解码器	详情	支持的文件类型
音频	AAC LC/LTP	●	●	组合标准比特率(最大 160kbps)和采样率(从 8 到 48kHz)的单声道/立体声内容	3GPP(.3gp) MPEG-4(.mp4、.m4a) ADTS 原始 AAC(aac、在 Android 3.1+中解码、在 Android 4.0+中编码、不支持 ADIF) MPEG-TS(.ts、不可查找、Android 3.0+)
	HE-AACv1 (AAC+)		●		
	HE-AACv2 (增强的 AAC+)		●		
	AMR-NB	●	●	4.75~12.2 kbps,采样@8kHz	3GPP(.3gp)
	AMR-WB	●	●	9 个频率,从 6.60 kbit/s 到 23.85 kbit/s,采样@16kHz	3GPP(.3gp)
	FLAC		● (Android 3.1+)	单声道/立体声(无多通道)。采样率最大 48kHz(但是在具有 44.1kHz 输出的设备上推荐最大 48kHz 的采样率,因为 48 到 44.1kHz 降低取样频率取样器不包括低通滤波器)。推荐采用 16 位;没有高频振动适用于 24 位	仅限 FLAC(.flac)
	MP3		●	Mono/Stereo 8-320Kbps 固定比特率(CBR)或可变比特率(VBR)	MP3(.mp3)

---

① 参见 http://developer.android.com/guide/appendix/media-formats.html。

(续表)

类型	格式	编码器	解码器	详情	支持的文件类型
音频	MIDI		●	MIDI Type 0 和 1。DLS Version 1 和 2。XMF 和 Mobile XMF。支持铃声格式 RTTTL/RTX、OTA 和 iMelody	Type 0 和 1(.mid、.xmf、.mxmf)。以及 RTTTL/RTX(.rtttl、.rtx)、OTA(.ota)和 iMelody (.imy)
	Vorbis		●		Ogg(.ogg) Matroska(.mkv、Android 4.0+)
	PCM/WAVE		●	8 位和 16 位线性 PCM (频率最大不得超出硬件限制)	WAVE(.wav)
Image	JPEG	●	●	基础+渐进	JPEG(.jpg)
	GIF		●		GIF(.gif)
	PNG	●	●		PNG(.png)
	BMP		●		BMP(.bmp)
	WEBP	●(Android 4.0+)	●(Android 4.0+)		WebP(webp)
Video	H.263	●	●		3GPP(.3gp)和 MPEG-4 (.mp4)
	H.264 AVC	●(Android 3.0+)	●	Baseline Profile (BP)	● 3GPP(.3gp) ● MPEG-4(.mp4) ● MPEG-TS(.ts、仅限 AAC 音频、不可查找、Android 3.0+)
	MPEG-4 SP		●		3GPP(.3gp)
	VP8		●(Android 2.3.3+)	仅在 Android 4.0 及以上版本中可流式处理	● WebM(.webm) ● Matroska(.mkv、Android 4.0+)

Android 提供下列多媒体功能。

(1) 播放音频与视频文件：使用 MediaPlayer 类的功能可以播放音频与视频文件。如果想要直接有一个简易的播放器(具有画面与控制界面)，也可以通过整合 VideoView 与 MediaController 类所提供的功能来达到目的。

(2) 录制音频与视频文件：使用 MediaRecorder 类提供的功能可以录制音频与视频并转换成对应的影音文件。虽然模拟器没有可获取音频与视频的硬件设备，不过大部分的实机都具有这类设备供用户录制影音文件。

Android 应用程序播放的影音文件来源可以来自下列 3 处。

(1) 资源文件：可以播放应用程序本身的资源文件(raw resources)。
(2) 外部文件：可以播放移动设备外部存储媒体内的影音文件。
(3) 网络数据流：可以播放来自网络上(URL)的影音流。

11.2 播放 Audio 文件

11.2.1 播放资源文件

一般来说，播放资源文件可能是最常使用的方式，因为这种方式十分简单，步骤如下：

STEP 1 将音频文件放至项目的 res/raw 目录内①，如图 11-1 的 ring.mp3 所示。使用 aapt(Android Asset Packaging Tool)即可找到该文件，而且项目内的 R 类会自动产生资源 ID 并引用到该文件。

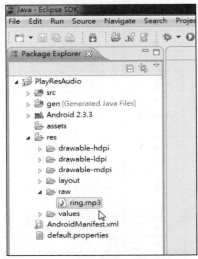

图 11-1

STEP 2 调用 MediaPlayer.create()来创建 MediaPlayer 的对象实体，然后再调用 start()即可播放音频文件。

```
//ring：音频文件名，不用加上扩展名。
MediaPlayer mp = MediaPlayer.create(context, R.raw.ring);
mp.start();
```

STEP 3 如欲暂停播放，可以调用 pause()；暂停后如欲继续播放，可以再次调用 start()，就会从上次暂停点继续播放。

STEP 4 如欲停止播放，调用 stop()。

STEP 5 如果不再播放，应该调用 release()，释放 MediaPlayer 所占用的资源。

---

① 文件名必须为小写字母。

MediaPlayer 类相关方法的说明如表 11-2 所示。

表 11-2

构 造 函 数
public MediaPlayer () 默认构造函数，用来创建 MediaPlayer 对象实体。如果不再使用 MediaPlayer，应该调用 release()释放所占用的资源；如果不释放资源，会因为产生过多的 MediaPlayer 对象实体而造成 Exception

方 法
public static MediaPlayer create(Context context, int resid) 根据资源 ID 创建 MediaPlayer 对象实体。如果创建成功会自动调用 prepare()，所以不需要再次调用。如果不再使用 MediaPlayer，应该调用 release()释放所占用的资源；如果不释放资源，会因为产生过多的 MediaPlayer 对象实体而造成 Exception ● context：Context 对象，通常是目前 Activity 对象 ● resid：资源 ID，用来指定欲播放的资源文件
public void start () 开始或继续播放；如果之前是暂停状态，会从上次暂停的位置继续播放；如果之前是停止状态或是从未播放过，会从头开始播放
public void pause () 暂停播放，调用 start()会继续播放
public void stop () 停止播放
public void release () 如果不再使用 MediaPlayer，应该调用 release()释放所占用的资源；如果不释放资源，会因为产生过多的 MediaPlayer 对象实体而造成 Exception
public void reset () 重置 MediaPlayer 到尚未初始化状态，调用此方法后必须重新调用 setDataSource()配置数据源与调用 prepare()
public void setDataSource (String path) 设置欲播放的数据源 path：可以是文件路径，也可以是 http/rtsp 等 URL 路径
public void setAudioStreamType (int streamtype) 指定欲播放的流类型。必须在 prepare()之前调用此方法 streamtype：流类型，可以在 AudioManager 类中找到流类型列表
public void prepare () 准备播放，等到播放器准备完成即可开始播放。调用的时机不对会产生 IllegalStateException；准备失败会产生 IOException
public void setOnBufferingUpdateListener (MediaPlayer.OnBufferingUpdateListener listener) 注册 MediaPlayer.OnBufferingUpdateListener 监听器，当网络数据下载的的缓冲量改变时会调用 MediaPlayer.OnBufferingUpdateListener 的 onBufferingUpdate()
public void setOnCompletionListener (MediaPlayer.OnCompletionListener listener) 注册 OnCompletionListener 监听器，当指定的影音文件播放完毕时会调用 OnCompletionListener 的 onCompletion ()

(续表)

方　　法
public void setOnPreparedListener (MediaPlayer.OnPreparedListener listener) 注册 OnPreparedListener 监听器，当播放器准备好要播放指定的影音文件时会调用 OnPreparedListener 的 onPrepared ()
public void setOnVideoSizeChangedListener (MediaPlayer.OnVideoSizeChangedListener listener) 注册 OnVideoSizeChangedListener 监听器，当视频的大小改变时会调用 OnVideoSizeChangedListener 的 onVideoSizeChanged ()
public void setDisplay (SurfaceHolder sh) 指定 SurfaceHolder 来呈现视频画面，如果不调用此方法会导致只有声音而没有图像 sh：用来呈现视频画面的 SurfaceHolder

 范例 PlayResAudio

范例(如图 11-2 所示)说明：
- 按下"播放"按钮，播放项目内的音频文件。
- 按下"暂停"按钮，暂停播放音频文件。
- 按下"停止"按钮，停止播放音频文件。
- TextView 组件会呈现目前播放文件的所在地址。

图 11-2

PlayResAudio/src/org/playResAudio/PlayResAudio.java

```
10.     public class PlayResAudio extends Activity{
11.         private MediaPlayer mp; //播放器
12.         private TextView tvInfo; //消息框
```

```
13.    private Button btnPlay; // "播放"按钮
14.    private Button btnPause; // "暂停"按钮
15.    private Button btnStop; // "停止"按钮
16.    private boolean isStoped = true; //true 代表处于完全停止播放状态
17.
18.    @Override
19.    public void onCreate(Bundle savedInstanceState) {
20.        super.onCreate(savedInstanceState);
21.        setContentView(R.layout.main);
22.        findViews();
23.    }
24.
25.    private void findViews() {
26.        tvInfo = (TextView)findViewById(R.id.tvInfo);
27.        btnPlay = (Button) findViewById(R.id.btnPlay);
28.        btnPause = (Button) findViewById(R.id.btnPause);
29.        btnStop = (Button) findViewById(R.id.btnStop);
30.
31.        //按下"播放"按钮
32.        btnPlay.setOnClickListener(new View.OnClickListener() {
33.            public void onClick(View view) {
34.                if(mp == null || isStoped){
35.                    mp = MediaPlayer.create(PlayResAudio.this, R.raw.ring);
36.                    isStoped = false;
37.                }
38.
39.                tvInfo.setText(getString(R.string.audioSource) +
40.                        getString(R.raw.ring));
41.                mp.start();
42.            }
43.        });
44.
45.        //按下"暂停"按钮
46.        btnPause.setOnClickListener(new View.OnClickListener() {
47.            public void onClick(View view) {
48.                if(mp == null || isStoped)
49.                    return;
50.                mp.pause();
51.            }
52.        });
```

```
53.
54.          //按下"停止"按钮
55.          btnStop.setOnClickListener(new View.OnClickListener() {
56.              public void onClick(View view) {
57.                  if(mp == null || isStoped)
58.                      return;
59.                  mp.stop();
60.                  isStoped = true;
61.              }
62.          });
63.      }
64.
65.      @Override
66.      protected void onPause() {
67.          super.onPause();
68.          if (mp != null) {
69.              mp.release();
70.              mp = null;
71.          }
72.      }
73. }
```

34~37 行：按下"播放"按钮时检查 MediaPlayer 对象是否为 null 或是完全处于停止播放状态。如果是，创建 MediaPlayer 对象实体方便之后播放，并将 isStoped 状态设为 false，代表处于非停止播放状态。

39~40 行：TextView 组件会呈现目前播放文件的所在地址。

41 行：调用 start()即可播放指定的音频文件。

48~50 行：按下"暂停"按钮时检查 MediaPlayer 对象是否为 null 或是完全处于停止播放状态。如果是，则结束方法执行，不需要再调用 pause()暂停播放；如果不是，则调用 pause()暂停播放。

57~60 行：按下"停止"按钮时检查 MediaPlayer 对象是否为 null 或是已经完全处于停止播放状态。如果是，则结束方法执行，不需要再调用 stop()停止播放；如果不是，则调用 stop()停止播放并将 isStoped 设为 true，代表处于完全停止播放状态。

69~70 行：Activity 在 onPause 状态代表暂停，应该调用 release()释放 MediaPlayer 所占用的资源。

11.2.2 播放外部文件或网络数据流

移动设备播放的音频文件可以来自外部文件(例如：SD 卡)或是网络①，播放这类文件的步

---

① 支持的通讯协议为 RTSP (RTP, SDP)、HTTP。

多媒体与相机功能

骤如下(MediaPlayer 相关方法说明参见表 11-2)：

STEP 1 使用 MediaPlayer 类的构造函数来创建 MediaPlayer 的对象实体。

> MediaPlayer mp = *new MediaPlayer*();

STEP 2 调用 setDataSource()并搭配路径来指定欲播放的文件。

> // path 代表欲播放的文件地址，或网络数据流的来源 URL。
> mp.setDataSource(*path*);

STEP 3 先调用 prepare()，然后调用 start()开始播放。

> mp.prepare();
> mp.*start*();

STEP 4 如果调用 stop()停止播放后还想继续播放，必须调用 reset()与 prepare()，方可再调用 start()开始播放。

 范例 MP_Audio

范例(如图 11-3 所示)说明：

- 按"播放>>SD card Audio"按钮，播放 SD 卡内的音频文件(范例音频文件的路径为 Example\CH11\mediaFiles\ring.mp3，请自行复制至模拟器的 sdcard 目录)[①]。
- 按"播放>>网络 Audio"按钮，播放网络数据流。

图 11-3

---

① 复制文件至模拟器的方式参见 3.3 节。

- 按"停止"按钮，停止播放。
- TextView 组件会呈现目前播放的音频文件来源地址。

MP_Audio/src/org/mP_Audio/MP_Audio.java

```java
13.     public class MP_Audio extends Activity {
14.         private MediaPlayer mp; //播放器
15.         private TextView tvInfo; //消息框
16.         private Button btnSDAudio; // "播放>>SD card Audio" 按钮
17.         private Button btnURLAudio; // "播放>>网络 Audio" 按钮
18.         private Button btnStop; // "停止" 按钮
19.
20.         @Override
21.         public void onCreate(Bundle savedInstanceState) {
22.             super.onCreate(savedInstanceState);
23.             setContentView(R.layout.main);
24.             findViews();
25.         }
26.
27.         private void findViews() {
28.             tvInfo = (TextView)findViewById(R.id.tvInfo);
29.             btnSDAudio = (Button)findViewById(R.id.btnSDAudio);
30.             btnURLAudio = (Button)findViewById(R.id.btnURLAudio);
31.             btnStop = (Button)findViewById(R.id.btnStop);
32.
33.             //按下 "播放>>SD card Audio" 按钮
34.             btnSDAudio.setOnClickListener(new OnClickListener() {
35.                 @Override
36.                 public void onClick(View v) {
37.                     String path = "/sdcard/ring.mp3";
38.                     playAudio(path);
39.                 }
40.             });
41.
42.             //按下 "播放>>网络 Audio" 按钮
43.             btnURLAudio.setOnClickListener(new OnClickListener() {
44.                 @Override
45.                 public void onClick(View v) {
46.                     String path =
47.                         "http://sites.google.com/site/ronforwork/Home/android-2/ring.mp3";
48.                     playAudio(path);
49.                 }
50.             });
51.
52.             //按下 "停止" 按钮
53.             btnStop.setOnClickListener(new OnClickListener() {
54.                 @Override
```

多媒体与相机功能

```
55.            public void onClick(View v) {
56.                if (mp != null) {
57.                    mp.stop();
58.                }
59.            }
60.        });
61.    }
```

36~39 行：按下"播放>>SD card Audio"按钮，指定播放来源为/sdcard/ring.mp3，并调用 63 行的 playAudio() 将路径信息传入。

45~49 行：按下"播放>>网络 Audio"按钮，指定播放来源为 http://sites.google.com/site/ronforwork/Home/android-2/ring.mp3，并调用 playAudio() 将路径信息传入。

55~59 行：按下"停止"按钮，只要 MediaPlayer 对象不为 null，即调用 stop() 停止播放。

```
63.    private void playAudio(String path) {
64.        tvInfo.setText(getString(R.string.audioSource) + path);
65.        if(mp == null)
66.            mp = new MediaPlayer();
67.        try {
68.            mp.reset();
69.            mp.setDataSource(path);
70.            mp.setAudioStreamType(AudioManager.STREAM_MUSIC);
71.            mp.prepare();
72.            mp.start();
73.        } catch (Exception e) {
74.            Log.e("MP_Audio", e.toString());
75.        }
76.    }
77.
78.    @Override
79.    protected void onPause() {
80.        super.onPause();
81.        if (mp != null) {
82.            mp.release();
83.            mp = null;
84.        }
85.    }
86. }
```

63 行：调用 playAudio() 可以播放外部文件或网络数据流。

64 行：TextView 组件会呈现目前播放的音频文件来源地址。

65~66 行：只要 MediaPlayer 对象为 null，即创建一个新的 MediaPlayer。

68 行：重置 MediaPlayer。

69 行：设置欲播放文件的来源地址。

70 行：调用 setAudioStreamType() 指定欲播放的流类型。

71 行：调用 prepare() 准备播放，等到播放器准备完成即可开始播放。
72 行：调用 start() 开始播放。

11.3 播放视频文件

11.3.1 简易视频播放器

一个视频播放器除了要能呈现画面外，至少要具备播放、暂停、向前或向后快进等基本功能，如果开发人员不想为播放内容的呈现与播放功能而费神，不妨结合 VideoView 与 MediaController 类的功能制作出一个简易的多媒体播放器。这样的播放器不仅可以播放视频文件，也同样支持音频文件的播放，说明如下。

- VideoView：专门用来加载各种视频文件内容，与其他窗口组件一样，VideoView 可以直接采用布局文件来设置。
- MediaController：这是一个简易的播放界面，提供基本的影音播放操作功能。

范例 VideoViewEx

范例(如图 11-4 所示)说明：
- 视频文件属于 oopsdigital 公司所有。
- VideoView 呈现视频文件内容(范例视频文件路径为 Example\CH11\mediaFiles\little-Monster.3gp，请自行复制至模拟器的 sdcard 目录)。

图 11-4

- MediaController 提供简易播放界面，让用户可以播放、暂停、向前或向后快进视频或音频文件。
- 以模拟器播放视频文件较容易发生延迟状况，在实机上播放则会十分顺畅。

VideoViewEx/res/layout/main.xml

```xml
1.  <?xml version="1.0" encoding="utf-8"?>
2.  <LinearLayout xmlns:android="http://schemas.android.com/apk/res/android"
3.      android:orientation="vertical"
4.      android:layout_width="match_parent"
5.      android:layout_height="match_parent" >
6.  
7.      <VideoView
8.          android:id="@+id/vvScreen"
9.          android:layout_width="wrap_content"
10.         android:layout_height="wrap_content" />
11. </LinearLayout>
```

7 行：在布局文件中设置 VideoView 组件。

VideoViewEx/src/org/videoViewEx/VideoViewEx.java

```java
8.  public class VideoViewEx extends Activity {
9.      //播放 SD 卡中的视频文件
10.     private String path = "/sdcard/littleMonster.3gp";
11.     //播放网络视频
12.     //private String path =
13.     //    "http://sites.google.com/site/ronforwork/Home/android-2/littleMonster.3gp";
14.     private VideoView vvScreen;
15.     
16.     @Override
17.     public void onCreate(Bundle savedInstanceState) {
18.         super.onCreate(savedInstanceState);
19.         setContentView(R.layout.main);
20.         findViews();
21.         playVideo();
22.     }
23.     
24.     private void findViews() {
25.         vvScreen = (VideoView) findViewById(R.id.vvScreen);
26.     }
27.     
28.     private void playVideo() {
29.         vvScreen.setVideoPath(path);
30.         MediaController mController = new MediaController(this);
31.         vvScreen.setMediaController(mController);
32.         vvScreen.requestFocus();
33.         vvScreen.start();
```

```
34.      }
35.  }
```

10~13 行：视频文件可以来自 SD 卡，也可以来自网络。
29 行：调用 setVideoPath()设置视频文件路径。
30 行：创建 MediaController。
31 行：指定 VideoView 要使用的 MediaController。
32 行：调用 requestFocus()取得停驻点。
33 行：开始播放视频文件。

11.3.2　MediaPlayer 播放视频文件

MediaPlayer 类不仅提供播放音频功能，也提供播放视频功能，而且两者原理大致相同；

比较大的差异是在播放视频文件时必须先准备一个播放画面，让视频文件内容得以呈现。SurfaceView 致力于画面的呈现，支持高速绘图，也就是支持较高的 FPS(Frames Per Second，画面更新率)，所以很适合用来呈现视频的播放、拍照前的预览与游戏画面的切换。所以在下面的范例中将会以 SurfaceView 呈现视频文件的内容，至于 MediaPlayer 相关方法说明，则参见之前的表 11-2。

SurfaceView 提供专门用来快速绘图的界面，该界面内嵌在 SurfaceView 里面，而 SurfaceView 则负责配置与正确显示界面内容。虽然 SurfaceView 是主要呈现画面的地方，但控制界面运作的却是 SurfaceHolder，可以调用 SurfaceView 的 getHolder()来取得 SurfaceHolder 对象。SurfaceHolder 相关方法说明如表 11-3 所示。

表 11-3

SurfaceHolder 界面
public abstract void addCallback (SurfaceHolder.Callback callback) SurfaceHolder 向 Callback 对象注册，当 surface 状态改变时会自动调用该 Callback 对象所实现的方法[1] callback：实现 SurfaceHolder.Callback 的对象
public abstract void setType (int type)[2] 设置界面种类 type：参见 SurfaceHolder 的 surface type 常量
public abstract void setFixedSize (int width, int height) 将界面大小固定 ● width：界面宽度 ● height：界面高度

SurfaceHolder.Callback 接口定义了与界面生命周期有关的 3 个重要方法，实现这 3 个方法，就可以根据界面所处状态来运行对应的程序；换句话说，就可以适当地响应界面的状态改变。Callback 方法的说明如表 11-4 所示。

---

[1] 与第 4 章中 Button 组件的事件处理机制非常相似。
[2] 此方法在 Android 3.0 中被标记为逐渐淘汰，API 文件说明改由系统自动设置，而不需要开发人员设置。但是为了兼容性，以下范例仍会调用此方法，否则在 Android 2.3 版系统上会无法呈现画面。

表 11-4

SurfaceHolder.Callback 类

public abstract void surfaceCreated (SurfaceHolder holder)
当界面第一次被创建时会调用此方法
holder：新创建 surface 所对应的 SurfaceHolder

public abstract void surfaceChanged (SurfaceHolder holder, int format, int width, int height)
当界面改变时(例如：格式或大小的改变)会调用此方法。调用 surfaceCreated()后，至少会调用一次此方法
- holder：发生改变的界面所对应的 SurfaceHolder
- format：界面新的像素格式(PixelFormat)
- width：界面新的宽度
- height：界面新的高度

public abstract void surfaceDestroyed (SurfaceHolder holder)
当界面被破坏时(例如：切换到其他画面，则原画面的界面就会被破坏)会调用此方法
holder：被破坏的界面所对应的 SurfaceHolder

 范例 MP_Video

范例(如图 11-5 所示)说明：
- 视频文件来自 www.oopsdigital.com。
- 以 SurfaceView 内嵌的 surface 呈现视频内容(范例视频文件的路径为 Example\CH11\mediaFiles\littleMonster.3gp，请自行复制至模拟器的 sdcard 目录)。
- 一旦取得视频的大小，即将界面大小设置与视频大小相同。
- 切换到其他画面后再切换回来，会重新播放视频。
- 以模拟器播放视频文件较容易发生延迟状况，在实机上播放则会十分顺畅。

图 11-5

MP_Video/src/org/mP_Video/MP_Video.java

```java
15.     public class MP_Video extends Activity {
16.         private final String tag = getClass().getName(); //取得类名称当作日志使用的标签
17.         private int videoWidth; //视频宽度
18.         private int videoHeight; //视频高度
19.         private MediaPlayer mp; //播放器
20.         private SurfaceView svScreen; //SurfaceView 组件，用来呈现视频画面
21.         private SurfaceHolder sHolder; //控制界面的 SurfaceHolder 组件
22.         private String path = "/sdcard/littleMonster.3gp"; //视频路径
23.         private boolean isVideoSizeKnown = false; //是否取得视频大小
24.         private boolean isVideoReady = false; //视频是否准备妥当
25.
26.         class SurfaceCallback implements SurfaceHolder.Callback{
27.             @Override
28.             public void surfaceCreated(SurfaceHolder holder) {
29.                 initMediaPlayer();
30.             }
31.
32.             @Override
33.             public void surfaceChanged(SurfaceHolder holder,
34.                     int format, int width, int height) {
35.                 Log.d(tag, "surfaceChanged called");
36.             }
37.
38.             @Override
39.             public void surfaceDestroyed(SurfaceHolder surfaceholder) {
40.                 releaseMediaPlayer();
41.                 doCleanUp();
42.             }
43.         }
44.
45.         @Override
46.         public void onCreate(Bundle savedInstanceState) {
47.             super.onCreate(savedInstanceState);
48.             setContentView(R.layout.main);
49.             findViews();
50.         }
51.
52.         private void findViews() {
53.             svScreen = (SurfaceView) findViewById(R.id.svScreen);
54.             sHolder = svScreen.getHolder();
55.             sHolder.addCallback(new SurfaceCallback());
```

```
56.        }
57.    }
```

26 行：SurfaceCallback 在 28、33、39 行实现 Callback 接口的 3 个方法，方法定义的说明请自行参见表 11-4，不再赘述。

28~30 行：当创建界面时，会调用 initMediaPlayer()，初始化 MediaPlayer。

33 行：当界面改变时会调用此方法。

39~42 行：当画面切换时，会调用 releaseMediaPlayer() 释放 MediaPlayer 所占用的资源，以及调用 doCleanUp() 清除相关变量的值。

54 行：调用 getHolder() 取得 SurfaceHolder 对象。

55 行：调用 addCallback() 向 SurfaceCallback 对象注册，当界面状态改变时会自动调用 26 行中的 SurfaceCallback 类所实现的对应方法。

```
59.    private void initMediaPlayer() {
60.        doCleanUp();
61.        if(mp == null)
62.            mp = new MediaPlayer();
63.        try {
64.            mp = new MediaPlayer();
65.            mp.setDataSource(path);
66.            mp.setDisplay(sHolder);
67.            mp.setAudioStreamType(AudioManager.STREAM_MUSIC);
68.            mp.prepare();
69.            mp.setOnBufferingUpdateListener(new OnBufferingUpdateListener() {
70.                @Override
71.                public void onBufferingUpdate(MediaPlayer mp, int percent) {
72.                    Log.d(tag, "onBufferingUpdate percent:" + percent);
73.                }
74.            });
75.            mp.setOnCompletionListener(new OnCompletionListener() {
76.                @Override
77.                public void onCompletion(MediaPlayer mp) {
78.                    Log.d(tag, "onCompletion called");
79.                }
80.            });
81.            mp.setOnPreparedListener(new OnPreparedListener() {
82.                @Override
83.                public void onPrepared(MediaPlayer mp) {
84.                    isVideoReady = true;
85.                    if (isVideoSizeKnown) {
86.                        playVideo();
87.                    }
88.                }
89.            });
90.            mp.setOnVideoSizeChangedListener(new OnVideoSizeChangedListener()
```

```
91.         @Override
92.         public void onVideoSizeChanged(
93.                 MediaPlayer mp, int width, int height) {
94.             if (width == 0 || height == 0) {
95.                 Log.e(tag,
96.                     "invalid video width(" + width + ") or height(" + height
97.                     + ")");
98.                 return;
99.             }
100.            isVideoSizeKnown = true;
101.            videoWidth = width;
102.            videoHeight = height;
103.            if (isVideoReady) {
104.                playVideo();
105.            }
106.        }
107.    });
108. } catch (Exception e) {
109.     Log.e(tag, e.toString());
110. }
111. }
```

60 行：调用 doCleanUp() 会清除相关变量的值。

66 行：调用 setDisplay() 指定 SurfaceHolder 来呈现视频画面。

69、75、81、90 行的 setOnXXXListener 方法请自行参见表 11-2 对应方法的说明，不再赘述。

83~88 行：当播放器准备好要播放时会调用 onPrepared()，将 isVideoReady 设为 true，并检查 isVideoSizeKnown 来确定是否已经取得视频的大小。必须在播放器准备好且取得视频的大小时才能调用 playVideo() 开始播放。

92~106 行：当视频的大小改变时会调用 onVideoSizeChanged()，先检查视频的宽或高是否为 0。如果为 0，代表没有正确取得视频的大小，执行 return 结束方法执行。如果正确取得视频大小，除了将 isVideoSizeKnown 设置为 true，同时将取得的宽与高分别设置给 video- Width 与 videoHeight。最后检查 isVideoReady 来确定视频是否准备妥当来决定是否调用 playVideo() 开始播放。

```
113. private void playVideo() {
114.     sHolder.setFixedSize(videoWidth, videoHeight);
115.     mp.start();
116. }
117.
118. private void releaseMediaPlayer() {
119.     if (mp != null) {
120.         mp.release();
121.         mp = null;
122.     }
```

```
123.        }
124.
125.        private void doCleanUp() {
126.            videoWidth = 0;
127.            videoHeight = 0;
128.            isVideoReady = false;
129.            isVideoSizeKnown = false;
130.        }
131.    }
```

113~116 行：调用 playVideo()会先调用 setFixedSize()将界面大小固定，然后再调用 start()开始播放。

118~123 行：调用 releaseMediaPlayer()会先检查 MediaPlayer 是否为 null，如果不为 null，则调用 release()释放 MediaPlayer 所占用的资源，然后再将其设为 null。

125~130 行：调用 doCleanUp()会重设 videoWidth、videoHeight、isVideoReady、isVideoSizeKnown 等变量，使其恢复到初始值。

11.4 录制音频文件

需要录音的情况十分常见，例如演讲或上课内容录音、开会内容录音等。最早期的录音方式为随身听加磁带，之后录音笔问世便成为主流，现在我们可以将录音功能整合到 Android 移动设备上。要编写录音程序，必须使用 MediaRecorder 类的功能，此类也提供录像的功能。虽然 Android developers 官方网站说明模拟器没有录音或录像的设备，所以暂时还不支持此类功能[①]。但笔者装上麦克风后测试的结果，证实至少在 Android 2.2 版以后的模拟器已经支持录音功能，只是效果不太好。

要编写具有录音功能的应用程序，就必须允许应用程序使用录音功能；如果将录音文件存放在 SD 卡内，则必须允许应用程序将文件存放在移动设备外部的存储媒体，所以必须在清单文件中作以下设置：

<uses-permission android:name="*android.permission.RECORD_AUDIO*" />
<uses-permission android:name="android.permission.WRITE_EXTERNAL_STORAGE" />

录音功能会比播放功能复杂一些，但也不会太难，请按照下列步骤执行：

STEP 1 使用 MediaRecorder 类的默认构造函数(default constructor)创建 MediaRecorder 对象实体。

MediaRecorder mr = *new MediaRecorder()*;

---

① 参见 http://developer.android.com/guide/topics/media/index.html 的 Audio and Video 部分。

STEP 2 调用 setAudioSource()指定录音来源。如果是麦克风，可以使用 MediaRecorder.AudioSource.MIC。

mr.*setAudioSource*(MediaRecorder.AudioSource.MIC);

STEP 3 调用 setOutputFormat()设置录音文件的输出格式(例如：3GPP)。

mr.*setOutputFormat*(MediaRecorder.OutputFormat.THREE_GPP);

STEP 4 调用 MediaRecorder 的 setAudioEncoder()设置录音的编码方式(例如：AMR_NB)。

mr.*setAudioEncoder*(MediaRecorder.AudioEncoder.AMR_NB);

STEP 5 调用 setOutputFile()设置录音文件的存放位置(例如存放在 SD 卡上)。

mr.*setOutputFile*("/sdcard/audio01.3gp");

STEP 6 调用 prepare()准备录音。

```
mr.prepare();
```

STEP 7 调用 start()开始录音。

mr.*start*();

STEP 8 调用 stop()结束录音。

mr.*stop*();

STEP 9 调用 reset()可以重复使用已经产生的 MediaRecorder 对象，但必须回到 Step 2 重新设置。

mr.*reset*();

STEP 10 调用 release()会立即释放 MediaRecorder 占用的资源。之后如果想要再使用 MediaRecorder 功能，必须重新创建 MediaRecorder 对象实体而无法重复使用已经被释放的 MediaRecorder 对象。

mr.release();

以上 MediaRecorder 类的常用方法详细说明如表 11-5 所示。

表 11-5

构 造 函 数
public MediaRecorder()
默认构造函数，用来创建 MediaRecorder 对象实体，可以用来录音或录像

(续表)

方法
public void setAudioSource(int audio_source) 指定录音来源设备。调用此方法必须在 setOutputFormat()之前 audio_source：录音来源；一般会设置成麦克风(MediaRecorder.AudioSource.MIC)
public void setVideoSource(int video_source) 指定录像来源设备。调用此方法必须在 setOutputFormat()之前 video_source：录像来源；一般会设置成 Camera(MediaRecorder.VideoSource.CAMERA)
public void setOutputFormat(int output_format) 设置录音/录像文件的输出格式。调用此方法必须在 setAudioSource()/setVideoSource()之后，但在 prepare()之前 output_format：录音/录像文件的输出格式。无论使用 AMR 对音频文件编码或是使用 H.263 对视频文件编码，都建议使用 3GP 格式。如果使用 MPEG-4 格式，可能会让一般 PC 的播放器无法播放
public void setAudioEncoder(int audio_encoder) 设置录音的编码方式。调用此方法必须在 setOutputFormat()之后，但在 prepare()之前 audio_encoder：录音使用的编码方式，目前可选择的编码方式有：AAC、AMR_NB、AMR_WB 共 3 种；其中 AAC 与 AMR_WB 编码在 API Level 10(Android 2.3.3)中才开始支持
public void setOutputFile(String path) 设置录音/录像文件的存放位置。调用此方法必须在 setOutputFormat()之后，但在 prepare()之前 path：文件所在路径
public void prepare() 准备录音/录像。调用此方法必须在 setAudioSource()、setOutputFormat()、setAudioEncoder()、setOutputFile()等方法之后，但在 start()之前。在 start()之后或 setOutputFormat()之前调用会产生 IllegalStateException；准备失败会产生 IOException
public void start() 开始录音/录像。调用完 prepare()才可调用此方法
public void stop() 结束录音/录像；若想再次启动录音功能，可以调用 reset()
public void reset() 将已停止的录音/录像状态重新启动，调用此方法可以重复利用原有的 MediaRecorder 对象，但必须像产生新的 MediaRecorder 对象一般重新设置。如果已调用 release()，则不可调用 reset()，因为已无法重复利用原有的 MediaRecorder 对象
public void release() 确定不再使用 MediaRecorder 对象来录音/录像，可以调用此方法释放 MediaRecorder 对象占用的资源，此时不可重复使用该 MediaRecorder 对象
public void setCamera (Camera c) 设置 Camera 为可以录像状态，调用此方法必须在 prepare()之前 c：准备用来录像的 Camera 对象
public void setProfile (CamcorderProfile profile) 应用录像配置文件来录像。调用此方法必须在 setVideoSource()/setAudioSource()之后，但在 setOutputFile()之前 profile：CamcorderProfile 配置文件

(续表)

方　　法
public void setMaxDuration (int max_duration_ms) 限制最长的录制时间 max_duration_ms：持续时间，单位为毫秒。如果为 0 或负数，代表不设限制
public void setMaxFileSize (long max_filesize_bytes) 限制录像文件的大小 max_filesize_bytes：文件大小，单位为字节。如果为 0 或负数，代表不设限制

 范例 MR_Audio

范例(如图 11-6 和图 11-7 所示)说明：

- 按下"录音"按钮即可开始录音，此时会显示"录音中..."，而且"录音"按钮会呈现无法使用状态(disable)，确保用户无法再次按下"录音"按钮直到按下"停止录音"按钮结束录音为止，如图 11-6 所示。
- 按下"停止录音"按钮会结束录音，并显示录音存储路径，而且会将"录音"按钮恢复成可使用状态(enable)，让用户可以再次录音，如图 11-7 所示。
- 录音完成后，可以按下"播放"按钮播放录音文件。

图 11-6

图 11-7

MR_Audio/src/org/mR_Audio/MR_Audio.java

```
16.     public class MR_Audio extends Activity {
17.         private final String tag = getClass().getName();
18.         private MediaRecorder mr; //录音器
19.         private TextView tvInfo; //消息框
20.         private Button btnRecord; // "录音"按钮
21.         private Button btnStopRecord; // "停止录音"按钮
22.         private Button btnPlay; // "播放"按钮
23.         private Button btnStopPlay; // "停止播放"按钮
```

```java
24.     private String path = "/sdcard/audio01.3gp"; //录音存储路径
25.     private MediaPlayer mp; //播放器
26.
27.     @Override
28.     public void onCreate(Bundle savedInstanceState) {
29.         super.onCreate(savedInstanceState);
30.         setContentView(R.layout.main);
31.         findViews();
32.     }
33.
34.     private void findViews() {
35.         tvInfo = (TextView)findViewById(R.id.tvInfo);
36.         btnRecord = (Button) findViewById(R.id.btnRecord);
37.         btnStopRecord = (Button) findViewById(R.id.btnStopRecord);
38.         btnPlay = (Button) findViewById(R.id.btnPlay);
39.         btnStopPlay = (Button) findViewById(R.id.btnStopPlay);
40.
41.         //按下"录音"按钮
42.         btnRecord.setOnClickListener(new OnClickListener() {
43.             @Override
44.             public void onClick(View view) {
45.                 if(!isSDExist()){
46.                     Toast.makeText(MR_Audio.this,
47.                         R.string.SDCardNotFound, Toast.LENGTH_LONG).show();
48.                     return;
49.                 }
50.                 btnRecord.setEnabled(false);
51.                 recordAudio();
52.             }
53.         });
54.
55.         //按下"停止录音"按钮
56.         btnStopRecord.setOnClickListener(new OnClickListener() {
57.             @Override
58.             public void onClick(View view) {
59.                 if(mr != null){
60.                     mr.stop();
61.                     mr.release();
62.                     mr = null;
63.                     tvInfo.setText(getString(R.string.filePath) + path);
64.                     btnRecord.setEnabled(true);
65.                 }
66.             }
67.         });
68.
69.         //按下"播放"按钮
```

```java
70.         btnPlay.setOnClickListener(new OnClickListener() {
71.             @Override
72.             public void onClick(View v) {
73.                 playAudio(path);
74.             }
75.         });
76.
77.         //按下"停止播放"按钮
78.         btnStopPlay.setOnClickListener(new OnClickListener() {
79.             @Override
80.             public void onClick(View v) {
81.                 if (mp != null) {
82.                     mp.stop();
83.                 }
84.             }
85.         });
86.     }
```

44~52 行：按下"录音"按钮先检查是否可以正常读取 SD 卡，如果可以，便禁用"录音"按钮，防止用户再次按下，接着调用 recordAudio()开始录音。

58~66 行：按下"停止录音"按钮，调用 stop()停止录音，接着调用 release()释放 MediaRecorder 所占用的资源，然后将录音文件所存放的位置显示在文本框上，并且启用"录音"按钮，代表用户可以再次录音。

72~74 行：按下"播放"按钮会调用 playAudio()开始播放录制好的音频文件。

80~84 行：按下"停止播放"按钮会调用 stop()停止播放。

```java
88.     private void recordAudio() {
89.         if(mr == null)
90.             mr = new MediaRecorder();
91.         else
92.             mr.reset();
93.         try {
94.             mr.setAudioSource(MediaRecorder.AudioSource.MIC);
95.             mr.setOutputFormat(MediaRecorder.OutputFormat.THREE_GPP);
96.             mr.setAudioEncoder(MediaRecorder.AudioEncoder.AMR_NB);
97.             mr.setOutputFile(path);
98.             mr.prepare();
99.             mr.start();
100.        } catch (Exception e) {
101.            Log.e(tag, e.toString());
102.        }
103.        tvInfo.setText(R.string.recording);
104.    }
105.
```

```
106.    private boolean isSDExist() {
107.        String state = Environment.getExternalStorageState();
108.        if (state.equals(Environment.MEDIA_MOUNTED))
109.            return true;
110.        else
111.            return false;
112.    }
113.
114.    private void playAudio(String path) {
115.        tvInfo.setText(getString(R.string.audioSource) + path);
116.        if(mp == null)
117.            mp = new MediaPlayer();
118.        try {
119.            mp.reset();
120.            mp.setDataSource(path);
121.            mp.setAudioStreamType(AudioManager.STREAM_MUSIC);
122.            mp.prepare();
123.            mp.start();
124.        } catch (Exception e) {
125.            Log.e(tag, e.toString());
126.        }
127.    }
128.
129.    @Override
130.    protected void onPause() {
131.        super.onPause();
132.        if (mp != null) {
133.            mp.release();
134.            mp = null;
135.        }
136.    }
137. }
```

88 行：调用 recordAudio() 会打开录音功能。

89~92 行：如果 MediaRecorder 不为 null，创建一个全新的 MediaRecorder；否则调用 reset() 重复利用已存在的 MediaRecorder。

94 行：设置麦克风为录音来源设备。

95 行：设置录音文件的输出格式为 3GPP。

96 行：设置录音的编码方式为 AMR_NB。

97 行：将录音文件存放在指定位置。

98 行：准备录音。

99 行：开始录音。

103 行：显示"录音中..."。

106~112 行：调用 isSDExist()会检查 SD 卡是否处于可访问状态，如果是，返回 true，否则返回 false。

114~127 行：调用 playAudio()会按照指定的路径播放录制好的音频文件。播放音频文件功能已于 11.2 节中说明，不再赘述。

11.5 拍照功能

Android 实机几乎都配有摄像头，Android 2.3 版更开始支持访问多个摄像头(multiple cameras)的功能。可以自行编写应用程序来操控摄像头拍照，也可以直接利用移动设备本身内置的相机应用程序来获取图像。Android 4.0 版以后的模拟器开始可以通过 webcam 的镜头来模拟拍照功能。

自行编写相机应用程序需要花费相当多的时间在编写程序上，因为要将功能写得完整并不简单；而且需要在各种实机上测试，以确保可以顺利操控各实机的摄像头。使用移动设备内置的相机应用程序来获取图像就没有上述缺点，而且因为几乎所有 Android 移动设备都有内置相机应用程序，所以不用担心无法拍照。

如果应用程序需要依赖摄像头，建议在清单文件内添加<uses-feature android:name= "android.hardware.camera" />[①]，这样 Google Play 才会发挥过滤功能，限制只有具备摄像头的移动设备才可以看见以及安装该应用程序。

```
<manifest ... >
    <uses-feature android:name="android.hardware.camera" />
    ...
</manifest ... >
```

创建一个方法来检查有没有应用程序可以运行拍照操作，如果有，则数量会大于 0。

```
public boolean isIntentAvailable(Context context, String action) {
    PackageManager packageManager = context.getPackageManager();
    Intent intent = new Intent(action);
    List<ResolveInfo> list = packageManager.queryIntentActivities(intent,
        PackageManager.MATCH_DEFAULT_ONLY);
    return list.size() > 0;
}
```

要利用内置相机应用程序拍照，就必须设置 Intent 做拍照操作——ACTION_IMAGE_CAPTURE；调用 startActivityForResult()打开 Activity 执行拍照并传送请求码(request code)。

[①] 关于<uses-feature>的设置列表，参见 http://developer.android.com/guide/topics/manifest/uses-feature-element.html#features-reference。

多媒体与相机功能

```
Intent intent = new Intent(MediaStore.ACTION_IMAGE_CAPTURE);
startActivityForResult(intent, requestCode);
```

执行完毕 startActivityForResult()会自动调用 onActivityResult()，此时可以调用 Bundle 的 get ("data")取得所拍的照片，以 ImageView 显示并自行编写存储程序。

```
protected void onActivityResult(int requestCode, int resultCode, Intent data) {
    super.onActivityResult(requestCode, resultCode, data);
    if (resultCode == RESULT_OK) {
        switch (requestCode) {
            case TAKE_PICTURE:
                Bitmap image = (Bitmap) data.getExtras().get("data");
                imageView.setImageBitmap(image);
                saveImage(image);
        }
    } else {
        return;
    }
}
```

 范例 TakePhotoEx

范例说明：

- 按下图 11-8 右上角的"拍照"按钮(Android 3.0 之前版本中则必须按 Menu 键)会启动移动设备内置的相机应用程序，如图 11-9 所示。

图 11-8

- 拍完照后回到图 11-8，会在 TextView 上显示存储位置，并在其下的 ImageView 呈现该图片[①]。

[①] 使用模拟器拍照可能会发生预览画面与呈现在 ImageView 中的相反的情况，但在实机测试中却没有发生这种情况。

347

图 11-9

范例创建步骤如下：

STEP 1　在 res/layout/main.xml 文件内新增 TextView 与 ImageView，分别用来呈现拍照后的文件存储路径与照片。

```
<TextView
    android:id="@+id/textView"
    android:layout_width="wrap_content"
    android:layout_height="wrap_content"
    android:layout_gravity="center"
    android:text="@string/filePath" />

<ImageView
    android:id="@+id/imageView"
    android:layout_width="match_parent"
    android:layout_height="match_parent"
    android:layout_gravity="center"
    android:src="@drawable/ic_launcher" />
```

STEP 2　创建 res/menu/cameramenu.xml，设置 ActionBar 的按钮。

```
<menu xmlns:android="http://schemas.android.com/apk/res/android" >

    <item
        android:id="@+id/takePicture"
        android:showAsAction="ifRoom|withText"
        android:icon="@drawable/ic_action_photo"
        android:title="@string/takePicture"/>

</menu>
```

STEP 3　在 AndroidManifest.xml 文件内添加下列程序代码。

多媒体与相机功能

```xml
<!-- 因为会访问 SD 卡，所以要设置 WRITE_EXTERNAL_STORAGE 权限 -->
<uses-permission android:name="android.permission.WRITE_EXTERNAL_STORAGE" />

<!-- 标识此应用程序需要使用相机设备 -->
<uses-feature android:name="android.hardware.camera" />
```

STEP 4 在 TakePhotoActivity.java 文件内添加下列程序代码。

```java
public class TakePhotoActivity extends Activity {
    private ImageView imageView;
    private TextView textView;
    private static final int TAKE_PICTURE = 1;

    @Override
    public void onCreate(Bundle savedInstanceState) {
        super.onCreate(savedInstanceState);
        setContentView(R.layout.main);
        findViews();
    }

    private void findViews() {
        textView = (TextView)findViewById(R.id.textView);
        imageView = (ImageView)findViewById(R.id.imageView);
    }

    @Override
    public boolean onCreateOptionsMenu(Menu menu) {
        MenuInflater inflater = getMenuInflater();
        inflater.inflate(R.menu.cameramenu, menu);
        return true;
    }

    @Override
    public boolean onOptionsItemSelected(MenuItem item) {
        switch (item.getItemId()) {
        case R.id.takePicture:
            if (isIntentAvailable(this, MediaStore.ACTION_IMAGE_CAPTURE)) {
                /* 如果要调用其他应用程序来做事，必须通过 Intent 并指定要做的事情(例如：
                拍照) */
                Intent intent = new Intent(MediaStore.ACTION_IMAGE_CAPTURE);
                /* 调用此方法并设置 requestCode 会打开一个 Activity 执行 Intent 所指定
                的操作，执行完毕会自动调用 onActivityResult() */
                startActivityForResult(intent, TAKE_PICTURE);
            } else {
                Toast.makeText(this, R.string.noCameraApp, Toast.LENGTH_SHORT).show();
            }
```

349

```java
            return true;
        default:
            return super.onOptionsItemSelected(item);
    }
}

/* 检查是否有默认应用程序可以处理指定的操作。在此主要检查是否有默认的相机应用程序 */
public boolean isIntentAvailable(Context context, String action) {
    PackageManager packageManager = context.getPackageManager();
    Intent intent = new Intent(action);
    /* 在此检查是否有默认应用程序可以执行拍照操作，如果有，则数量会大于 0 */
    List<ResolveInfo> list = packageManager.queryIntentActivities(intent,
            PackageManager.MATCH_DEFAULT_ONLY);
    return list.size() > 0;
}

@Override
protected void onActivityResult(int requestCode, int resultCode, Intent data) {
    super.onActivityResult(requestCode, resultCode, data);
    /* 操作成功，则 resultCode 为 RESULT_OK */
    if (resultCode == RESULT_OK) {
        switch (requestCode) {
            case TAKE_PICTURE:
                /* 利用内置相机程序拍照，调用 get()并搭配 data 这个键即可取得照
                片(Bitmap 类型) */
                Bitmap image = (Bitmap) data.getExtras().get("data");
                /* 使用 ImageView 呈现照片 */
                imageView.setImageBitmap(image);
                saveImage(image);
        }
    } else {
        return;
    }
}

/* 存储 Bitmap 图形 */
private void saveImage(Bitmap image) {
    if (!isSDExist()) {
        Toast.makeText(this, R.string.SDCardNotFound, Toast.LENGTH_LONG)
                .show();
        return;
    }

    /* 取得外部公开目录——pictures 的路径 */
```

```java
        File imageDir = Environment
                .getExternalStoragePublicDirectory(Environment.DIRECTORY_PICTURES);
        if (!imageDir.exists()) {
            imageDir.mkdirs();
        }

        File imageFile = new File(imageDir, "image01.jpg");
        try {
            /* BufferedOutputStream 使用缓冲区功能会加快写入速度 */
            BufferedOutputStream bos = new BufferedOutputStream(
                    new FileOutputStream(imageFile));
            /* 采用 JPEG 压缩，质量为 90% */
            image.compress(Bitmap.CompressFormat.JPEG, 90, bos);
            bos.flush();
            bos.close();
            textView.setText(getString(R.string.filePath) + imageFile.toString());
        } catch (IOException ioe) {
            Log.e(getPackageName(), ioe.toString());
        }
        String[] paths = { imageFile.toString() };
        callMediaScanner(paths);
    }

    /* 检查 SD 卡的状态。MEDIA_MOUNTED 代表可以访问 SD 卡，返回 true；否则返回 false */
    private boolean isSDExist() {
        String state = Environment.getExternalStorageState();
        if (state.equals(Environment.MEDIA_MOUNTED))
            return true;
        else
            return false;
    }

    /* 启动媒体扫描仪，扫描 paths 所代表的路径是否有多媒体文件，如果有，则创建该文件的连接，
     * 让其他应用程序(如相片集(Gallery))可以知道有新文件添加 */
    private void callMediaScanner(String[] paths) {
        MediaScannerConnection.scanFile(this, paths, null,
                new MediaScannerConnection.OnScanCompletedListener() {
                    public void onScanCompleted(String path, Uri uri) {
                        Log.i("ExternalStorageEx", "Scanned " + path + ":");
                        Log.i("ExternalStorageEx", "-> uri=" + uri);
                    }
                });
    }
}
```

11.6 录像功能

除了可以使用移动设备内置的相机应用程序拍照外,也可以用来录像。因为录像也需要依赖摄像头,所以建议在清单文件内也添加<uses-feature android:name="android.hardware.camera" />设置。其他录像步骤大致与拍照步骤相同。

创建一个方法以检查有没有应用程序可以运行录像操作,如果有,则数量会大于 0。

```java
public boolean isIntentAvailable(Context context, String action) {
    PackageManager packageManager = context.getPackageManager();
    Intent intent = new Intent(action);
    List<ResolveInfo> list = packageManager.queryIntentActivities(intent,
        PackageManager.MATCH_DEFAULT_ONLY);
    return list.size() > 0;
}
```

要利用内置相机应用程序录像,就必须设置 Intent 做录像操作——ACTION_VIDEO_CAPTURE;调用 startActivityForResult()打开 Activity 执行录像并传送请求码(request code)。

```java
Intent intent = new Intent(MediaStore.ACTION_VIDEO_CAPTURE);
startActivityForResult(intent, requestCode);
```

执行完毕 startActivityForResult()会自动调用 onActivityResult(),此时可以调用 Intent 的 getData()取得录像文件的 Uri,以 VideoView 显示并自行编写存储程序。

```java
protected void onActivityResult(int requestCode, int resultCode, Intent data) {
    super.onActivityResult(requestCode, resultCode, data);
    if (resultCode == RESULT_OK) {
        switch (requestCode) {
            case TAKE_VIDEO:
                Uri uri = data.getData();
                MediaController controller = new MediaController(this);
                videoView.setMediaController(controller);
                videoView.setVideoURI(uri);
                videoView.start();
                saveVideo(uri);
        }
    } else {
        return;
    }
}
```

范例 TakeVideoEx

范例说明：
- 按下图 11-10 右上角的"录像"按钮(在 Android 3.0 之前版本中则必须按 Menu 键)会启动移动设备内置的相机应用程序。
- 录像完毕[①]后会回到图 11-10，在 TextView 上显示存储位置，并在其下的 VideoView 中播放该影片。

图 11-10

范例创建步骤如下：

STEP 1 在 res/layout/main.xml 文件内新增 TextView 呈现录像后的文件存储路径；VideoView 则播放录像结果。

```
<TextView
    android:id="@+id/textView"
    android:layout_width="wrap_content"
    android:layout_height="wrap_content"
    android:layout_gravity="center"
    android:text="@string/filePath" />

<VideoView
    android:id="@+id/videoView"
    android:layout_width="wrap_content"
    android:layout_height="wrap_content"
    android:layout_gravity="center"/>
```

STEP 2 创建 res/menu/cameramenu.xml，设置 ActionBar 的按钮。

```
<menu xmlns:android="http://schemas.android.com/apk/res/android" >
```

---

① 使用模拟器录像可能会发生无法录像的情况，但实机测试却没有发生这种情况。

```xml
<item
    android:id="@+id/takeVideo"
    android:showAsAction="ifRoom|withText"
    android:icon="@drawable/ic_action_video"
    android:title="@string/takeVideo"/>
</menu>
```

STEP 3 在 AndroidManifest.xml 文件内添加下列程序代码。

```xml
<!-- 因为会访问 SD 卡，所以要设置 WRITE_EXTERNAL_STORAGE 权限 -->
<uses-permission android:name="android.permission.WRITE_EXTERNAL_STORAGE" />

<!-- 标识此应用程序需要使用相机设备 -->
<uses-feature android:name="android.hardware.camera" />
```

STEP 4 在 TakeVideoActivity.java 文件内添加下列程序代码。

```java
public class TakeVideoActivity extends Activity {
    private VideoView videoView;
    private TextView textView;
    private static final int TAKE_VIDEO = 2;

    @Override
    public void onCreate(Bundle savedInstanceState) {
        super.onCreate(savedInstanceState);
        setContentView(R.layout.main);
        findViews();
    }

    private void findViews() {
        textView = (TextView) findViewById(R.id.textView);
        videoView = (VideoView) findViewById(R.id.videoView);
    }

    @Override
    public boolean onCreateOptionsMenu(Menu menu) {
        MenuInflater inflater = getMenuInflater();
        inflater.inflate(R.menu.cameramenu, menu);
        return true;
    }

    @Override
    public boolean onOptionsItemSelected(MenuItem item) {
        switch (item.getItemId()) {
            case R.id.takeVideo:
                if (isIntentAvailable(this, MediaStore.ACTION_VIDEO_CAPTURE)) {
```

```java
            /* 如果要调用其他应用程序来做事，必须通过 Intent 并指定要做的事情(例如：
            录像) */
            Intent intent = new Intent(MediaStore.ACTION_VIDEO_CAPTURE);
            /* 调用此方法并设置 requestCode 会打开一个 Activity，执行 Intent 所指定
            的操作，
             * 执行完毕会自动调用 onActivityResult() */
            startActivityForResult(intent, TAKE_VIDEO);
        } else {
            Toast.makeText(this, R.string.noCameraApp, Toast.LENGTH_SHORT)
                    .show();
        }
        return true;

    default:
        return super.onOptionsItemSelected(item);
    }

}

/* 检查是否有默认应用程序可以处理指定的操作。在此主要检查是否有默认的相机应用程序 */
public boolean isIntentAvailable(Context context, String action) {
    PackageManager packageManager = context.getPackageManager();
    Intent intent = new Intent(action);
    /* 在此检查是否有默认应用程序可以执行录像操作，如果有，则数量会大于 0 */
    List<ResolveInfo> list = packageManager.queryIntentActivities(intent,
            PackageManager.MATCH_DEFAULT_ONLY);
    return list.size() > 0;
}

@Override
protected void onActivityResult(int requestCode, int resultCode, Intent data) {
    super.onActivityResult(requestCode, resultCode, data);
    /* 操作成功，则 resultCode 为 RESULT_OK */
    if (resultCode == RESULT_OK) {
        switch (requestCode) {
        case TAKE_VIDEO:
            /* 利用内置相机程序录像，调用 getData()取得录像文件的 Uri */
            Uri uri = data.getData();
            MediaController controller = new MediaController(this);
            videoView.setMediaController(controller);
            /* 使用 VideoView 播放影片 */
            videoView.setVideoURI(uri);
            videoView.start();
            saveVideo(uri);
        }
```

```java
    } else {
        return;
    }
}

/* 存储录像 */
private void saveVideo(Uri uri) {
    if (!isSDExist()) {
        Toast.makeText(this, R.string.SDCardNotFound, Toast.LENGTH_LONG)
                .show();
        return;
    }
    /* 取得外部公开目录——影片的路径 */
    File videoDir = Environment
            .getExternalStoragePublicDirectory(Environment.DIRECTORY_MOVIES);
    if (!videoDir.exists()) {
        videoDir.mkdirs();
    }

    /* 通常内置相机应用程序录像完毕后会自动存成录像文件并产生该文件的 Uri,
     * 下面的程序就是要读出 Uri 所指定文件的内容, 再写入 videoFile 所指定的文件内 */
    String videoPathFromUri = getRealPathFromURI(uri);
    File videoFile = new File(videoDir, "video01.3gp");
    try {
        FileInputStream fis = new FileInputStream(videoPathFromUri);
        FileOutputStream fos = new FileOutputStream(videoFile);
        byte[] buffer = new byte[1024];
        int byteCount;
        while ((byteCount = fis.read(buffer)) != -1) {
            fos.write(buffer, 0, byteCount);
        }
        fis.close();
        fos.close();
        textView.setText(getString(R.string.filePath)
                + videoFile.toString());
    } catch (IOException ioe) {
        Log.e(getPackageName(), ioe.toString());
    }
    String[] paths = { videoFile.toString() };
    callMediaScanner(paths);
}

/* 检查 SD 卡的状态。MEDIA_MOUNTED 代表可以访问 SD 卡, 返回 true; 否则返回 false */
private boolean isSDExist() {
```

```java
        String state = Environment.getExternalStorageState();
        if (state.equals(Environment.MEDIA_MOUNTED))
            return true;
        else
            return false;
    }

    /* 按照 Uri 取得录像存储路径 */
    public String getRealPathFromURI(Uri uri) {
        /* DATA 字段存储内置相机程序存储录像的路径 */
        String[] columns = { MediaStore.Images.Media.DATA };
        ContentResolver contentResolver = getContentResolver();
        /* 提供 Uri 与域名向系统查询对应数据，会返回 Cursor 以方便取得结果集合 */
        Cursor cursor = contentResolver.query(uri, columns, null, null, null);
        /* 按照域名取得对应字段索引，如果该字段不存在，会产生 IllegalArgumentException */
        int column_index = cursor
                .getColumnIndexOrThrow(MediaStore.Images.Media.DATA);
        cursor.moveToFirst();
        return cursor.getString(column_index);
    }

    /* 启动媒体扫描仪，扫描 paths 所代表的路径是否有多媒体文件，
     * 如果有，则创建该文件的连接 */
    private void callMediaScanner(String[] paths) {
        MediaScannerConnection.scanFile(this, paths, null,
                new MediaScannerConnection.OnScanCompletedListener() {
                    public void onScanCompleted(String path, Uri uri) {
                        Log.i("ExternalStorageEx", "Scanned " + path + ":");
                        Log.i("ExternalStorageEx", "-> uri=" + uri);
                    }
                });
    }
}
```

第 12 章
手机实用功能开发

本章学习目标：

- 手机铃声设置
- 手机音量与振动的设置
- 来短信与来电处理
- 查询联系人数据

12.1 手机铃声设置

手机铃声(ringtones)可以说是个性的表现,相信每个拥有手机的人都思考过应该选择怎样的铃声来代表自己。而手机铃声泛指手机上的各种铃声,但最重要的不外乎是电话铃声(来电时电话的响铃声音)、通知铃声(接收到信息时发出的通知声音)与闹钟铃声(闹钟时间一到所发出的声音)。用户若想选择自己首选的铃声或改变音量,甚至设置静音、振动,都可以通过下列步骤来实现(以模拟器为例):在主画面中按 Menu 按键,选择"设置">"音效",会显示音效设置列表,如图 12-1 所示。这一节会探讨各种手机铃声的选择;下一节则说明各种音量的调节,甚至包括静音与振动的设置。

一般而言,手机铃声会放在手机内存的 system/media/audio 目录内;电话铃声、通知铃声与闹钟铃声会分别存放在 audio 目录的 ringtones、notifications、alarms 这三个子目录中。使用实机测试铃声不会有问题,因为手机的内存内已经存放了对应的铃声,不过模拟器的内存内却没有存放任何铃声,所以无法测试。有两种解决方式:

(1) 在 SD 卡内置创建相同路径的目录(sdcard/media/audio/ringtones、notifications、alarms),并将铃声文件复制至对应目录中,如图 12-2 所示。

图 12-1

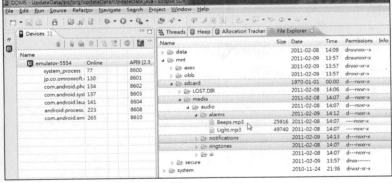

图 12-2

要达到此目的,最快速的方式就是将 PC 上完整的 media 目录直接拖动到模拟器的 SD 卡上,如图 12-3 所示。虽然复制到模拟器的 SD 卡而非内存内[①],Android 系统仍然可以加载铃声,但是建议复制完毕后重新启动模拟器,以便 Android 系统可以加载新的铃声文件。

(2) 将铃声复制至 SD 卡的任一目录内。使用 Android 系统内置的音乐播放程序,在"播放列表"内找到铃声文件后久按该文件会弹出选项,选择"设成来电铃声",如图 12-4 所示。

---

① 无法将文件直接复制至设备(包含模拟器)的内存内,因为该空间属于只读状态。

手机实用功能开发

图 12-3　　　　　　　　　　　　　　　　　　图 12-4

要让用户选择铃声，必须在画面上显示对应的铃声选择器(ringtone picker)以方便用户选择，创建铃声选择器的步骤如下：

STEP 1　创建 Intent 并将 action 设置成 ACTION_RINGTONE_PICKER。

Intent intent = new Intent(ACTION_RINGTONE_PICKER);

STEP 2　完成铃声选择器的选项详细设置[①]，采用默认与个性化的差异可参见图 12-5 和图 12-6。

- 设置铃声选择器的标题文本：默认标题为"铃声"，可更改成"电话铃声"。

intent.putExtra(EXTRA_RINGTONE_TITLE, "电话铃声");

- 取消"静音"选项：默认有"静音"选项，可以取消。

intent.putExtra(EXTRA_RINGTONE_SHOW_SILENT, false);

- 设置当前铃声(或称为默认铃声)为默认状态：未设置前，"静音"选项为默认状态；设置后，当前铃声(例如：Alpha 铃声)成为默认状态，并且多了"默认铃声"选项。

intent.putExtra(EXTRA_RINGTONE_EXISTING_URI,
　　getActualDefaultRingtoneUri(RingtoneEx.this, TYPE_RINGTONE));
// getActualDefaultRingtoneUri()会取得当前默认铃声的实际位置

图 12-5　　　　　　　　　　　　　图 12-6

---

① 如果想要了解还有哪些选项详细设置，参见 API 文件中关于 RingtoneManager.ACTION_RINGTONE_PICKER 的详细说明。

STEP 3 调用 startActivityForResult()打开指定的铃声选择器。

startActivityForResult(intent, requestCode);

STEP 4 用户选完铃声按下"确定"按钮会自动调用 onActivityResult()，所以必须改写 onActivityResult()以取得用户选择的铃声，并调用 setActualDefaultRingtoneUri()将该铃声设置成当前铃声。

```
@Override
protected void onActivityResult(int requestCode, int resultCode, Intent intent) {
    Uri uri = intent.getParcelableExtra(EXTRA_RINGTONE_PICKED_URI);
    RingtoneManager.setActualDefaultRingtoneUri(context, TYPE_RINGTONE,
        uri);
}
```

由此可知，startActivityForResult()与 onActivityResult()是一组交互的方法，startActivityForResult()可说是调用者(caller)，而 onActivityResult()是一个接收者(receiver)。

STEP 5 调用 setActualDefaultRingtoneUri()设置铃声必须在清单文件内添加 WRITE_SETTINGS 权限。

```
<uses-permission android:name="android.permission.WRITE_SETTINGS" />
```

创建铃声选择器所需使用到的相关方法说明如表 12-1 所示。

表 12-1

RingtoneManager 类
常　　量
public static final String ACTION_RINGTONE_PICKER Activity 的 action 设置：在画面上显示铃声选择器
public static final int TYPE_RINGTONE 指定铃声种类为电话铃声
public static final int TYPE_NOTIFICATION 指定铃声种类为通知铃声
public static final int TYPE_ALARM 指定铃声种类为闹钟铃声
public static final String EXTRA_RINGTONE_TITLE 铃声选择器的标题
public static final String EXTRA_RINGTONE_SHOW_SILENT 铃声选择器是否显示"静音"选项
public static final String EXTRA_RINGTONE_EXISTING_URI 在铃声选择器上设置当前铃声为默认状态，并且添加"默认铃声"选项
public static final String EXTRA_RINGTONE_TYPE 设置铃声选择器显示的铃声种类，若不设置，则默认为电话铃声

(续表)

RingtoneManager 类

常　量

public static final String EXTRA_RINGTONE_DEFAULT_URI
设置铃声选择器的"默认铃声"选项实际对应的铃声 URI，若不设置，当用户选择"默认铃声"选项时，一律为当前的电话铃声

public static final String EXTRA_RINGTONE_PICKED_URI
返回用户选定铃声的 URI

构 造 函 数

public RingtoneManager (Activity activity)
创建 RingtoneManager 对象
activity：当前 Activity

方　法

public static Ringtone getRingtone (Context context, Uri ringtoneUri)
根据指定的 URI 返回 Ringtone 对象
context：Context 对象，通常为当前 Activity
ringtoneUri：铃声的 URI

public static Uri getActualDefaultRingtoneUri (Context context, int type)
根据指定的铃声种类取得该铃声种类的当前铃声
context：Context 对象，通常为当前 Activity
type：铃声种类

public static void setActualDefaultRingtoneUri (Context context, int type, Uri ringtoneUri)
将指定铃声设置为默认铃声。需要 WRITE_SETTINGS 权限
- context：Context 对象，通常为当前 Activity
- type：铃声种类
- ringtoneUri：铃声的 URI

public void stopPreviousRingtone ()
停止最后播放的铃声

Ringtone 类

public int getStreamType ()
取得铃声的流类型

public String getTitle (Context context)
取得铃声的标题文本
context：Context 对象，通常为当前 Activity

public boolean isPlaying ()
检查铃声是否在播放中。如果是，则返回 true；否则返回 false

public void play ()
播放铃声

public void stop ()
停止播放铃声

(续表)

Ringtone 类
Settings.System 类
public static final Uri DEFAULT_ALARM_ALERT_URI 取得当前闹钟铃声的 URI
public static final Uri DEFAULT_NOTIFICATION_URI 取得当前通知铃声的 URI
public static final Uri DEFAULT_RINGTONE_URI 取得当前电话铃声的 URI
Activity 类
常　　量
public static final int RESULT_OK 属于 Activity 标准结果，代表操作成功。例如：选好铃声后按下"确定"按钮就会传递 RESULT_OK 结果
public static final int RESULT_CANCELED 属于 Activity 标准结果，代表取消操作。例如：按下"取消"按钮就会传递 RESULT_CANCELED 结果
方　　法
public void startActivityForResult (Intent intent, int requestCode) 打开一个 Activity，该 Activity 结束后会自动调用 onActivityResult()，并传递请求码(request code)以方便 onActivityResult()接收到之后可以作为识别的根据 ● intent：打开指定 Activity 所需使用的 Intent ● requestCode：请求码
protected void onActivityResult (int requestCode, int resultCode, Intent data) 调用 startActivityForResult()打开 Activity，当 Activity 结束后会自动调用 onActivityResult()，onActivity-Result()会接收到传递过来的请求码、结束码与 Intent 对象 ● requestCode：传递过来的请求码 ● resultCode：传递过来的结束码 ● data：传递过来的 Intent 对象

 范例 RingtoneEx

范例(如图 12-7 所示)说明：
- TextView 文本框显示各种铃声的当前铃声。
- 按下"电话铃声"按钮，会弹出电话铃声选择器，列出可以选择的电话铃声，并将当前电话铃声设置为默认状态。
- 按下"通知铃声"按钮，会弹出通知铃声选择器，列出可以选择的通知铃声，并将当前通知铃声设置为默认状态。
- 按下"闹钟铃声"按钮，会弹出闹钟铃声选择器，列出可以选择的闹钟铃声，并将当前闹钟铃声设置为默认状态。

手机实用功能开发

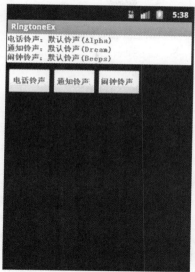

图 12-7

RingtoneEx/src/org/ringtoneEx/RingtoneEx.java

```
3.    import static android.media.RingtoneManager.*;
4.    import static android.provider.Settings.System.*;
5.    import android.app.Activity;
6.    import android.content.Intent;
7.    import android.media.Ringtone;
8.    import android.media.RingtoneManager;
9.    import android.net.Uri;
10.   import android.os.Bundle;
11.   import android.view.View;
12.   import android.view.View.OnClickListener;
13.   import android.widget.Button;
14.   import android.widget.TextView;
```

3~4 行：因为本范例会使用许多 RingtoneManager、Settings.System 的类常量，所以使用 import static 功能，之后使用这些类常量时可省略类名称。

```
17.   private TextView tvInfo; //显示各种类的当前铃声
18.   private Button btnRingtone; // "电话铃声"按钮
19.   private Button btnNotification; // "通知铃声"按钮
20.   private Button btnAlarm; // "闹钟铃声"按钮
21.   private final static int req_ringtone = 0; //代表电话铃声选择器
22.   private final static int req_notification = 1; //代表通知铃声选择器
23.   private final static int req_alarm = 2; //代表闹钟铃声选择器
24.   private RingtoneManager ringtoneMgr;
25.
```

365

```
26.    @Override
27.    public void onCreate(Bundle savedInstanceState) {
28.        super.onCreate(savedInstanceState);
29.        setContentView(R.layout.main);
30.        ringtoneMgr = new RingtoneManager(this);
31.        findViews();
32.    }
```

30 行：创建 RingtoneManager 对象。

```
46.    private void showRingtone() {
47.        Ringtone phone = getRingtone(this, DEFAULT_RINGTONE_URI);
48.        Ringtone notification = getRingtone(this, DEFAULT_NOTIFICATION_URI);
49.        Ringtone alarm = getRingtone(this, DEFAULT_ALARM_ALERT_URI);
50.        StringBuilder sbInfo = new StringBuilder();
51.        sbInfo.append(getString(R.string.ringtone_title) + ": ");
52.        if(phone != null)
53.            sbInfo.append(phone.getTitle(this));
54.        sbInfo.append("\n");
55.        sbInfo.append(getString(R.string.notification_title) + ": ");
56.        if(notification != null)
57.            sbInfo.append(notification.getTitle(this));
58.        sbInfo.append("\n");
59.        sbInfo.append(getString(R.string.alarm_title) + ": ");
60.        if(alarm != null)
61.            sbInfo.append(alarm.getTitle(this));
62.        tvInfo.setText(sbInfo);
63.    }
```

46 行：执行此方法会在 TextView 组件上显示各类铃声信息。

47~49 行：分别取得各个种类的当前铃声。

51~62 行：如果有铃声被设置成静音，则该铃声对应的 Ringtone 对象会为 null，为了避免调用 getTitle()时产生 NullPointerException，先检查 Ringtone 对象是否为 null，如果不为 null，再调用 getTitle()取得该铃声的标题(铃声名称)。最后再将铃声信息显示在 tvInfo 所代表的 TextView 组件上。

```
65.    class OnClickHandler implements OnClickListener{
66.        @Override
67.        public void onClick(View v) {
68.            Intent intent = new Intent(ACTION_RINGTONE_PICKER);
69.            //按下"电话铃声"按钮
70.            if(v == btnRingtone){
71.                intent.putExtra(EXTRA_RINGTONE_TITLE,
72.                    getString(R.string.ringtone_title));
```

```
73.            intent.putExtra(EXTRA_RINGTONE_SHOW_SILENT, false);
74.            intent.putExtra(EXTRA_RINGTONE_EXISTING_URI,
75.                    getActualDefaultRingtoneUri(
76.                            RingtoneEx.this, TYPE_RINGTONE));
77.            startActivityForResult(intent, req_ringtone);
78.        }
79.        //按下"通知铃声"按钮
80.        else if(v == btnNotification){
81.            intent.putExtra(EXTRA_RINGTONE_TITLE,
82.                    getString(R.string.notification_title));
83.            intent.putExtra(EXTRA_RINGTONE_TYPE, TYPE_NOTIFICATION);
84.            intent.putExtra(EXTRA_RINGTONE_DEFAULT_URI,
85.                    DEFAULT_NOTIFICATION_URI);
86.            intent.putExtra(EXTRA_RINGTONE_SHOW_SILENT, false);
87.            intent.putExtra(EXTRA_RINGTONE_EXISTING_URI,
88.                    getActualDefaultRingtoneUri(
89.                            RingtoneEx.this, TYPE_NOTIFICATION));
90.            startActivityForResult(intent, req_notification);
91.        }
92.        //按下"闹钟铃声"按钮
93.        else{
94.            intent.putExtra(EXTRA_RINGTONE_TITLE,
95.                    getString(R.string.alarm_title));
96.            intent.putExtra(EXTRA_RINGTONE_TYPE, TYPE_ALARM);
97.            intent.putExtra(EXTRA_RINGTONE_DEFAULT_URI,
98.                    DEFAULT_ALARM_ALERT_URI);
99.            intent.putExtra(EXTRA_RINGTONE_SHOW_SILENT, false);
100.           intent.putExtra(EXTRA_RINGTONE_EXISTING_URI,
101.                   getActualDefaultRingtoneUri(
102.                           RingtoneEx.this, TYPE_ALARM));
103.           startActivityForResult(intent, req_alarm);
104.       }
105.   }
106. }
```

68 行：创建 Intent 并将 action 设置成 ACTION_RINGTONE_PICKER，代表将在画面上显示铃声选择器。

71~72 行：设置铃声选择器的标题文本。

73 行：取消"静音"选项。

74~76 行：在铃声选择器上将当前电话铃声设置成默认状态，并且添加"默认铃声"选项。

77 行：打开指定的铃声选择器等待用户选择，用户选择完毕后会自动调用下面 109 行中的 onActivityResult()。

83 行：设置铃声选择器显示的铃声种类为通知铃声，若不设置，则默认为电话铃声。

84~85 行：将通知铃声的"默认铃声"选项设置为当前通知铃声，若不设置，则"默认铃声"选项代表的是当前电话铃声而非通知铃声，这样一来用户就会觉得奇怪，因为现在选择的种类明明是通知铃声，但单击"默认铃声"出来的声音却是电话铃声，这样不就变成张冠李戴了吗？可以将此两行删除后试试看便知问题出在哪里。

77、90、103 行：按照按下的按钮不同，传递不同的请求码，方便之后执行 onActivityResult() 时作为识别之用。

```
108.    @Override
109.    protected void onActivityResult(int requestCode, int resultCode, Intent intent) {
110.        if(resultCode != RESULT_OK) return;
111.        Uri uri = intent.getParcelableExtra(EXTRA_RINGTONE_PICKED_URI);
112.        if(uri != null){
113.            switch(requestCode){
114.                case req_ringtone:
115.                    setActualDefaultRingtoneUri(this, TYPE_RINGTONE, uri);
116.                    break;
117.                case req_notification:
118.                    setActualDefaultRingtoneUri(this, TYPE_NOTIFICATION, uri);
119.                    break;
120.                case req_alarm:
121.                    setActualDefaultRingtoneUri(this, TYPE_ALARM, uri);
122.                    break;
123.            }
124.            showRingtone();
125.        }
126.    }
127.
128.    @Override
129.    protected void onPause() {
130.        super.onPause();
131.        ringtoneMgr.stopPreviousRingtone();
132.    }
```

110 行：resultCode != RESULT_OK 代表用户没有成功选择铃声，所以执行 return;直接结束 onActivityResult()。

111 行：返回用户所选择铃声的 URI，方便之后调用 setActualDefaultRingtoneUri()将该铃声设置成当前铃声。

112~125 行：如果铃声的 URI 不为 null(代表铃声选择成功)，则使用 switch-case 比较请求码来判断当初用户选择哪一种类的铃声，并调用 setActualDefaultRingtoneUri()将该铃声设置成指定种类的默认铃声。例如用户在电话铃声选择器中选了 A 铃声，则该铃声就会成为电话铃声的当前铃声。

131 行：停止正在播放的铃声。

12.2 手机音量与振动的设置

除了前一节提及的铃声外,手机还有音乐、系统以及来电声音(电话中对方讲话的声音)。不同种类的声音都有自己专属的流类型;如果要调整这些声音的音量,必须在调用 AudioManager 类的 adjustStreamVolume() 同时指定流类型[1]与音量大小[2];除此之外,还可以指定标志(flag)[3]来达到特殊要求。如果要取得当前音量与最大音量,可以分别调用 AudioManager 的 getStreamVolume()与 getStreamMaxVolume()。

手机有 3 种铃声模式(ringer mode),分别是正常模式、静音模式与振动模式[4],可以调用 AudioManager 类的 setRingerMode()来设置铃声模式。如果要取得当前铃声模式,可以调用 AudioManager 的 getRingerMode()。也可以直接让手机振动,只要调用 Vibrator 类的 vibrate()并指定振动持续的时间,即可立即让手机振动。要打开手机振动器的功能,必须在清单文件内增加<uses-permission android:name="android.permission.VIBRATE" />设置。

以上相关类的说明如表 12-2 所示。

表 12-2

AudioManager 类
常　　量
public static final int STREAM_RING 电话铃声流
public static final int STREAM_NOTIFICATION 通知铃声流
public static final int STREAM_ALARM 闹钟铃声流
public static final int STREAM_MUSIC 音乐流
public static final int STREAM_SYSTEM 系统声音流
public static final int STREAM_VOICE_CALL 来电声音流
public static final int ADJUST_LOWER 减小音量

---

① 例如:STREAM_RING(电话铃声流)、STREAM_MUSIC(音乐流)、STREAM_SYSTEM(系统声音流)等,参见 API 文件中关于 AudioManager 类常量的说明。
② 例如:ADJUST_LOWER(减小音量)、ADJUST_RAISE(加大音量),参见 API 文件中关于 AudioManager 类常量的说明。
③ 例如:FLAG_PLAY_SOUND(调整音量时会播放声音)、FLAG_SHOW_UI(调整音量时会以 Toast 来显示现在的音量)、FLAG_VIBRATE(进入震动模式时手机会震动一下),参见 API 文件中关于 AudioManager 类常量的说明。
④ 正常模式(RINGER_MODE_NORMAL)、静音模式(RINGER_MODE_SILENT)、震动模式(RINGER_MODE_VIBRATE),参见 API 文件中关于 AudioManager 类常量的说明。

(续表)

AudioManager 类
常　　量
public static final int ADJUST_RAISE 加大音量
public static final int FLAG_PLAY_SOUND 调节音量时会播放内置音效，让用户知道现在的声音大小
public static final int FLAG_SHOW_UI 弹出 Toast 消息框来显示现在的音量大小
public static final int FLAG_VIBRATE 进入振动模式时手机会振动一下，让用户知道现在为振动状态。这个标志设置仅适用于铃声流，而不适用于音乐与系统声音流，这是因为用户应该不会希望音乐是以振动来呈现的
public static final int RINGER_MODE_NORMAL 正常模式
public static final int RINGER_MODE_SILENT 静音模式
public static final int RINGER_MODE_VIBRATE 振动模式
方　　法
public void adjustStreamVolume (int streamType, int direction, int flags) 根据指定的流类型与音量大小来调整音量 ● streamType：流类型 ● direction：调整音量的大小 ● flags：标志，可以设置特殊功能如 FLAG_PLAY_SOUND(调节音量时会播放声音)、FLAG_SHOW_UI(调整音量时会以 Toast 来显示现在音量)。如果想要应用两个以上标志的功能，可以使用加法运算，例如： FLAG_PLAY_SOUND + FLAG_SHOW_UI
public int getStreamVolume (int streamType) 取得特定流类型的当前音量 streamType：流类型
public int getStreamMaxVolume (int streamType) 取得特定流类型的最大音量 streamType：流类型
public void setRingerMode (int ringerMode) 设置铃声模式 ringerMode：欲设置的铃声模式
public int getRingerMode () 取得当前铃声模式
Vibrator 类
public void vibrate (long milliseconds) 让手机振动 milliseconds：振动持续的时间，单位为毫秒

范例 AudioMgrEx

范例(如图 12-8 所示)说明：
- TextView 文本框显示各种类声音的音量以及铃声模式。
- 选择声音后按下"大声"或"小声"按钮，可以调整该声音的音量。在调整过程中会弹出 Toast 显示音量，并播放对应的声音，让用户了解当前音量大小。
- 最下面 3 个图标单选按钮分别代表铃声模式的正常、静音与振动。选择"振动"时，手机会振动一下。

图 12-8

AudioMgrEx/src/org/audioMgrEx/AudioMgrEx.java

```
22.      private TextView tvInfo; //显示音量与静音、振动状态
23.      private RadioGroup rgVolume, rgVibrate;
24.      private RadioButton rbRingtone; //铃声
25.      private RadioButton rbMusic; //音乐
26.      private RadioButton rbSystem; //系统声音
27.      private RadioButton rbNormal; //正常
28.      private RadioButton rbSilent; //静音
29.      private RadioButton rbVibrate; //振动
30.      private Button btnRaise, btnLower; // "大声"、"小声"按钮
31.      private Ringtone testTone;
32.      private int streamType, direction;
33.      private AudioManager audioMgr;
34.      private Vibrator vibrator;
35.      private MediaPlayer mp;
36.
37.      @Override
```

```
38.     public void onCreate(Bundle savedInstanceState) {
39.         super.onCreate(savedInstanceState);
40.         setContentView(R.layout.main);
41.         audioMgr = (AudioManager)getSystemService(Context.AUDIO_SERVICE);
42.         vibrator = (Vibrator)getSystemService(Context.VIBRATOR_SERVICE);
43.         findViews();
44.     }
```

42 行：取得控制手机振动器的 Vibrator 对象。

```
46.     private void findViews() {
47.         tvInfo = (TextView)findViewById(R.id.tvInfo);
48.         showInfo();
49.
50.         OnCheckedChangeListener rgVolumeListener = new OnCheckedChangeListener() {
51.             @Override
52.             public void onCheckedChanged(RadioGroup group, int checkedId) {
53.                 if(testTone != null && testTone.isPlaying())
54.                     testTone.stop();
55.                 if(mp != null && mp.isPlaying())
56.                     mp.stop();
57.
58.                 if(rbRingtone.isChecked()){
59.                     testTone = getRingtone(
60.                         AudioMgrEx.this, DEFAULT_RINGTONE_URI);
61.                     streamType = STREAM_RING;
62.                 }
63.                 else if(rbMusic.isChecked()){
64.                     mp = MediaPlayer.create(AudioMgrEx.this, R.raw.ring);
65.                     streamType = STREAM_MUSIC;
66.                 }
67.                 else{
68.                     streamType = STREAM_SYSTEM;
69.                 }
70.             }
71.         };
72.         rgVolume = (RadioGroup)findViewById(R.id.rgVolume);
73.         rgVolume.setOnCheckedChangeListener(rgVolumeListener);
74.         rbRingtone = (RadioButton)findViewById(R.id.rbRingtone);
75.         rbMusic = (RadioButton)findViewById(R.id.rbMusic);
76.         rbSystem = (RadioButton)findViewById(R.id.rbSystem);
77.         rbRingtone.setChecked(true);
78.
79.         OnClickListener btnListener = new OnClickListener() {
```

```
80.        @Override
81.        public void onClick(View v) {
82.            //按下"大声"按钮
83.            if(v == btnRaise)
84.                direction = ADJUST_RAISE;
85.
86.            //按下"小声"按钮
87.            else
88.                direction = ADJUST_LOWER;
89.
90.            if(rbRingtone.isChecked()){
91.                audioMgr.adjustStreamVolume(streamType, direction,
92.                    FLAG_SHOW_UI + FLAG_VIBRATE);
93.                if(testTone != null && !testTone.isPlaying())
94.                    testTone.play();
95.            }
96.            else if(rbMusic.isChecked()){
97.                audioMgr.adjustStreamVolume(streamType, direction,
98.                    FLAG_SHOW_UI);
99.                if(mp != null && !mp.isPlaying())
100.                   mp.start();
101.           }
102.           else{
103.               audioMgr.adjustStreamVolume(streamType, direction,
104.                   FLAG_SHOW_UI + FLAG_PLAY_SOUND);
105.           }
106.           showInfo();
107.       }
108.   };
109.   btnRaise = (Button)findViewById(R.id.btnRaise);
110.   btnRaise.setOnClickListener(btnListener);
111.   btnLower = (Button)findViewById(R.id.btnLower);
112.   btnLower.setOnClickListener(btnListener);
```

50~52 行：以匿名内部类实现 OnCheckedChangeListener 的 onCheckedChanged()，当同一组的 RadioButton 选项改变时会自动调用此方法。

53~56 行：一旦改变选项，便停止之前正在播放的铃声或音乐。

58~62 行：如果选择"铃声"选项，准备好要播放的当前铃声，并设置声音流为铃声。

63~66 行：如果选择"音乐"选项，准备好要播放的音乐，并设置声音流为音乐。

67~69 行：如果选择"系统声音"选项，设置声音流为系统声音。不准备要播放的声音文件是因为要使用系统内置声音。

73 行：RadioGroup 对象——rgVolume 注册 OnCheckedChangeListener，当所属 RadioButton 选项改变时会自动调用 52 行中的 onCheckedChanged()。

77 行：将"铃声"选项设置为选择状态，这会触发选项改变事件而自动调用 52 行中的 onCheckedChanged()。

83~88 行：按下"大声"按钮就加大音量；按下"小声"按钮就减小音量。

91 行：调用 adjustStreamVolume()调整音量。

92 行：FLAG_SHOW_UI 代表调整音量时会弹出 Toast 显示当前音量；FLAG_VIBRATE 代表进入振动模式时手机会振动一下。

93~94 行：如果未播放铃声，开始播放。

99~100 行：如果未播放音乐，开始播放。

104 行：FLAG_SHOW_UI 代表调整音量时会弹出 Toast 显示当前音量；FLAG_PLAY_SOUND 代表会播放内置音效。

```
114.        rbNormal = (RadioButton)findViewById(R.id.rbNormal);
115.        rbSilent = (RadioButton)findViewById(R.id.rbSilent);
116.        rbVibrate = (RadioButton)findViewById(R.id.rbVibrate);
117.        switch(audioMgr.getRingerMode()){
118.            case RINGER_MODE_NORMAL:
119.                rbNormal.setChecked(true);
120.                break;
121.            case RINGER_MODE_SILENT:
122.                rbSilent.setChecked(true);
123.                break;
124.            case RINGER_MODE_VIBRATE:
125.                rbVibrate.setChecked(true);
126.                break;
127.        }
128.
129.        OnCheckedChangeListener rgVibrateListener = new OnCheckedChangeListener() {
130.            @Override
131.            public void onCheckedChanged(RadioGroup group, int checkedId) {
132.                //选择"正常"选项
133.                if(rbNormal.isChecked()){
134.                    audioMgr.setRingerMode(RINGER_MODE_NORMAL);
135.                }
136.                //选择"静音"选项
137.                else if(rbSilent.isChecked()){
138.                    audioMgr.setRingerMode(RINGER_MODE_SILENT);
139.                }
140.                //选择"振动"选项
141.                else{
142.                    audioMgr.setRingerMode(RINGER_MODE_VIBRATE);
143.                    vibrator.vibrate(200);
144.                }
145.                showInfo();
146.            }
147.        };
```

```
148.        rgVibrate = (RadioGroup)findViewById(R.id.rgVibrate);
149.        rgVibrate.setOnCheckedChangeListener(rgVibrateListener);
150.    }
```

117~127 行：利用 switch-case 检查当前铃声模式，并将对应的选项设置为默认状态。例如：铃声状态为静音，就自动选择"静音"选项。

129~131 行：以匿名内部类实现 OnCheckedChangeListener 的 onCheckedChanged()，当同一组的 RadioButton 选项改变时会自动调用此方法，其概念与 50～52 行相同。

133~144 行：根据选择的铃声模式，将手机设置成对应的模式。

143 行：会让手机振动 200 毫秒。

```
152.    private void showInfo() {
153.        int volume_ringtone = audioMgr.getStreamVolume(STREAM_RING);
154.        int maxVolume_ringtone = audioMgr.getStreamMaxVolume(STREAM_RING);
155.        int volume_music = audioMgr.getStreamVolume(STREAM_MUSIC);
156.        int maxVolume_music = audioMgr.getStreamMaxVolume(STREAM_MUSIC);
157.        int volume_system = audioMgr.getStreamVolume(STREAM_SYSTEM);
158.        int maxVolume_system = audioMgr.getStreamMaxVolume(STREAM_SYSTEM);
159.        StringBuilder sbInfo = new StringBuilder();
160.        sbInfo.append(getString(R.string.curVol_maxVol) + "\n" +
161.            getString(R.string.ringtone) + ": " +
162.                volume_ringtone + "/" + maxVolume_ringtone + "\n" +
163.            getString(R.string.music) + ": " +
164.                volume_music + "/" + maxVolume_music + "\n" +
165.            getString(R.string.system) + ": " +
166.                volume_system + "/" + maxVolume_system + "\n");
167.
168.        sbInfo.append(getString(R.string.ringerMode) + ": ");
169.        switch(audioMgr.getRingerMode()){
170.            case RINGER_MODE_NORMAL:
171.                sbInfo.append(getString(R.string.normal));
172.                break;
173.            case RINGER_MODE_SILENT:
174.                sbInfo.append(getString(R.string.silent));
175.                break;
176.            case RINGER_MODE_VIBRATE:
177.                sbInfo.append(getString(R.string.vibrate));
178.                break;
179.        }
180.        tvInfo.setText(sbInfo);
181.    }
```

153~158 行：取得指定流类型的当前音量与最大音量。

169~180 行：根据 switch-case 比较结果将当前铃声模式显示在 TextView 组件上。

12.3 来短信与来电处理

手机收到短信或有来电时，都会发出 Broadcast(广播)，如果想要对该 Broadcast 做出响应，必须拦截来短信与来电对应的 action(来短信的 action 为"android.provider.Telephony.SMS_RECEIVED"；来电的 action 为"android.intent.action.PHONE_STATE")，并注册指定的 BroadcastReceiver，将响应置于改写的 onReceive()内。这些观念在 6.4 节中已经说明，不在此赘述。在此要说明的是如何在接收到来短信与来电时取出相关信息(例如：来短信内容与来电号码)。发出的 Broadcast 内含有 Intent 对象，而 Intent 上的 Bundle 对象就存储着来短信或来电的相关信息，分别说明如下：

1. 来短信内容

一封短信如果是 8 位字符(1 个字符以 8 位存储，半角字即属此种字符)，最长可为 140 个字符；16 位字符(1 个字符以 16 位存储，全角字即属此种字符)最长可为 70 个字符。短信内容大都以 PDU(Protocol Description Unit)格式存储，而该格式并非仅有短信内容，还包含发送者(sender)和时间戳(time stamp)等信息。取得 PDU 格式的数据后解析成各个相关信息的方式说明如下：

(1) 取得 bundle 对象内存储的 PDU 数据并调用 SmsMessage.createFromPdu()，将 PDU 数据转换成 SmsMessage 对象。

```
/* 返回数组是因为短信内容过长时会变成两封以上短信，所以需要数组来存储 */
Object[] pdus = (Object[])bundle.get("pdus");
//将 PDU 数据转换成 SmsMessage 对象
SmsMessage[] smsMsgs = new SmsMessage[pdus.length];
for (int i = 0; i < pdus.length; i++)
    smsMsgs[i] = SmsMessage.createFromPdu((byte[])pdus[i]);
```

(2) 调用 SmsMessage 的 getDisplayMessageBody()取得短信内容[①]；调用 getDisplayOriginatingAddress()取得发信地址；调用 getTimestampMillis()取得信息时间，但单位为毫秒，所以常利用 Date 构造函数转换成 Date 对象。

```
String msgBody = smsMsgs[0].getDisplayMessageBody();
String sender = smsMsgs[0].getDisplayOriginatingAddress();
Date time = new Date(smsMsgs[0].getTimestampMillis());
```

2. 来电内容

来电比来短信简单，因为少了短信的内容。来电时 Broadcast 的 Intent action 为"android.intent.action.PHONE_STATE"[②]，从来电中可以取得 2 种信息：手机状态与来电号码，说明如下：

(1) 在清单文件内注册 BroadcastReceiver，并拦截手机状态改变的 action。

[①] 短信内容若有中文，在模拟器上会呈现乱码，实机则无此问题。
[②] TelephonyManager 的常量 ACTION_PHONE_STATE_CHANGED 值为"android.intent.action.PHONE_STATE"。

手机实用功能开发

```xml
<receiver android:name="PhoneReceiver">
    <intent-filter>
        <action android:name="android.intent.action.PHONE_STATE" />
    </intent-filter>
</receiver>
```

(2) 要取得手机状态，必须在清单文件内添加 READ_PHONE_STATE 权限。

```xml
<uses-permission android:name="android.permission.READ_PHONE_STATE" />
```

(3) 调用 Bundle 的 getString() 并搭配 TelephonyManager.EXTRA_STATE 常量取得手机状态，状态值有 3 种：EXTRA_STATE_IDLE(待机中)、EXTRA_STATE_OFFHOOK(拨号或通信中)、EXTRA_STATE_RINGING(来电中)。

```java
Bundle bundle = intent.getExtras();
String phoneState = bundle.getString(TelephonyManager.EXTRA_STATE);
```

(4) 如果手机状态为来电中，即可调用 Bundle 的 getString() 并搭配 TelephonyManager.EXTRA_INCOMING_NUMBER 常量取得来电号码。

```java
if(phoneState.equals(EXTRA_STATE_RINGING)){ //手机在来电状态
    String phoneNo = bundle.getString(TelephonyManager.EXTRA_INCOMING_NUMBER);
```

处理来短信或来电所需使用到的方法说明如表 12-3 所示。

表 12-3

SmsMessage 类
public static SmsMessage createFromPdu (byte[] pdu)
将 PDU 数据转换成 SmsMessage 对象
public String getDisplayMessageBody ()
取得短信或电子邮件的信息内容
public String getDisplayOriginatingAddress ()
取得短信或电子邮件的发信地址，如果是由手机发送的短信，则发信地址大多为电话号码
public long getTimestampMillis ()
取得信息时间，这个时间是由短信服务中心(service centre)所标记的时间戳
TelephonyManager 类
常　　量
public static final String ACTION_PHONE_STATE_CHANGED
Broadcast 的 Intent action，代表手机状态改变。需要设置 READ_PHONE_STATE 权限
public static final String EXTRA_STATE
取得手机状态。当 Broadcast 的 Intent action 为 ACTION_PHONE_STATE_CHANGED 时，可以使用 EXTRA_STATE 常量取得对应的手机状态。状态值有 3 种：EXTRA_STATE_IDLE(待机中)、EXTRA_STATE_OFFHOOK(拨号或通信中)、EXTRA_STATE_RINGING(来电中)
public static final String EXTRA_INCOMING_NUMBER
取得来电号码。当 Broadcast 的 Intent action 为 ACTION_PHONE_STATE_CHANGED，而且状态值为 EXTRA_STATE_RINGING 时，才可以使用 EXTRA_INCOMING_NUMBER 常量取得对应的来电号码。换句话说，就是一定要在来电状态才可获取来电号码，其他状态就无法取得来电号码

范例 TelephonyEx

范例说明：
- 在第一页显示正在等待短信或来电的信息，如图 12-9 所示。
- 如果有短信，就会显示发信地址、时间与短信内容，如图 12-10 所示。
- 如果有来电，就会显示来电号码，如图 12-11 所示。

图 12-9

图 12-10

图 12-11

TelephonyEx/AndroidManifest.xml

```xml
<manifest xmlns:android="http://schemas.android.com/apk/res/android"
    package="org.telephonyEx"
    android:versionCode="1"
    android:versionName="1.0" >

    <uses-sdk android:minSdkVersion="8" />

    <!-- 欲拦截短信，必须添加 RECEIVE_SMS 的权限设置；
         欲取得手机状态，则必须添加 READ_PHONE_STATE 的权限设置 -->
    <uses-permission android:name="android.permission.RECEIVE_SMS" />
    <uses-permission android:name="android.permission.READ_PHONE_STATE" />

    <application
        android:icon="@drawable/icon"
        android:label="@string/app_name" >
        <activity
            android:name=".TelephonyEx"
            android:label="@string/app_name" >
            <intent-filter>
                <action android:name="android.intent.action.MAIN" />
                <category android:name="android.intent.category.LAUNCHER" />
```

```xml
        </intent-filter>
    </activity>
    <activity android:name=".Result" />

    <!-- 设置短信与来电的 BroadcastReceiver，并指定欲拦截的 Intent action -->
    <receiver android:name="SMSReceiver" >
        <intent-filter>
            <action android:name="android.provider.Telephony.SMS_RECEIVED" />
        </intent-filter>
    </receiver>
    <receiver android:name="PhoneReceiver" >
        <intent-filter>
            <action android:name="android.intent.action.PHONE_STATE" />
        </intent-filter>
    </receiver>
</application>
</manifest>
```

TelephonyEx/src/org/telephonyEx/SMSReceiver.java

```
10.     public class SMSReceiver extends BroadcastReceiver {
11.         private final static int requestCode = 0; //来短信请求码设置为 0
12.         @Override
13.         public void onReceive(Context context, Intent intent){
14.             Bundle bundle = intent.getExtras();
15.             String msgBody = "";
16.             String sendAddr = "";
17.             Date time = new Date(0);
18.
19.             if (bundle != null) {
20.                 Object[] pdus = (Object[])bundle.get("pdus");
21.                 SmsMessage[] smsMsgs = new SmsMessage[pdus.length];
22.                 for (int i = 0; i < pdus.length; i++){
23.                     smsMsgs[i] = SmsMessage.createFromPdu((byte[])pdus[i]);
24.                     msgBody += smsMsgs[i].getDisplayMessageBody();
25.                 }
26.                 sendAddr = smsMsgs[0].getDisplayOriginatingAddress();
27.                 time = new Date(smsMsgs[0].getTimestampMillis());
28.             }
29.             Intent i = new Intent(context, Result.class);
30.             Bundle b = new Bundle();
31.             b.putInt("requestCode", requestCode);
32.             b.putString("sendAddr", sendAddr);
```

```
33.            b.putString("time", time.toLocaleString());
34.            b.putString("msgBody", msgBody);
35.            i.putExtras(b);
36.            i.addFlags(Intent.FLAG_ACTIVITY_NEW_TASK);
37.            context.startActivity(i);
38.        }
39.    }
```

10~13 行：清单文件已经设置 SMSReceiver 拦截来短信，所以来短信时会自动调用 13 行的 onReceive()。

20 行：取得 bundle 对象内存储的 PDU 数据。

23 行：调用 SmsMessage.createFromPdu()将 PDU 数据转换成 SmsMessage 对象。

24 行：取得短信内容。

26 行：取得发信地址。

27 行：取得发信时间并转换成 Date 对象。

31~34 行：将短信相关信息存储在 Bundle 对象内。

36~37 行：如果 startActivity()在 Activitty 外部被调用，Intent 对象就必须添加 FLAG_ACTIVITY_NEW_TASK 标志以打开新的任务，这是因为没有当前 Activity 就不会有任务打开另一个新的 Activitty。未添加 FLAG_ACTIVITY_NEW_TASK 标志会产生 Exception。

TelephonyEx/src/org/telephonyEx/PhoneReceiver.java

```
9.     public class PhoneReceiver extends BroadcastReceiver {
10.        private final static int requestCode = 1; //来电请求码设置为 1
11.        @Override
12.        public void onReceive(Context context, Intent intent){
13.            Bundle bundle = intent.getExtras();
14.            String phoneNo = "";
15.            String phoneState = "";
16.            if(bundle != null)
17.                phoneState = bundle.getString(EXTRA_STATE);
18.
19.            if(phoneState.equals(EXTRA_STATE_RINGING)){
20.                phoneNo = bundle.getString(EXTRA_INCOMING_NUMBER);
21.                Intent i = new Intent(context, Result.class);
22.                Bundle b = new Bundle();
23.                b.putInt("requestCode", requestCode);
24.                b.putString("phoneNo", phoneNo);
25.                i.putExtras(b);
26.                i.addFlags(Intent.FLAG_ACTIVITY_NEW_TASK);
27.                context.startActivity(i);
28.            }
29.        }
30.    }
```

9~12 行：清单文件已经设置 PhoneReceiver 拦截来电，所以有来电时，会自动调用 12 行的 onReceive()。

17 行：取得手机状态。

19~20 行：如果手机状态属于来电中，就取得来电号码。

23~24 行：将来电相关信息存储在 Bundle 对象内。

TelephonyEx/src/org/telephonyEx/Result.java

```
6.      public class Result extends Activity {
7.          TextView tvInfo; //用来显示相关信息
8.          @Override
9.          public void onCreate(Bundle savedInstanceState) {
10.             super.onCreate(savedInstanceState);
11.             setContentView(R.layout.result);
12.             findViews();
13.             showResults();
14.         }
15.         private void findViews(){
16.             tvInfo = (TextView)findViewById(R.id.tvInfo);
17.         }
18.         private void showResults(){
19.             Bundle bundle = this.getIntent().getExtras();
20.             int requestCode = bundle.getInt("requestCode");
21.             StringBuilder sb = new StringBuilder();
22.             switch(requestCode){
23.                 case 0: //代表来短信
24.                     sb.append(getString(R.string.sendAddr) +
25.                             bundle.getString("sendAddr") + "\n");
26.                     sb.append(getString(R.string.time) +
27.                             bundle.getString("time") + "\n");
28.                     sb.append(getString(R.string.msgBody)+ "\n" +
29.                             bundle.getString("msgBody")+ "\n");
30.                     tvInfo.setText(sb);
31.                     break;
32.                 case 1: //代表有来电
33.                     sb.append(getString(R.string.caller) +
34.                             bundle.getString("phoneNo") + "\n");
35.                     tvInfo.setText(sb);
36.                     break;
37.             }
38.         }
39.     }
```

20~37 行：先取得请求码，利用 switch-case 判断请求码代表的是来短信(值为 0)还是来电(值为 1)。如果是来短信，就取得发信地址、发信时间与短信内容后呈现；如果手机状态属于来电中，就取得来电号码后呈现。

12.4 查询联系人数据

从 Android 2.0(API Level 5)开始，Android 提供新的 Contacts API 以方便管理和整合多个账户内的联系人数据。为了处理不同账户间的联系人重复的问题，Android 会将相同联系人的数据汇总(aggregate)起来成为同一个联系人的数据。汇总的规则如下：
有两个以上联系人符合下列其中一种情况，就会被汇总成同一个联系人。
- 姓名完全相同。
- 姓名顺序虽然颠倒，但是相同，例如：Ron Huang 与 Huang Ron。
- 虽然只有姓或名相同，但是还有其他信息(如电话号码或电子邮件地址)相同。
- 有联系人完全没有姓名，但是有相同的电话号码或电子邮件地址。

当比较姓名时，会忽略大小写；比较电话时会忽略空格符及特殊字符，如：*、#、(、)；如果两个联系人电话号码相同，但一个有国家代码，另一个没有国家代码，也会被认为相同[①]。

汇总后的联系人数据分别存储在 contacts 数据表(联系人主要数据，例如：联系人 ID、联系人姓名)[②]与 data 数据表(联系人详细数据，例如：电话号码、电子邮件地址)[③]，但不论是 contacts 或 data 数据表内容，都引用原始联系人(raw contacts)[④]的数据，而且原始联系人可能有多个，如图 12-12 所示。

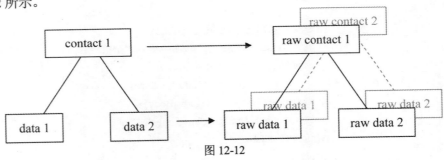

图 12-12

关系 contacts 与 data 这两个数据表的就是_ID 字段，_ID 可以当作是联系人 ID；但这个联系人 ID 与原始联系人的 ID 并不相同，这是因为 contacts 数据表可能由多个原始联系人汇总而成。

查询联系人的数据可以使用 Activity 的 managedQuery()或 ContentResolver 的 query()，两个方法的参数一样，也都是返回 Cursor 对象。一般建议使用 managedQuery()，因为会将 Activity

---

① 日本国家代码例外，不会被认为相同。
② contacts 数据表的字段及其说明参见 ContactsContract.Contacts 类的 API 文件。
③ data 数据表的字段及其说明参见 ContactsContract.Data 类的 API 文件。
④ 参见 ContactsContract.RawContacts 类的 API 文件。

的生命周期考虑进去，例如：当 Activity 暂停时会将查询上传；重启 Activity 时会再查询一次[①]；不过 Android 3.0(Honeycomb)会把 managedQuery()定义为逐渐淘汰，为了实现向后兼容性，范例会使用 query()。不论 managedQuery()或 query()都与第 8 章 SQLiteDatabase 的 query()非常相似，您应该已经熟练掌握。查询联系人使用到的相关方法说明如表 12-4 所示。

表 12-4

ContentResolver 类

public final Cursor query (Uri uri, String[] projection, String selection, String[] selectionArgs, String sortOrder)
对指定的 URI 作查询，并取得 Cursor 对象(存储着查询结果)

- uri：欲查询数据表的 URI
- projection：以数组存储欲查询的域名。设为 null 会查询所有字段的数据，性能较差，所以建议只设置欲查询的字段
- selection：设置查询条件，效果与 SQL 语法 WHERE 条件式的功能相同(此参数内容不需要再放入 WHERE 字符串)，如果值为 null，代表不设置条件，所以会返回所有数据列
- selectionArgs：可以在参数 selection 内容加上 "?" 并使用 selectionArgs 内的值替换该 "?"
- sortOrder：排序，效果与 SQL 语法 ORDER BY 的功能相同

 范例 ContactEx

范例(如图 12-13 所示)说明：
- 以 ListView 呈现联系人 ID 与姓名。
- 单击联系人会弹出 Toast 消息框呈现联系人姓名与各类型的电话号码。

图 12-13

ContactEx/src/org/contactEx/ContactEx.java

18.　　public final class ContactEx extends Activity {

[①] 详细说明参见 http://developer.android.com/guide/topics/providers/content-providers.html 的 Making the query。

```java
19.     private ListView lvContacts;
20.
21.     @Override
22.     public void onCreate(Bundle savedInstanceState) {
23.         super.onCreate(savedInstanceState);
24.         setContentView(R.layout.main);
25.         findViews();
26.         createContactList();
27.     }
28.
29.     private void createContactList() {
30.         Cursor cursor = getContacts();
31.         String[] fields = { Contacts._ID, Contacts.DISPLAY_NAME };
32.         int[] tvResID = {R.id.contactID, R.id.contactName};
33.
34.         SimpleCursorAdapter adapter = new SimpleCursorAdapter(
35.             this, R.layout.contactlist, cursor,
36.             fields, tvResID, CursorAdapter.FLAG_REGISTER_CONTENT_OBSERVER );
37.         lvContacts.setAdapter(adapter);
38.     }
39.
40.     private Cursor getContacts() {
41.         Uri uri = Contacts.CONTENT_URI;
42.         String[] fields = { Contacts._ID, Contacts.DISPLAY_NAME };
43.         String selection = null;
44.         String[] selectionArgs = null;
45.         String sortOrder = Contacts.DISPLAY_NAME;
46.         return getContentResolver().query(
47.             uri, fields, selection, selectionArgs, sortOrder);
48.     }
49.
50.     private Cursor getPhones(long id) {
51.         Uri uri = Phone.CONTENT_URI;
52.         String[] fields = { Phone.TYPE, Phone.NUMBER };
53.         String selection = Phone.CONTACT_ID + " = "+ id;
54.         String[] selectionArgs = null;
55.         String sortOrder = null;
56.         return getContentResolver().query(
57.             uri, fields, selection, selectionArgs, sortOrder);
58.     }
```

29 行：createContactList()会在 ListView 上呈现联系人 ID 与姓名。

30 行：取得联系人数据并指派给 cursor 变量。

31 行：指定呈现 cursor 内容中 _ID 与 DISPLAY_NAME 这两个字段的数据。

32 行：将欲呈现的数据放在指定的 TextView 组件上(定义在 contactlist.xml layout 文件内，参见 35~36 行说明)。

34 行：调用 SimpleCursorAdapter 构造函数将 cursor 对象内容转换成 ListView 所呈现的选项。

35~36 行：以 fields 参数指定要从 cursor 内取出哪些字段的数据，并将取出的数据指定以 R.layout.contactlist 文件所定义的 TextView 组件(以 tvResID 参数指定是哪些 TextView 组件)来呈现数据。简单地说，就是要将 cursor 内的 _ID 与 DISPLAY_NAME 这两个字段的数据取出后分别呈现在 R.id.rowID、R.id.contactName 组件上。注意：cursor 内容如果没有 _id 字段，SimpleCursorAdapter 构造函数会出现以下错误：IllegalArgumentException: column '_id' does not exist。FLAG_REGISTER_CONTENT_OBSERVER 代表在 cursor 上注册数据内容监控器，一旦数据有改变并发出通知，会调用 CursorAdapter 的 onContentChanged()。

37 行：调用 setAdapter()将选项数据添加到 ListView 组件。

40 行：调用 getContacts()可以取得联系人数据。

41 行：contacts 数据表的 URI。

42 行：欲查询 _ID 与 DISPLAY_NAME(联系人姓名)这两个字段的数据。

43 行：值为 null，代表不设置条件，所以会返回所有数据列。

44 行：值为 null，代表不设置条件参数。

45 行：将 DISPLAY_NAME 字段排序，默认为升序排序，若要降序排序，则加上 DESC。

46 行：调用 getContentResolver()返回 ContentResolver 对象，再调用 query()对指定的 URI 作查询，会取得 Cursor 对象(存储查询结果)。

50 行：调用 getPhones()可以取得联系人电话号码等数据。

51 行：电话号码相关信息的 URI。

52 行：欲查询 TYPE(电话类型)与 NUMBER(电话号码)这两个字段的数据。

53 行：指定查询 id 所代表的联系人数据。

```
60.     private void findViews() {
61.         lvContacts = (ListView) findViewById(R.id.contactList);
62.         lvContacts.setOnItemClickListener(new OnItemClickListener() {
63.             @Override //单击 ListView 上面的项
64.             public void onItemClick(AdapterView<?> parent,
65.                 View view, int position, long id) {
66.                 LinearLayout linear = (LinearLayout)view;
67.                 CharSequence contactName =
68.                     ((TextView)linear.getChildAt(1)).getText();
69.                 StringBuilder text = new StringBuilder();
70.                 Cursor phones = getPhones(id);
71.                 text.append(contactName);
72.
73.                 if(phones.getCount()<=0){
74.                     text.append("\n" + getString(R.string.phoneNo_not_found));
```

```
75.                    Toast.makeText(ContactEx.this, text, Toast.LENGTH_LONG).show();
76.                    return;
77.                }
78.
79.                while(phones.moveToNext()){
80.                    int typeID =
81.                        phones.getInt(phones.getColumnIndex(Phone.TYPE));
82.                    String type =
83.                        getString(Phone.getTypeLabelResource(typeID));
84.                    String phoneNo =
85.                        phones.getString(phones.getColumnIndex(Phone.NUMBER));
86.                    text.append("\n" + type + ":  " + phoneNo);
87.                }
88.                Toast.makeText(ContactEx.this, text, Toast.LENGTH_LONG).show();
89.            }
90.        });
91.    }
92. }
```

62~64 行：ListView 组件调用 setOnItemClickListener() 向 OnItemClickListener 监听器注册，接下来利用匿名内部类实现 OnItemClickListener 的 onItemClick()。当 ListView 组件上面的选项被单击时，onItemClick() 会自动被调用并执行。

66 行：view 代表的是 contactlist.xml layout 文件内定义的 LinearLayout 组件。

68 行：取得 LinearLayout 组件上 index 为 1 的子组件，也就是 R.id.contactName 所代表的 TextView 组件(存储着 DISPLAY_NAME(联系人姓名)数据)。之后再调用 getText() 将该 TextView 组件上的文本取出。

70 行：参数 id 其实就是联系人 ID。详细说明：70 行的 id 就是 65 行的 id，来自于前面 35 行中的 SimpleCursorAdapter 对象所使用 cursor 对象内 _ID 字段所存储的数据。

73~77 行：如果查无电话数据，就显示尚未输入该联系人的电话号码并结束。

79~88 行：如果联系人有电话号码，就以 Toast 消息框呈现联系人姓名与各类型的电话号码。83 行调用 getTypeLabelResource() 并指定电话类型，即可返回该电话类型对应的描述文本；例如 2 代表移动设备①。

不可不知

访问联系人数据必须在清单文件内添加 READ_CONTACTS 权限。

```
<uses-permission android:name="android.permission.READ_CONTACTS" />
```

---

① 经笔者测试，getTypeLabelResource()支持本地化；换句话说，切换成英文时，移动设备会变成 Mobile。

API 文件建议的查询方法[①]：

- 查询个别联系人(individual contact)，请改用 CONTENT_LOOKUP_URI 而非 CONTENT_URI。
- 根据电话号码来查询联系人，请使用 PhoneLookup.CONTENT_FILTER_URI，性能会更好。
- 根据用户输入的局部姓名来作查询(最常用在 AutoCompleteTextView 组件上)，请使用 CONTENT_FILTER_URI。
- 根据联系人其他数据如：电子邮件地址、昵称等，请查询 ContactsContract.Data 数据表(就是前述 data 数据表)，因为查询结果会包含联系人 ID 与联系人姓名等数据。

---

① 参见 API 文件中关于 ContactsContract.Contacts 类的 Query 部分。

第13章
AdMob广告制作

本章学习目标：

- AdMob 简介
- 注册 AdMob 账户
- AdMob 广告实现

13.1 AdMob 简介

AdMob 于 2006 年创建,是一家提供移动广告的公司,向其他需要打广告的客户收费,然后将其广告显示在移动设备上。为了能让广告大量曝光,AdMob 让开发人员可以在移动设备应用程序(例如 Android phone 或 iPhone 的应用程序)或网页上(在此专指供手机或平板电脑浏览的移动版网页,而非传统网页)植入 AdMob 的广告。至于会播放何种广告由该公司控制。提供页面空间放置广告的人称为广告发布商,只要移动设备的用户单击该广告,发布商就可获利。这让移动设备应用程序或网页开发人员趋之若鹜,竞相摆放广告,形成了一个庞大的广告联播网。Google 看中移动设备的商机,于 2009 年 11 月 9 日宣布以 7 亿 5 千万美元并购 AdMob 公司[①]。Google 旗下其实已经有专门负责网页广告的 AdSense®,它与 AdMob 移动网页广告业务部分重叠,所以 Google 宣布 2012 年 5 月 1 日起将移动网页广告部分移至 AdSense,而 AdMob 则专注于移动设备应用程序的广告,参见图 13-1 中的深色部分。

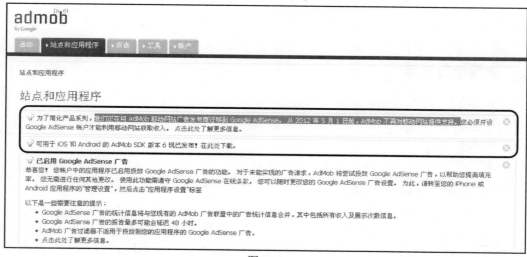

图 13-1

Android、iPhone、iPad、Windows Phone 的应用程序都可以内嵌 AdMob 广告,如图 13-2 右上角所示[③]。移动设备平台很可能成为 Google 广告联播网的新成员,而且可能成为最具份量的一员,因为人手一部智能手机的时代不久将会来临。

---

① 参见 http://en.wikipedia.org/wiki/AdMob。
② "Google AdSense 是一种免费的广告计划,可让各种规模的网站发布商在自己的网站上显示相关的 Google 广告",这是 Google AdSense 官网对 Google AdSense 的解释,参见 http://support.google.com/adsense/bin/answer.py?hl=zh-Hant&ctx=as2&answer=9712&rd=3。
③ 该图为移动设备上当红的游戏"愤怒的小鸟"。

AdMob 广告制作

图 13-2

在 Android Market 销售编写好的应用程序是一种获利方式,但有越来越多的应用程序不收取费用[①]而只靠内嵌广告来获利。只要应用程序功能不错,再加上免付费即可安装,就可能迅速受到广大移动设备用户群的青睐,获利不一定差。在本书第 1 章的实际案例中,在 2010 年 7 月,该应用程序免费版(内嵌 AdMob 广告)的获利为 6 200 美元,即超过付费版的 4 400 美元。

要想在 Android 应用程序放置 AdMob 广告,需要完成下列两大步骤。

(1) 注册 AdMob 账户:13.2 节会详加说明。

(2) 利用 AdView 显示广告内容:在 Android 应用程序内放置 AdView,用户运行应用程序时,即可在 AdView 上看到广告。完整设置方式参见 13.3 节。

13.2 注册 AdMob 账户

成为 AdMob 广告发布商不需要支付任何费用,只要注册账户即可,等到广告收益达到 20 美元(国际电汇者则是 100 美元)[②],AdMob 就会通过当初在账户设置的付款方式来支付款项。因为 AdMob 已成为 Google 大家庭中的一员,所以如同申请 Google 其他服务一样,可以直接使用 Google 账户来开通 AdMob 功能即可。注册 AdMob 账户的步骤如下:

STEP 1　进入 AdMob 简体中文服务首页,网址为 http://zhcn.admob.com/,按下右边的"注册 AdMob"按钮,如图 13-3 所示。

[①] Android 应用程序的不收费比例高达 67%,参见 http://en.wikipedia.org/wiki/Android_Market#Comparisons_to_competitors。

[②] 参见 http://support.google.com/admob/bin/answer.py?hl=zh-Hant&answer=1307281&topic=1307280&ctx=topic。

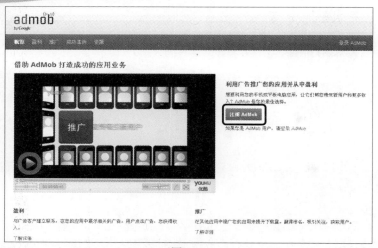

图 13-3

STEP 2 如果已经有 Google 账户，直接单击"在此登录"的链接即可，如图 13-4 所示。如果没有 Google 账户，就必须填写相关信息来创建一个新的账户。

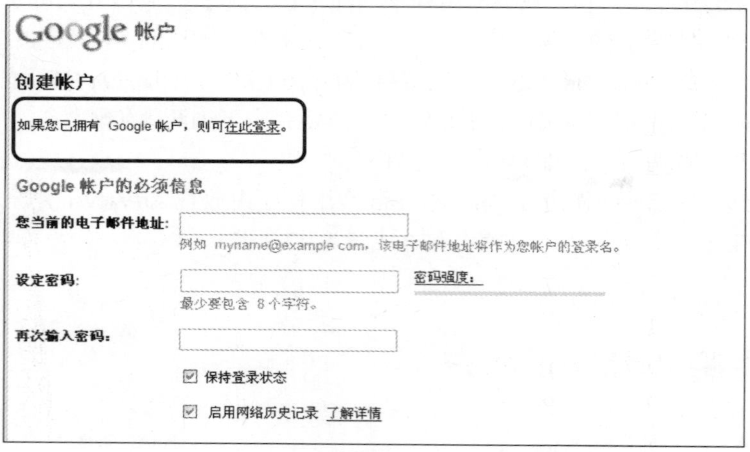

图 13-4

STEP 3 账户创建完毕后会启动账户，如图 13-5 所示，按下"添加网站/应用程序"按钮。

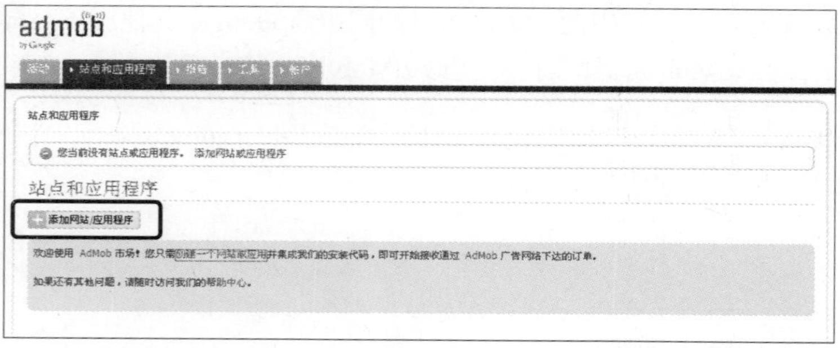

图 13-5

AdMob 广告制作

STEP 4　在做其他设置之前,系统会要求先完成填写付款详细信息,粗体字部分都是必填字段,如图 13-6 所示。

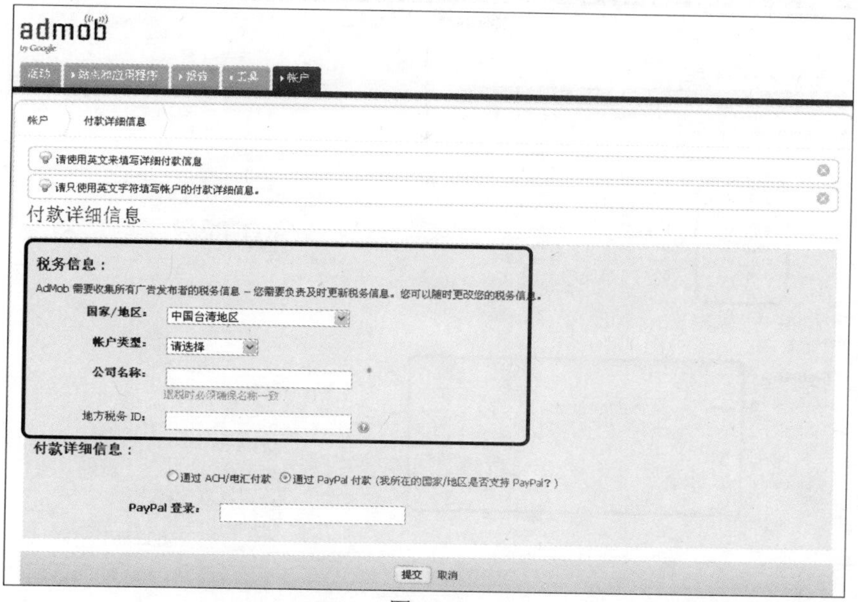

图 13-6

(1) 国家或地区:选择"中国台湾地区"。
(2) 账户类型:可选择"个别"。
(3) 企业名称:填写个人英文名称。
(4) 付款详细数据[1]

- 通过 ACH/电汇付款:收款人姓名与银行相关数据都要以英文填写,最好先询问银行各字段应如何填写。使用这种方式,银行一般都会收手续费[2]。因为属于国际电汇付款,所以广告收益必须达到 100 美元,AdMob 才会汇款。
- 通过 PayPal 付款:输入 PayPal 登录账号即可。这是最简单的收款方式,而且广告收益只要达到 20 美元,AdMob 就会汇款。PayPal 的每笔手续费约 2~3% +若干人民币。因为 PayPal 的运作方式按理并不算是银行,因此银行相关法律对 PayPal 并不适用,用户存在 Paypal 账户中的资金无法得到法律上的保护[3]。PayPal 注册步骤参考 PayPal 官方网站的注册教学,写得非常详细,网址为 http://www.paytaiwan.com/paypal_signup_guide.htm。

STEP 5　单击画面上方主菜单的"站点与应用程序"选项卡后选择"Android 应用程序";并于下方详细数据部分填写已经开发好且想要放广告的应用程序名称等相关信息,如图 13-7 所示。按最下方的"继续"按钮继续下一步骤。

---

[1] 参见 http://support.google.com/admob/bin/answer.py?hl=zh-Hant&answer=1307283&topic=1307280&ctx=topic。
[2] 经过询问,无论汇进来金额多少,银行都会收取手续费!
[3] 参见 http://zh.wikipedia.org/zh-tw/PayPal#.E6.B3.95.E5.BE.8B.E5.8F.8A.E9.A2.A8.E9.9A.AA.E5.95.8F.E9.A1.8C。

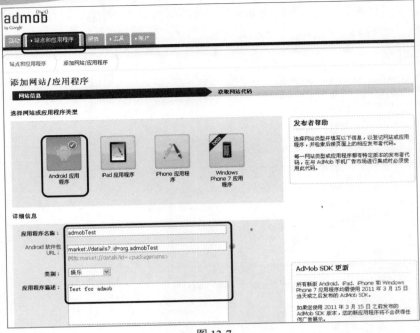

图 13-7

STEP 6　按下"下载 AdMob Android SDK"按钮即可下载 AdMob SDK，如图 13-8 所示。

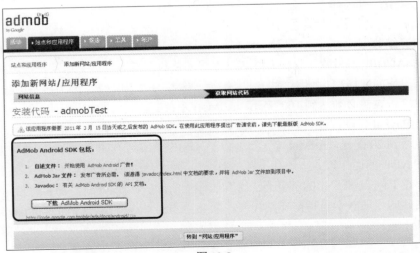

图 13-8

STEP 7　单击画面上方主菜单中的"站点与应用程序"选项卡后，将光标移至应用程序名称旁边，会显示"管理设置"按钮，如图 13-9 所示。按下该按钮继续下一个步骤。

AdMob 广告制作

图 13-9

STEP 8 图 13-10 上显示的发布商 ID 需要记录下来，因为之后必须将该 ID 填写在 Android layout 文件的 <AdView> 标签内。

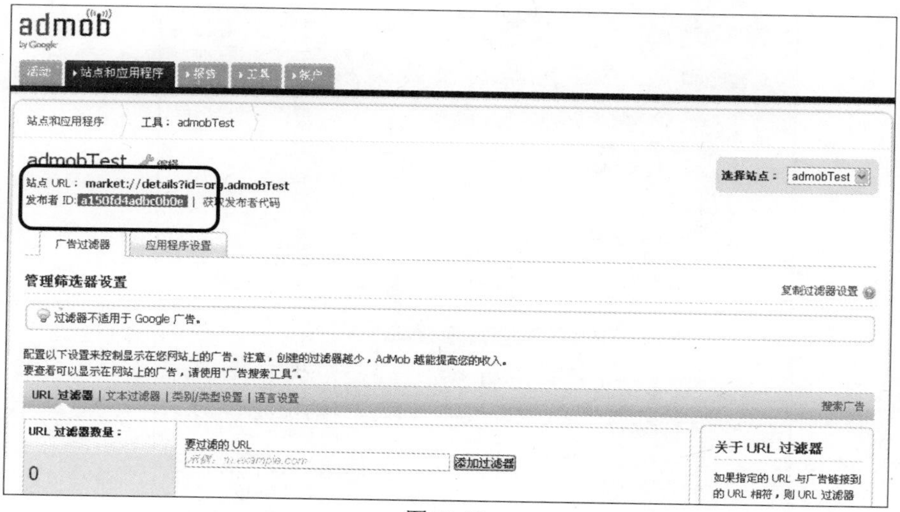

图 13-10

13.3 AdMob 广告实现

经过前面的步骤，已经取得 AdMob SDK 与发布商 ID，接下来就是让 Android 应用程序放置广告，可参见下面的范例说明。

 范例 admobTest

范例(如图 13-11 所示)说明：
如果设置成功，AdMob 广告会显示在 AdView 组件上。

图 13-11

范例创建步骤如下：

STEP 1 将之前下载的 AdMob SDK 解压缩，会得到 GoogleAdMobAdsSdk-xxx.jar(xxx 代表版本)。接下来将该 SDK 的 JAR 文件路径添加到应用程序内。打开 Eclipse，对着 Android 项目右击后选择 Build Path | Configure Build Path，如图 13-12 所示。

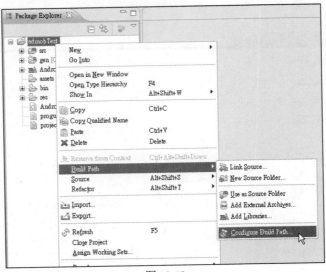

图 13-12

接下来会弹出如图 13-13 所示的窗口，按下 Add External JARs 按钮并指定 AdMob SDK 的 JAR 文件位置后，按下下方的 OK 按钮完成设置。

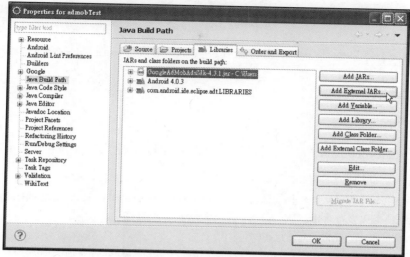

图 13-13

STEP 2 在清单文件内设置 INTERNET 与 ACCESS_NETWORK_STATE 使用权限，并声明 com.google.ads.AdActivity 与设置 configChanges 属性。

```
<manifest xmlns:android="http://schemas.android.com/apk/res/android"
    package="org.admobTest"
    android:versionCode="1"
    android:versionName="1.0" >

    <uses-sdk android:minSdkVersion="8" />

    <!-- 设置 INTERNET 权限是因为传送广告内容需要通过 Internet；设置 ACCESS_
NETWORK_STATE 权限是因为要能访问网络信息 -->
    <uses-permission android:name="android.permission.INTERNET" />
    <uses-permission android:name="android.permission.ACCESS_NETWORK_STATE" />

    <application
        android:icon="@drawable/ic_launcher"
        android:label="@string/app_name" >
        <activity
            android:name=".AdmobTestActivity"
            android:label="@string/app_name" >
            <intent-filter>
                <action android:name="android.intent.action.MAIN" />

                <category android:name="android.intent.category.LAUNCHER" />
            </intent-filter>
        </activity>
```

```xml
<!-- 需要声明 com.google.ads.AdActivity。设置 configChanges 属性，代表 Activity 会自行处
理所设置的改变，否则 Activity 会重启；如果不设置，运行时也会发生错误 -->
<activity
    android:name="com.google.ads.AdActivity"
    android:configChanges="keyboard|keyboardHidden|orientation|screenLayout|uiMode|screenSize|smallestScreenSize" />
</application>
</manifest>
```

STEP 3 在 res/layout/main.xml 文件内新建 AdView 组件[①]，之后广告内容会显示在上面。

```xml
<!--
adSize：设置广告显示的大小，BANNER 代表 MMA(Mobile Marketing Association)标准大小。
adUnitId：填入发布商 ID。
loadAdOnCreate：是否一开始就加载广告。
testDevices：设置测试用的设备，换句话说，测试完毕要移除此属性。如果是模拟器，要填入 TEST_
EMULATOR；如果是实机，则必须输入实机的 ID，就是 IMEI 码(International Mobile Equipment Identity)，
在移动设备的设备选项中即可找到。
-->
<com.google.ads.AdView
    android:id="@+id/adView"
    android:layout_width="wrap_content"
    android:layout_height="wrap_content"
    ads:adSize="BANNER"
    ads:adUnitId="a14f4a2xxxxxxx"
    ads:loadAdOnCreate="true"
    ads:testDevices="TEST_EMULATOR" />
```

STEP 4 在 AdmobTestActivity.java 文件内添加下列程序代码。

```java
public class AdmobTestActivity extends Activity {

    @Override
    public void onCreate(Bundle savedInstanceState) {
        super.onCreate(savedInstanceState);
        setContentView(R.layout.main);
        /* AdView 与其他 View 一样可以通过 findViewById()找到定义在布局文件内对应的组件 */
        AdView adView = (AdView)this.findViewById(R.id.adView);
        /* 开始在背景加载广告 */
        adView.loadAd(new AdRequest());
    }
}
```

---

① AdView 其实是 RelativeLayout 的子类，可查看 AdMob SDK 的 javadoc 文件。

第 14 章

将应用程序发布至 Google Play

 本章学习目标：

- 如何将应用程序发布至 Google Play
- 产生并签名应用程序
- 申请 Android 开发人员账号
- 使用开发人员管理界面发布应用程序

14.1 如何将应用程序发布至 Google Play

如果要将应用程序发布至 Google Play(原名为 Android Market)，必须经历下列 3 大过程：
(1) 产生并签名应用程序。
(2) 申请 Android 开发人员账号。
(3) 登录开发人员应用程序管理界面(Android developer console)以发布应用程序。
下面各节将会一一说明如何完成这些过程。

14.2 产生并签名应用程序

开发好的项目必须先导出为 APK(Android Package)文件[1]，并且经过签名的操作才能发布至 Google Play 供 Android 移动设备的用户下载。APK 文件的产生与签名过程说明如下[2]。
(1) 创建私钥(private key)：使用 JDK 的 keytool 工具[3]产生私钥，等到 APK 文件产生之后必须以私钥签名[4]。
(2) 产生可发布至 Google Play 的应用程序：使用 Eclipse 的导出功能可以产生 APK 文件。
(3) 使用私钥签名 APK 文件：使用 JDK 的 jarsigner 工具[5]完成签名操作。
(4) 调试 APK 文件：最后使用 Android 的 zipalign 工具[6]调试 APK 文件，当 APK 文件安装在移动设备时可以达到性能优化。
上述 4 个步骤十分烦琐，其实可以使用 Eclipse + ADT 工具快速达成。

14.2.1 使用 Eclipse + ADT 产生并签名应用程序

一般开发 Android 应用程序都会安装 Eclipse 与 ADT 工具[7]，这样一来只要使用 Eclipse 的导出向导(export wizard)，并且按照指示操作就可以完成上述 APK 文件的产生与签名 4 个步骤的工作，非常方便。步骤说明如下：

STEP 1 对欲导出的项右击并选择 Export，如图 14-1 所示。

---

[1] APK 文件(附件名为 apk)其实就是 Android 的应用程序，非常类似 Java 的 JAR 文件，都是以 ZIP 格式压缩的文件。
[2] 各步骤详细说明参见 http://developer.android.com/guide/publishing/app-signing.html#releasemode。
[3] keytool 工具在 JDK 目录的 bin 子目录内，例如：C:\Program Files\Java\jdk1.6.0_24\bin\keytool.exe。
[4] 因为私钥专门用来签名欲发布至 Android Market 的应用程序，所以又称为发布密钥。
[5] jarsigner 工具在 JDK 目录的 bin 子目录内，例如：C:\Program Files\Java\jdk1.6.0_24\bin\jarsigner.exe。
[6] zipalign 工具在 JDK 目录的 bin 子目录内，例如：C:\android-sdk-windows\tools\zipalign.exe。
[7] 下载与安装说明参见本书的第 2 章。

将应用程序发布至 Google Play

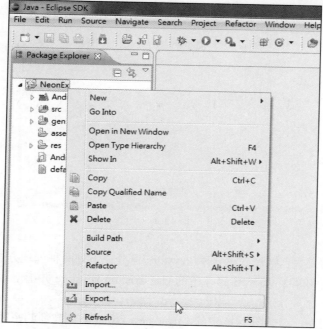

图 14-1

STEP 2 选择 Android | Export Android Application 选项后按下 Next 按钮，如图 14-2 所示。

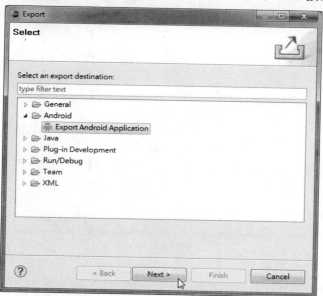

图 14-2

STEP 3 如欲更改导出的项目，请按下 Browse 按钮，如不更改，则直接按下 Next 按钮继续，如图 14-3 所示。

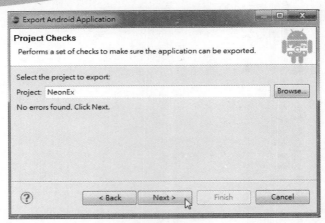

图 14-3

STEP 4 根据之前是否已经产生密钥库(keystore)可分成下列两种处理方式。

- 已经有密钥库[①]：选择图 14-4 中的 Use existing keystore 并指定该密钥库的路径位置(Location 字段)与输入密码(Password 字段)。
- 没有密钥库：选择图 14-4 中的 Create new keystore 创建新的密钥库，指定新创建密钥库的路径位置(Location 字段)与输入密码(Password 字段)，并再次确认密码(Confirm 字段)。

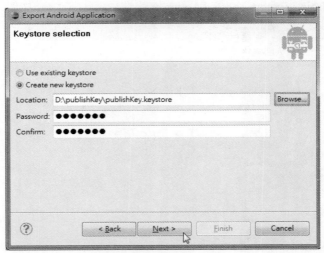

图 14-4

STEP 5 在密钥库内置创建密钥，需填写的字段说明如下(参见图 14-5)。

- Alias：密钥的别名。
- Password：密码，可以与密钥库的密码相同。
- Confirm：确认密码。

---

① 要发布至 Google Play 的应用程序不能使用调试密钥(debug key)签名；第 9 章提及的 debug.keystore 密钥库内的密钥就是调试密钥，无法使用在欲发布至 Google Play 的应用程序上。

- Validity：有效年限，可填入 1~1000 年[①]。
- First and Last Name 以下字段：只要挑选其中一个字段填写即可。

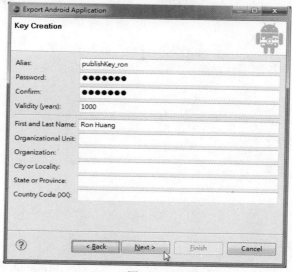

图 14-5

STEP 6 指定 APK 文件导出后所存放的路径，如图 14-6 所示。按下 Finish 按钮后即可产生 APK 文件。

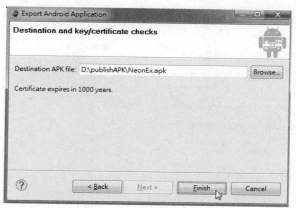

图 14-6

14.2.2　签名应用程序注意事项

在签名应用程序的过程中，必须注意下列重要事项。

1. 应用程序的版本

为了控制应用程序的版本，会在清单文件内使用 android:versionCode 与 android:version-

---

① 要求至少要填 25 年以上。

Name 这两个属性。

- android:versionCode：内部管理控制的版本号码，必须为整数值，用户不会看到此版本号码。虽然可以使用任何整数，但每次改版时，新版本的 versionCode 都必须比前一个版本大，否则会无法发布[①]。
- android:versionName：对外发布的版本名称，值为字符串，用户会看到此名称。

以下是上述两个版本控制属性在清单文件内设置的范例：

```
<?xml version="1.0" encoding="utf-8"?>
<manifest xmlns:android="http://schemas.android.com/apk/res/android"
    package="org.neonEx"
    android:versionCode="2"
    android:versionName="1.1">
    ...
</manifest>
```

2. 应用程序的名称与图标

在清单文件内，<application>标签的 android:icon 与 android:label 两个属性非常重要，因为它们分别代表应用程序的图标与名称。如果用户有安装应用程序，就可以在移动设备上看到该应用程序的图标与名称。以下是这两个属性在清单文件内设置的范例：

```
<?xml version="1.0" encoding="utf-8"?>
<manifest xmlns:android="http://schemas.android.com/apk/res/android"
    package="org.neonEx"
    android:versionCode="2"
    android:versionName="1.1">
    <application
        android:icon="@drawable/icon"
        android:label="@string/app_name"
        ...
    </application>
</manifest>
```

3. 移除 debuggable 设置

将清单文件的<application>标签属性 android:debuggable 设置为 true，代表可以在实机上运行调试模式，如下列程序代码所示。如果只是将应用程序直接安装在模拟器或自己的移动设备上而不通过 Google Play，无论是否添加 android:debuggable 属性，都可以正常地安装运行该应用程序。

```
<application
    android:icon="@drawable/icon"
```

[①] 参见 14.4.2 节。

```
            android:label="@string/app_name"
            android:debuggable="true">
            ...
</application>
```

但是如果要将应用程序发布至 Google Play，就必须将 android:debuggable 属性直接移除或设置为 false，否则将项导出时会显示警示信息，如图 14-7 所示。

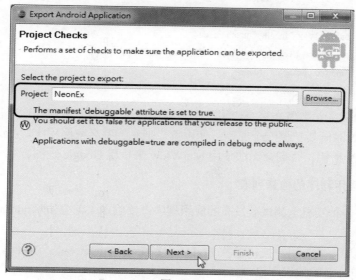

图 14-7

4. 使用私钥申请 Maps API 密钥

如果应用程序使用 MapView 组件来呈现 Google 地图，就必须申请 Maps API 密钥(Maps API Key)。在开发阶段，因为还不需要发布至 Google Play，所以暂时可以使用调试密钥来申请 Maps API 密钥。之前第 9 章使用的就是调试密钥。

要将应用程序发布至 Google Play，必须产生私钥[①]来签名应用程序，不能再使用调试密钥；而应用程序一旦使用到 MapView 组件的功能，也必须改为使用以私钥所申请的 Maps API 密钥，而不能再使用调试密钥所申请的 Maps API 密钥，否则会无法呈现 Google 地图。更改的步骤说明如下。

STEP 1 使用私钥申请 Maps API 密钥：找到私钥库(keystore)所在位置，然后按照 9.1 节的说明来申请，不再赘述。

STEP 2 修改 MapView 组件的 Maps API 密钥：在布局文件内，将 MapView 组件的 android:apiKey 属性值改填入以私钥申请的 Maps API 密钥号码后存储，如下面程序代码中的第 10 行所示。

---

① 产生私钥的方式参见 14.2.1 节。

```xml
1.  <?xml version="1.0" encoding="utf-8"?>
2.  <LinearLayout mlns:android="http://schemas.android.com/apk/res/android"
3.      android:orientation="vertical"
4.      android:layout_width="match_parent"
5.      android:layout_height="match_parent" >
6.      <com.google.android.maps.MapView
7.          android:id="@+id/mView"
8.          android:layout_width="match_parent"
9.          android:layout_height="match_parent"
10.         android:apiKey="填入以私钥申请的 Maps API 密钥号码"
11.         android:clickable="true" />
12. </LinearLayout>
```

假设某一个开发人员欲发布多个应用程序至 Google Play,每一个应用程序都可以使用不同的私钥来签名[1],这时候要把握一个原则:使用哪一把私钥签名应用程序,就必须使用同一把私钥申请 Maps API 密钥;否则会无法使用 MapView 来呈现 Google 地图。

5. 准备发布应用程序的检查列表[2]

(1) 尽可能在各种实机上测试欲发布的应用程序:以确定大部分的 Android 移动设备都能运行该应用程序。

(2) 在应用程序内增加用户授权条款(End User License Agreement):让用户了解其权限以保护开发人员个人或公司的知识产权。

(3) 考虑是否添加 Google Play 授权(Google Play Licensing):如果将应用程序设定为要收费,可以考虑是否添加 Google Play 授权机制,该项机制是当用户每次在 Google Play 对应用程序进行安装或更新时,都需要连接上 Google 官方服务器来确认该应用程序是否合法取得,其目的是阻止盗版等侵权行为。

(4) 确定应用程序的名称与图标:设置 android:icon 与 android:label 两个属性(如前所述)。

(5) 移除 debuggable 设置、关闭日志功能与移除不必要的文件与数据:

- 移除 android:debuggable="true"的设置(如前所述)。
- 移除应用程序内的日志文件、备份文件以及其他不必要的文件,如果原始代码内有用到日志功能,也将其关闭。

(6) 设置应用程序版本:设置 android:versionCode 与 android:versionName 两个属性(如前所述)。

(7) 产生私钥:用来申请 Maps API 密钥与签名欲发布的应用程序(如前所述)。

(8) 再次测试欲发布的应用程序:将已经编译且签名的应用程序再次测试,以确保应用程序能够正确运行。

---

[1] 不过为了管理方便,一般都建议用同一把密钥签名所有欲发布的应用程序。如果有太多把私钥,可能会弄混是用哪把密钥签名哪个应用程序,这样在更新应用程序版本时会发生错误,可参见 14.4.2 节。

[2] 参见 http://developer.android.com/guide/publishing/preparing.html。

14.3 申请 Android 开发人员账号

要发布应用程序至 Google Play，必须先申请开发人员账号，申请步骤如下。

STEP 1 打开 Google Play Android Developer Console 网页：(1)可以直接输入网址 https://play.google.com/apps/publish。(2)或至 Android 官方网站首页——http://developer.android.com/index.html，并单击标题栏中的 Distribute │ 左侧导航栏中的 Publishing │ Get Started │ 右侧的 Google Play Android Developer Console，如图 14-8 所示。

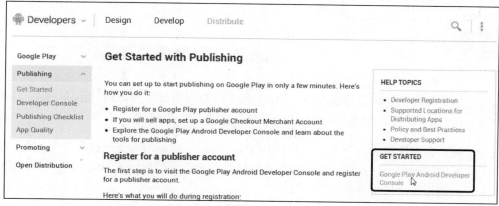

图 14-8

STEP 2 填入 Gmail 账号(就是 Google 账号)与密码并按下 Sign in 按钮，如图 14-9 所示。

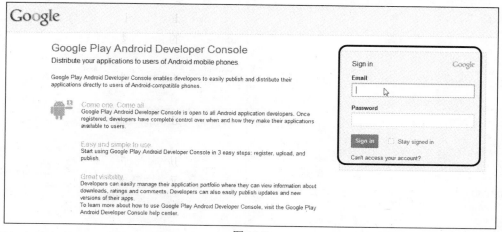

图 14-9

STEP 3 同意开发人员发布协议，如图 14-10 所示；选中"我同意并愿意将我的帐户注册信息与 Google Play 开发者分发协议相关联"之后单击"继续付款"链接。

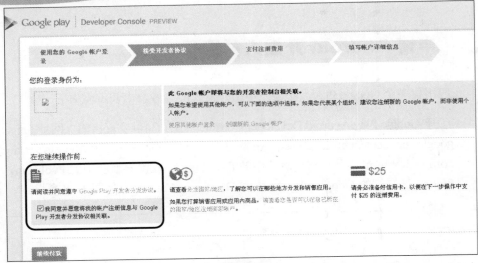

图 14-10

STEP 4 接下来将注册成为开发人员，完成之后才能将应用程序发布到 Google Play。注册费用为 25 美金，如图 14-11 所示。单击"立即启动"链接。

STEP 5 网页会将注册成为开发人员的费用列出，并要求以信用卡支付该笔费用。除了要填写信用卡相关信息外，还需填写账单地址与电话号码，如图 14-12 所示。填写完毕后请按下"接受并继续"按钮。

图 14-11　　　　　　　　　　　图 14-12

STEP 6 网页会再次列出注册费用为 25 美金，如图 14-13 所示。如果选中"让我的电子邮件地址保持为机密"，用户就无法看到开发人员的电子邮件地址。确定之后按下"立即下单"按钮。

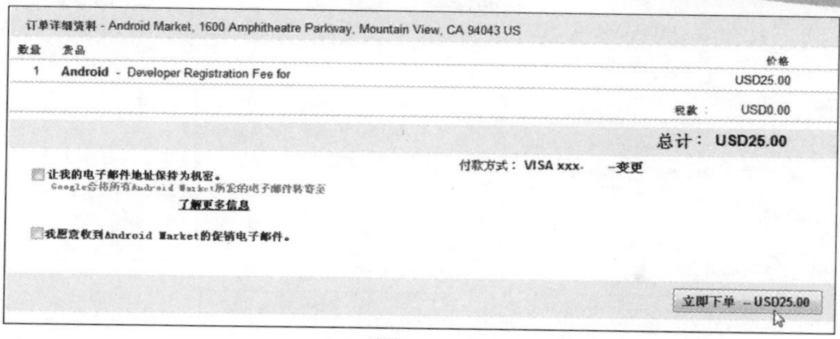

将应用程序发布至 Google Play

图 14-13

STEP 7 等待信用卡授权完毕后，就会完成付费手续。这个由一般用户升级成的开发人员账号不仅可以使用原本 Google 的相关服务，现在更具备发布 Android 应用程序的功能，所以请好好保存，因为其价值至少为 25 美金。

14.4 使用开发人员管理界面发布应用程序

开发人员如果想要将应用程序发布到 Google Play 上，必须使用发布专用的开发人员管理界面 (Android developer console)。进入该管理界面的方式与申请开发人员账号登录的方式相同①，不再赘述。进入管理界面后，根据应用程序是否首次发布，分别说明如下。

14.4.1 应用程序首次发布

首次发布的步骤说明如下：

STEP 1 在应用程序管理界面，单击"上传应用程序"链接，如图 14-14 所示。

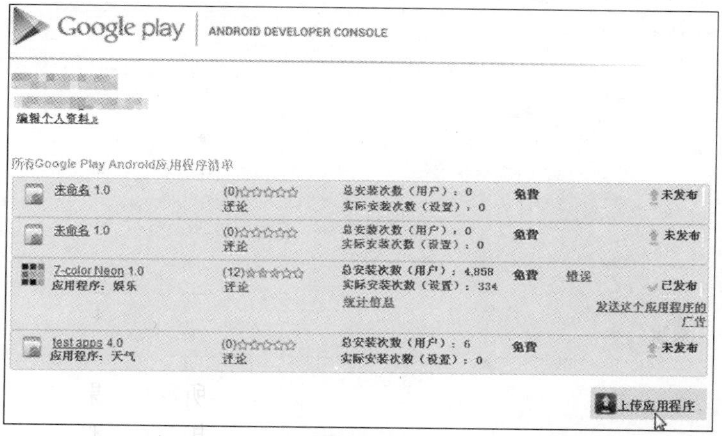

图 14-14

---

① 参见 14.3 节的 Step 1、2。

STEP 2　选择欲上传的应用程序后按下"上传"按钮，如图 14-15 所示。

图 14-15

STEP 3　要发布应用程序，必须完成 3 部分：上传资产、填写应用程序详细信息、填写发布选项；将在同一网页完成这些操作。

第 1 部分——上传资产：上传应用程序相关图形文件，如图 14-16 所示。各个字段说明如下。
- 屏幕截取画面：提供屏幕操作画面的截图。
- 高解析度应用程序图示：会用在 PC 浏览器版的 Google Play 上，此图标不会替换原来应用程序的启动图标(清单文件中的 android:icon 属性所指定的即为启动图标)[①]。
- 宣传图片：营销专用的图形文件，非必要字段。
- 主题图片：Google Play 特色区的图标[②]，非必要字段。
- 宣传影片：可以录制一段宣传用的 video 文件放在 YouTube 上供人浏览。非必要字段。
- 停止宣传：不勾选就代表同意应用程序促销到其他非 Google 官方的应用程序商城。

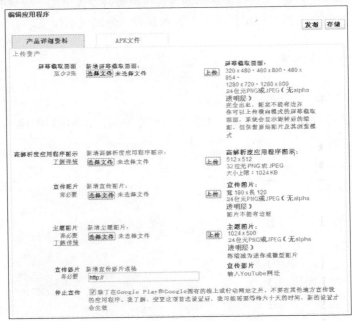

图 14-16

---

① 依据笔者测试结果，如果使用手机直接连至 Google Play，显示出来的是启动图标；如果使用 PC 浏览器进入 Google Play 网站浏览应用程序，显示出来的会是高分辨率的应用程序图标。这是因为手机屏幕小，所以只呈现小尺寸的启动图标。
② 使用 PC 浏览器查看应用程序的详细信息时才会显示此图标。

第 2 部分——应用程序详细数据：填写应用程序相关信息，如图 14-17 所示。各个字段说明如下。

- 语言：如果应用程序支持多国语言，建议提供应用程序在各个支持语言的说明。单击"新建语言"链接可以新建应用程序在指定语言的说明。
- Title：应用程序的标题，30 个字以内。
- Description：应用程序的描述，4000 个字以内。
- Recent Changes：应用程序的改版说明，只要编写最新一版的改版情况即可，500 个字以内。
- Promo Text：营销用的文本，80 个字以内。
- 应用程序类型：分成应用程序与游戏两大类。
- 类别：应用程序的详细分类。

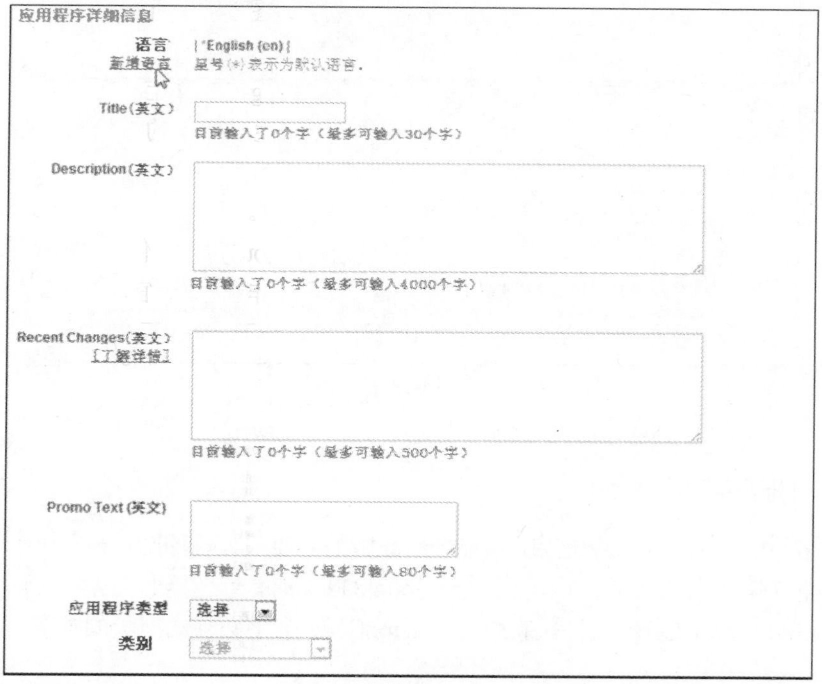

图 14-17

第 3 部分——发布选项、联系人信息与同意(事项)：如图 14-18 所示。各个字段说明如下。

(1) 发布选项
- 复制保护：是否启动防止复制机制，不过此机制不久之后即不再得到支持。
- 内容分级：应用程序内容的分级，分为心智成熟度高、中、低与所有人等 4 种[①]。
- 定价：默认为免费，如果要销售应用程序，就必须设置商家账户。可以选中欲将应用程序在哪些国家或地区上架。

(2) 联系人信息
- 网站：开发人员的网站。

---

① 详细分级方式参见 http://support.google.com/googleplay/android-developer/support/bin/answer.py?hl=zh-Hant&answer=188189。

- 电子邮件：开发人员的电子邮件地址。
- 电话号码：开发人员的电话。

(3) 同意(事项)

必须选中"这个应用程序符合《Android 内容指南》的规定"与"我了解不论我的所在位置或国籍为何，…"这两个选项，否则无法发布。

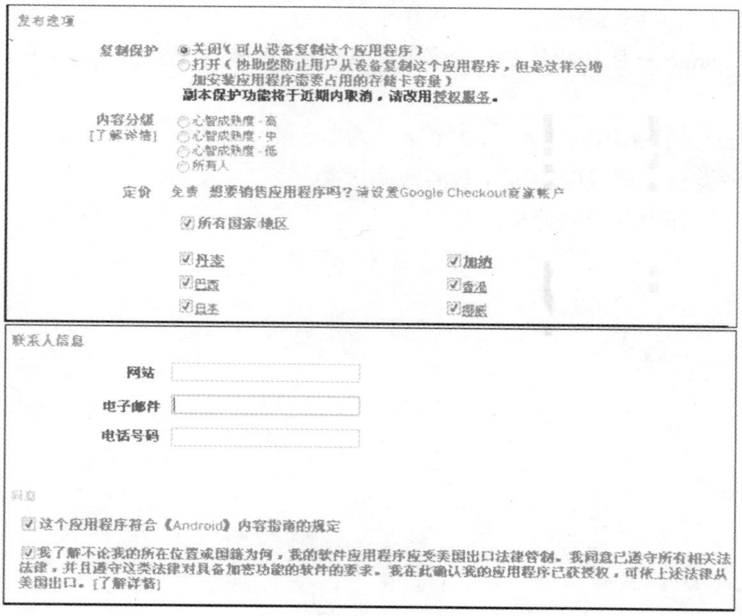

图 14-18

14.4.2 应用程序改版

应用程序发布到 Google Play 之后，可能会有改版的需要；如果要改版，必须注意下列事项。

(1) 版本控制：上传更新版的应用程序到 Google Play 时，管理界面会先检查应用程序清单文件内的 android:versionCode 属性。新版的 versionCode 数字必须比旧版的数字大，否则会无法更新。

(2) 必须使用相同私钥签名：更新应用程序的版本时，新版本的应用程序必须使用和旧版本相同的私钥来签名，否则会无法更新。所以一般建议使用同一把私钥签名所有欲发布的应用程序，避免多把密钥在管理上的麻烦。

(3) 填写改版信息：图 14-17 中的 Recent Changes 字段应该列出改版后的新功能与上一版功能的差异性，让用户了解新旧版本的差异，以决定是否要更新应用程序。

附 录

导入范例程序错误时的解决方法

下列几种情况都有可能造成将本书范例导入到 Eclipse 后发生错误，解决方法说明如下。

1. Android API 版本不符问题

本书范例程序采用 Android API 15(Android 4.0.3)，如果未安装该版本的 API，会导致所有 Java 源文件编译失败，有下列两种解决方法(可挑选任一种)。

- 安装 API 15：安装方式请参见 2.4 节。请选中 Android 4.0.3 (API 15)并安装。
- 编辑 default.properties 文件：导入范例项目后，打开 default.properties 文件内容，将 target 设置成已经安装的 Android API 版本，如图 F-1 所示。例如读者只有安装 API 8(Android 2.2)而没有安装 API 15，可以将 target=android-15 改成 target=android-8 即可。建议改完后重新启动 Eclipse。

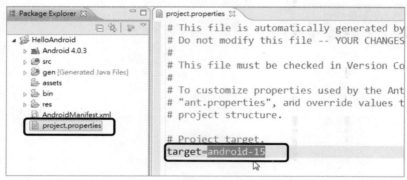

图 F-1

2. Eclipse 编码问题

Windows 版本的 Eclipse 默认编码为 MS950，而本书所有范例都是 UTF-8 编码格式。

如果未将 Eclipse 编码格式同样设置成 UTF-8，导入或打开本书范例时会产生乱码情况，这样很可能会导致编译失败。解决方法请参见 2.5 节。

3. 范例只读文件问题

因为范例程序为只读文件(可通过 www.tupwk.com.cn 下载)，建议将范例复制到硬盘驱动器，并改成非只读文件后再导入 Eclipse 内。步骤如下：

STEP 1 将范例目录 Example 复制到硬盘驱动器(例如 C:/)。

STEP 2 对 Example 目录右击并选择"属性"；在弹出的"Example 属性"窗口中，取消选择"只读"，取消后按下"确定"按钮即可，如图 F-2 所示。

导入范例程序错误时的解决方法　附录

图 F-2

4. Java 编译程序问题

以匿名类实现事件处理方法时会发生错误，如图 F-3 所示。

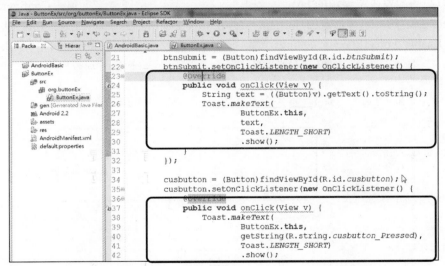

图 F-3

第一种解决方法，步骤如下。

STEP 1 会发生这样的问题是因为安装了 JDK 7 版，必须先将编译程序的版本改至 6.0 版。

STEP 2 更改整个 Eclipse 环境的 Java 编译程序版本：单击 Eclipse 菜单 Window | Preferences，会弹出如图 F-4 所示的窗口。在右边窗格中，在 Compiler compliance level 字段中单击 1.6(就是 6.0 版)，按 OK 按钮即可[①]。

---

① 如果仍有问题，则重启 Eclipse；如果重启后仍有问题，先移除有问题的 Android 项，然后再重新导入。

415

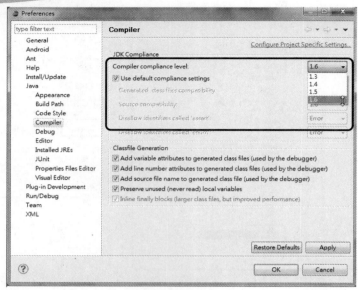

图 F-4

第二种解决方法如下。

也可以只更改发生错误项的 Java 编译程序版本：改为单击图 F-4 右上角的 Configure Project Specific Settings，然后选择指定的项，之后会弹出如图 F-5 所示的窗口(与图 F-4 的窗口几乎一样)，选择 Enable project specific settings，然后在 Compiler compliance level 字段中单击 1.6。

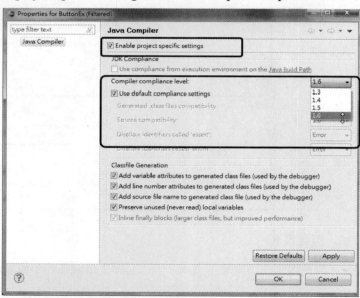

图 F-5